Handbook of
Behavioral Neurobiology

Volume 8
Developmental Psychobiology and
Developmental Neurobiology

HANDBOOK OF BEHAVIORAL NEUROBIOLOGY

General Editor:
Norman T. Adler
University of Pennsylvania, Philadelphia, Pennsylvania

Volume 1 Sensory Integration
Edited by R. Bruce Masterton

Volume 2 Neuropsychology
Edited by Michael S. Gazzaniga

Volume 3 Social Behavior and Communication
Edited by Peter Marler and J. G. Vandenbergh

Volume 4 Biological Rhythms
Edited by Jürgen Aschoff

Volume 5 Motor Coordination
Edited by Arnold L. Towe and Erich S. Luschei

Volume 6 Motivation
Edited by Evelyn Satinoff and Philip Teitelbaum

Volume 7 Reproduction
Edited by Norman T. Adler, Donald Pfaff, and Robert W. Goy

Volume 8 Developmental Psychobiology and Developmental Neurobiology
Edited by Elliott M. Blass

A Continuation Order Plan is available for this series. A continuation order will bring delivery of each new volume immediately upon publication. Volumes are billed only upon actual shipment. For further information please contact the publisher.

Handbook of
Behavioral Neurobiology

Volume 8
Developmental Psychobiology and Developmental Neurobiology

Edited by
Elliott M. Blass
Johns Hopkins University
Baltimore, Maryland

PLENUM PRESS • NEW YORK AND LONDON

Library of Congress Cataloging in Publication Data

Main entry under title:

Developmental psychobiology and developmental neurobiology.

  (Handbook of behavioral neurobiology; v. 8)
  Includes bibliographies and index.
  1. Developmental psychobiology. 2. Developmental neurobiology. I. Blass, Elliott M.,
1940–   . II. Series. [DNLM: 1. Adaptation, Psychological. 2. Behavior—physiology.
3. Nervous System—growth & development. W1 HA51I v.8/WL 103 D4893]
QP360.D49   1985   152   85-19276
ISBN 0-306-42034-1

© 1986 Plenum Press, New York
A Division of Plenum Publishing Corporation
233 Spring Street, New York, N.Y. 10013

Printed in the United States of America

This text is dedicated to
the memory of my father
Joseph H. Blass

Contributors

ARTHUR P. ARNOLD, *Department of Psychology and Brain Research Institute, University of California, Los Angeles, California*

ELLIOTT M. BLASS, *Department of Psychology, Johns Hopkins University, Baltimore, Maryland*

SARAH W. BOTTJER, *Department of Psychology and Brain Research Institute, University of California, Los Angeles, California*

ROBERT M. BRADLEY, *Department of Oral Biology, School of Dentistry; and Department of Physiology, Medical School, University of Michigan, Ann Arbor, Michigan*

JOHN C. FENTRESS, *Departments of Psychology and Biology, Dalhousie University, Halifax, Nova Scotia, Canada*

CHARLES A. GREER, *Sections of Neuroanatomy and Neurosurgery, Yale University School of Medicine, New Haven, Connecticut*

L. HAVERKAMP, *Department of Neurology, Baylor College of Medicine, Houston, Texas*

HELMUT V. B. HIRSCH, *Center for Neurobiology, State University of New York at Albany, Albany, New York*

MICHAEL LEON, *Department of Psychobiology, University of California, Irvine, California*

PETER J. MCLEOD, *Department of Psychology, Dalhousie University, Halifax, Nova Scotia, Canada*

CHARLOTTE M. MISTRETTA, *Center for Human Growth and Development; Center for Nursing Research, School of Nursing; and Department of Oral Biology, School of Dentistry, University of Michigan, Ann Arbor, Michigan*

TIMOTHY H. MORAN, *Department of Psychiatry and Behavioral Sciences, Johns Hopkins University School of Medicine, Baltimore, Maryland*

R. W. OPPENHEIM, *Department of Anatomy, Wake Forest University, Bowman Gray School of Medicine, Winston-Salem, North Carolina*

PATRICIA E. PEDERSEN, *Sections of Neuroanatomy and Neurosurgery, Yale University School of Medicine, New Haven, Connecticut*

GORDON M. SHEPHERD, *Sections of Neuroanatomy and Neurosurgery, Yale University School of Medicine, New Haven, Connecticut*

Preface

Previous volumes in this series have focused on adult phenotypic characteristics. They have described behavioral, neural, and neurophysiological processes in adult animals engaging in individual or social activities. In emphasizing adult characteristics, developmental processes that gave rise to adult forms of behavioral expression were essentially ignored. This was appropriate, for to do justice to the developmental landscape requires an extensive treatment of developmental theories, data bases, and their biological context. The current volume starts to address developmental issues, focusing on the interface between developmental psychobiology and developmental neurobiology. Developmental psychobiology's other natural interface, behavioral ecology, will be treated in a separate volume.

The present volume highlights two perspectives of neurobiologists and psychobiologists concerned with adaptation during development. One perspective represents the traditional approach, of taking a longitudinal, or life history, view of development, tracing morphological and physiological changes that accrue with maturation, growth, and differentiation. The model here is the adult "end point." The multivariegated pathways taken to these various end points can and have been traced and experiental influences on development have been assessed.

This approach has a long and rich history in embryology, developmental neurobiology, and developmental psychobiology. It has identified periods when neural, muscular, and behavioral attributes become progressively circumscribed through events that occur in specifiable time frames. This approach has also allowed the specification of epochs during development when phenotypic expression is relatively unaffected by particular classes of interactions with the biological environment. The longitudinal approach has permitted biologists to identify events that set the stage for the "critical" period, thereby placing it in a developmental continuum and not having it focused upon as a starting or pivotal point of specification.

A more recent perspective focuses on changing developmental processes as meeting different demands faced at a particular point in development. This approach complements the more traditional one. It is predicated on the idea that the young organism is not a miniature and incomplete adult, but has highly specialized needs of its own, commensurate with its station, and during its phylogenetic history has responded to these needs by developing behavioral and morphological specializations. This provides a view of a succession of anatomical and behavioral phenotypes, each selected for in its own right (theoretically), culminating in and influencing the more stable adult phenotype that deals with the rigors of its more consistent physical and social environment.

I have chosen to focus this volume on the interface between developmental psychobiology and developmental neurobiology. This focus represents a natural synthesis between two disciplines that, in the past, have often followed parallel tracks concerning ontogeny. Recently, however, the two areas have been realigned. The behaviorists, for their part, have asked more pointed questions, thereby uncovering phenomena that allow for more precise anatomical and neurophysiological characterizations that are cast within a behavioral framework. Many neurobiologists, for their part, have become responsive to the idea that their data base and theoretical positions did not have a behavioral referent and were therefore of limited validity for a coherent theory of neurobehavioral development. Biopsychological and ethological paradigms and tools are currently being used to explore the neurobiological mechanisms underlying the ontogeny of behavioral processes as they may naturally unfold.

The opening chapter by Oppenheim and Haverkamp sets the intellectual tone of the volume. It provides psychobiologists and neurobiologists with the perspective that, historically, the issues that many of us had grappled with (and still are grappling with) have been settled long ago by the embryologists. For much of this century, their perspective on development and developmental issues has been considerably broader than that of most psychobiologists. The embryologists have abandoned the raucous and divisive conflicts that have hampered developmental psychobiology and that, to this day, still plague developmental psychology. The embryologists were able to identify, as Oppenheim and Haverkamp demonstrate, specific processes by which the developmental course could be altered through influences outside the organism or the particular cell, tissue, or structure in question. In certain cases, the mechanism has been identified. This form of analysis was also able to exclude classes of events that were without apparent effect and has provided examples of normal development of portions of the nervous system in the absence of any obvious exogenous stimulation.

The power of an embryological frame of reference is seen in the number and diversity of issues that can be addressed through it. Although too soon to articulate the "laws" of neurobehavioral development, it is not too soon, indeed it is imperative, to seek from the embryological literature guidance for analyzing issues of neurobiological and psychobiological development. This type of approach provides us with the rare opportunity in our reductionist era to phrase questions at a "higher" behavioral level based on the concepts, principles, and approaches derived from "lower" levels of analysis.

This type of thinking is extended by Fentress and McLeod in their treatment of motor development. Their chapter is remarkably wide-ranging, addressing issues as diverse as the development of lobster and shrimp claw morphology to complex social interactions among mammals, to imitation by human infants, to speech development. The authors call our attention to the diversity of rules that are followed during development. Some attributes obey rules that are linear and sequential and do not appear to be affected in any obvious way by external events. These motor systems, at the point of maturation or earlier, come into the service of specific sensory stimuli and motivational states. Other rules appear to involve a reduction of complexity before new, additional complexity is added. Yet another set of rules applies to the development of a motor pattern at a time that precedes the functional expression of that pattern. That is, a silent period is interposed between acquiring and practicing the pattern in one context (e.g., the nest), not practicing the pattern for a protracted period of time, and, later as an adult, expressing the pattern once again. Sexual behavior is an obvious case in point.

The next chapter, by Moran, points to the strides made in certain areas of developmental psychobiology during the past decade. It brings to light the fact that newborn mammals, when in the nest, are remarkably competent to extract particular features of their environment. These features either predict or are contemporaneous with "reinforcing" and sedating characteristics of the mother—for example, her warmth, softness, teat, and milk. There is a second theme concerned with the highly organized behaviors of infant mammals that only occur under contrived circumstances of experimentation. Specifically, as Hall has demonstrated, infant rats under special circumstances of high activation and elevated ambient temperature can eat. Moreover, the feeding controls appear to be remarkably sophisticated, being highly similar in many ways to that of adult rats. Moreover, Moran has demonstrated that even 3-day-old infant rats, under circumstances of elevated ambient temperature and electrical stimulation of the medial forebrain bundle, express integrated behavioral acts. These acts can culminate in lordosis seen during sexual receptivity in adult rats.

Moran's chapter therefore speaks to the issues of infantile capacity and function during normal development, on the one hand, and to the circumstances under which the nascent or latent behavioral patterns that typify adult motivational behaviors can become expressed prematurely. Understanding how and when the potential patterns become actualized represents a major challenge to developmental psychobiologists and developmental neurobiologists. This chapter by Moran points to the importance of stimuli that acquire their biological significance during the nesting period and to the function of these stimuli as they become attached to, elicit, and direct motor patterns that become sequentially available during development.

This last issue represents a major focus of Bottjer and Arnold's chapter in which the circumstances underlying the acquisition of bird song and the putative neurobiological mechanisms that mediate this behavior are revealed. They touch upon a number of issues that are fundamental to all students of development (ones also raised by Oppenheim and Haverkamp in their introduction). Bottjer and Arnold attempt to specify the events that occur during the so-called critical period, including the interactions between the bird and its environment that foster song acquisition. Their exposition of mechanisms of bird song point out the inadequacy of traditional paradigms of learning that do not have as a referent point certain classes of social reinforcers. Theirs and other findings on memory and the circumstances of its expression in bird song demand that psychological theories of memory be modified to account for seasonal behavior that appears to be under hormonal control. It is clear to me that the literatures on bird song, on courtship behavior, on courtship calls, and on adult recognition of sexual signals must be addressed by psychologists interested in the behavioral and neurological mechanisms of conditioning and memory.

Bottjer and Arnold draw an intriguing parallel between the characteristics of the avian song system and that of the human speech system. While the parallel has certain limitations, it nonetheless provides the opportunity to ask very refined questions concerning the integration of the singing mechanism and to use the information acquired from such analyses to phrase questions that might not otherwise be obvious about the human speech mechanism. The implications of this system for better understanding neurobehavioral losses during aging or disease, recovery of function after central insult, and life-history plastic changes are obvious and compelling.

The contribution by Pedersen, Greer, and Shepherd represents a synthesis of developmental psychobiology and neurobiology at its very best. In it the authors relate their analyses of central olfactory mechanisms that are differentially labeled by (C14), 2-deoxy-D-glucose during the act of suckling. Success of this analysis reflects the authors' abilities to capitalize on a very specific behavior in the developing rat pup, suckling, and to bring to bear on this behavior state-of-the-art techniques available to the neuroanatomist and neurophysiologist. The experimental questions concerning neurobiology were guided by the behavior. That is, Pedersen's previous behavioral analyses had revealed different experiential and time components in determining infant rats suckling behavior. The anatomical analyses by the authors and their colleagues exploited the different aspects of the behavioral act to reveal neural structures and possible functions that had not been uncovered previously and that, in all likelihood, may have remained otherwise hidden in the near future.

This collaborative effort between the psychobiologist, Pedersen, and the neurobiologists, Greer and Shepherd, therefore represents a union of precise behavioral questions and application of sensitive neurobiological techniques to bring about a more realistic merging of the different levels of neurobiological analysis. The opportunity for mutual collaborations of scientists whose trainings and expertise are diverse yet complementary must be followed if progress is to be realized at the interface between these disciplines.

A similar note is struck in the chapter by Mistretta and Bradley, who summarize their own famous experiments on the development of gustatory senses in rats and sheep, animals whose developmental histories differ so vastly. The authors also report their efforts to alter physiologically the developmental, physiological, and behavioral characteristics of the sensory system. These efforts had a relatively minor effect on the expression of taste threshold or preference. I personally find this behavioral invariance very comforting, because the sensory system can maintain its fidelity even in the face of unusual historical and nutritional circumstances. Yet the gustatory system is extremely flexible, and this is expressed in changes in perception and motivation in preference and ingestive behavior following toxicity or certain rewards. This even holds for developing rats as young as 5 days of age.

From these reports by Mistretta and Bradley, one anticipates the next rounds of major breakthroughs in their work as they will soon be able to specify the changes in single and multiple units in the gustatory system at the level of the thalamus in animals that have had very different developmental gustatory histories. There is also the fascinating issue of functional equivalence over developmental time. Given that the peripheral coding properties of developing systems differ so vastly from those properties in the adult system, one marvels at the ability of the animal to keep straight the events associated with a particular taste. This represents another opportunity for reductionism to flow "the other way" with some precision. That is, the information obtained from the neurophysiological study can be utilized to pose informative questions on behavioral function and mechanisms of taste perception.

Hirsch's very thoughtful chapter on visual system development thoroughly reviews the extant neurophysiological and behavioral literatures on the development of the mammalian visual system. One is struck by the thoroughness of investigation by the neurophysiologists who have identified the morphological and neurophysiological characteristics of various visual subsystems and the characteristics of their components. They have also determined how these are affected by various classes of visual experience, ranging from total rearing in the dark to rearing with distorting lenses to rearing in unusual environments such as those with striped patterns. The behavioral data are less complete, and there are, unfortunately, very few parallels between behavioral and neurophysiological lines of inquiry. Without the larger framework offered by behavior, the neurophysiological studies run the risk of isolation and irrelevance. It would be very beneficial for the field if developmental psychobiologists could capitalize upon some of the major findings of the neurophysiologists, particularly those pertaining to the role of experience in depth perception and the importance of visual experience in sensory–motor integration, to elucidate further the mechanisms through which animals learn to use vision to locomote through space and to capture prey, for example. Such findings would set the stage for the neurophysiologist and neurobiologist to explore further the anatomical and physiological characteristics of these developing systems and mark their changes. Another fascinating and unexplored dimension in behavioral correlates of the visual system was pointed out by Hirsch in his discussion of visual seeking behavior by the kitten at a particular point in its development.

The final chapter of this volume is by Michael Leon, who discusses temperature regulation during development. Leon documents various strategies utilized by altricial animals during their thermally vulnerable developmental period, when they are at the mercy of exogenous sources for heat and sustenance. Leon makes it perfectly clear that the relationship is not unidirectional between the mother (or parent) and infant. That is, infants can exert certain control over the interaction by calling for help when cooled and/or isolated. The infant can also heat and cool itself by moving about in the nest, reaching the upper portions of the huddle to dissipate heat when warm, reaching the core of the huddle when cool and heat retention is called for.

In the intimate setting of the nest, infants learn about the consequences of their own behavior and discover stimuli that provide succor and the stimuli associated with such comfort. Not only does the infant learn about these relationships but this learning occurs within the context of considerable excitement that surely must carry on to future epochs in the animal's life, as discussed previously.

There are two major omissions in this text—chapters on psychopharmacology and behavioral endocrinology during development. The latter will be addressed in the next volume of the series on development. The former was omitted because the field has not advanced to the point where there is sufficient information available on neurochemical development and its behavioral consequences to make such a chapter beneficial at this time. We hope this situation will change within the next few years.

I wish to express thanks to Norman Adler, who is the overall editor of this series, and especially to the contributors to this volume. They were enormously responsible and receptive to suggestions that arose during the course of chapter preparation. Finally, on a more personal developmental note, I dedicate this volume to the memory of my father, Joseph H. Blass, who was very interested in many aspects of behavior, especially its development.

ELLIOTT M. BLASS

Contents

CHAPTER 1

Early Development of Behavior and the Nervous System: An Embryological
    Perspective   1

R. W. Oppenheim and L. Haverkamp

Introduction   1
Some Embryological Findings   3
Early Neural Development 11
Neurobehavioral Development: The Role of Neural Activity 17
Behavioral Embryology: Practice and Perceptual Experience 24
References 30

CHAPTER 2

Motor Patterns in Development 35

John C. Fentress and Peter J. McLeod

Introduction 35
    Preliminary Issues 35
    Approach Taken 36
Review of Issues 37
    Basic Dimensions of Movement 37
    Spontaneous and Elicited Movements 41
    Higher-Order Motor Patterns 50
    Socially Integrated Motor Patterns 53
Some Personal Explorations 60
    Rodent Motor Patterns 60

Conditioned Movement Sequences and Stereotypies 70
Canid Motor Patterns 72
Emergent Issues 76
Developmental Processes: A Motor Perspective 77
Capabilities and Strategies 77
Perspectives on Mechanism 79
Summary Themes and Conclusions 85
References 86

CHAPTER 3

Environmental and Neural Determinants of Behavior in Development 99

Timothy H. Moran

Introduction 99
Suckling as an Organized Behavior 100
External Controls 101
Internal Controls of Suckling 103
Suckling and Reinforcement 105
Huddling 106
Independent Ingestion 109
Water Intake 109
Nutrient Intake 110
Brain Stimulation and Behavioral Organization 112
Discussion 118
Concluding Remarks 123
References 123

CHAPTER 4

The Ontogeny of Vocal Learning in Songbirds 129

Sarah W. Bottjer and Arthur P. Arnold

Introduction 129
Background 130
Behavioral Data 130
Theoretical Framework 134
Neuroanatomical Data 135
Sexual Dimorphisms and Brain–Behavior Correlations 137
Neural Correlates of Song Learning 140
The Auditory Template 140
Auditory–Motor Integration: The Matching Process 141
The Central Motor Program 149
Possible Mechanisms of Song Learning 151
Development of the Central Motor Program 151

Sensitive Periods for Song Learning 152
Song Learning and Brain Volume Changes 154
Relations of Avian and Human Vocal Learning 155
Similarities Between Vocal Learning in Birds and Humans 155
Relationships Between Vocal Perception and Production 157
References 158

CHAPTER 5

Early Development of Olfactory Function 163

Patricia E. Pedersen, Charles A. Greer, and Gordon M. Shepherd

Introduction 163
Main Olfactory Epithelium 164
Vomeronasal Organ 167
Main Olfactory Bulb 169
Accessory Olfactory Bulb 175
Central Olfactory Projections 175
Modified Glomerular Complex 178
The Ontogeny of Olfactory Function: Overview 180
Prenatal Activity in the Accessory Olfactory Bulb 182
Early Postnatal Activity in the MGC 187
Postnatal Activity in the Main Olfactory Bulb 190
Discussion 192
References 197

CHAPTER 6

Development of the Sense of Taste 205

Charlotte M. Mistretta and Robert M. Bradley

Introduction 205
Development of the Rat Gustatory System 207
Changes in Electrophysiological Responses from Peripheral Taste Nerves .. 207
Anatomical Correlates of Response Changes in Peripheral Taste Nerves ... 213
Changes in Electrophysiological Responses from Central Nervous System
Taste Neurons and Behavioral Implications 214
Attempts to Modify the Rat Gustatory System through Early Experience ... 218
Development of the Sheep Gustatory System and Applications to
Development of Taste Responses in Humans 220
Changes in Electrophysiological Responses from Peripheral Taste Nerves .. 221
Anatomical Correlates of Response Changes in Peripheral Taste Nerves ... 224
Changes in Electrophysiological Responses from Central Nervous System
Taste Neurons and Implications for Human Behavioral Responses 227

Proposed Membrane Changes Underlying Development of Taste
  Responses 230
Summary. 232
References 233

CHAPTER 7

The Tunable Seer: Activity-Dependent Development of Vision 237

Helmut V. B. Hirsch

Experience-Dependent Developmental Programs 237
The Retino-Geniculo-Cortical Pathway in Adult Cats 239
  Functional Specialization in the Retina 241
  Parallel Pathways in the Visual System 242
  Orientation Selectivity of Area 17 Cells 243
  Distribution of Preferred Orientations in Retina, Lgn, and Area 17 245
Development of the Retino-Geniculo-Cortical Pathway in Normal Cats . 246
  Optics and Alignment of the Eyes 247
  Retinal Development 247
  Lgn Development 250
  Development of Area 17 254
Activity-Dependent Development of the Visual System 258
  Neuronal Activity and Subcortical Development 259
  Neuronal Activity and Cortical Development 261
  Factors Controlling Differences in Experience Sensitivity of Cortical Cells . 266
Extraretinal Factors in Visual System Development 270
Behavioral Significance of Experience-Dependent Changes 272
  Development of Visually Guided Behavior in the Kitten 272
  Correlating Behavioral and Visual System Development 275
  Effects of Stripe-Rearing on Visual Development 277
  Experience-Dependent Changes in "Attention". 280
Summary: The Tunable Seer 281
  Developmental Changes 281
  Experience-Dependent Changes in Area 17 282
  Behavioral Significance of Experience-Dependent Development 283
  Integration of Genetic and Environmental Information 284
  Implications of Experience-Dependent Development 284
References 285

CHAPTER 8

Development of Thermoregulation 297

Michael Leon

Introduction 297
  Thermal Aspects of Embryonic Development 298
  Sexual Differentiation 300

Development of Thermoregulation 301
    Strategies for Thermoregulation 301
    Maternal Defense of Neonatal Temperature 304
    Adaptive Features of Hypothermia 307
Mechanisms of Thermoregulation in Young Mammals 307
    Physiological Mechanisms 307
    Behavioral Mechanisms 308
Plasticity in the Development of Thermoregulatory Capacity 311
    Effects of Being Reared in Different Ambient Conditions 311
    Thermal Mediation of Early Experience 313
    Thermal Activation 314
Summary 316
References 317

Index 323

Early Development of Behavior and the Nervous System

An Embryological Perspective

R. W. OPPENHEIM AND L. HAVERKAMP

> Dedicated to the memory of G. E. Coghill (1872–1941) whose writings on related issues have been an inspiration to our own thoughts on these matters.

I. INTRODUCTION

For much of the period between 1950 and 1970 the Canadian psychologist D. O. Hebb was a leading exponent of a developmental approach to neural and behavioral problems. Perhaps his best-known contribution in this regard was his now classic book *Organization of Behavior* (1949). In this work Hebb not only outlined a plausible and testable conceptual framework for understanding the role of sensory and perceptual experience in the development of the nervous system and behavior, but also provided empirical support for the model. By virtue of this, Hebb's book is considered to be an important milestone in the age-old debate over the role of nature versus nuture, heredity versus environment, or preformation versus epigenesis in the development of behavioral and psychological processes.

Although we have always been greatly impressed with Hebb's thoughtfulness on these issues, we have never understood why he failed to draw on the rich literature in embryology to strengthen his conceptual arguments. Hebb was not alone in this shortcoming, however, for most psychologists of his era either were unaware of or failed to appreciate the contributions that embryology could make to understanding principles of both neural and behavioral development. Yet there were

R. W. OPPENHEIM   Department of Anatomy, Wake Forest University, Bowman Gray School of Medicine, Winston-Salem, North Carolina 27103.   L. HAVERKAMP   Department of Neurology, Baylor College of Medicine, Houston, Texas 77004.

important exceptions, Karl Lashley, for instance, was very much interested in the ways in which embryology might help elucidate neurobehavioral development (see, e.g, Lashley, 1938).[1] Other leading psychologists of this period with equally strong interests in embryology were L. Carmichael (1946), A. Gesell (1945), M. McGraw (1935), and J. Piaget (1969). Even if one attributes Hebb's failure on this score to the general lack of influence of embryology on psychological thought during this period, it does not explain why, in another book written many years later (Hebb, 1980), but which deals with similar conceptual issues, the same oversight is repeated.

In his more recent book a chapter devoted to the issue of heredity versus environment contains a tabular list of factors that Hebb claims cover all the various influences that control behavioral development (Table 1). However, only two of these categories deal at all with factors that might conceivably be studied by embryologists: number II, which includes "nutritive and toxic" influences, and number IV, which covers sensory experience. Although these are obviously important factors in neurobehavioral development, they do not begin to reflect the rich variety of normally occurring interactions that are the focus of most neuroembryological investigations. Indeed, to an embryologist, Hebb's categories ignore or exclude many of the most interesting and relevant phenomena involved in early development of the nervous system (and by implication in the earliest structural basis for behavior).

We have singled out Hebb in this context not because he is necessarily more vulnerable to criticism on this issue, but because it seems apparent that if someone as thoughtful and knowledgeable as Hebb could seemingly miss the relevance of embryology for helping to elucidate these issues, then there must certainly be many others in developmental psychology or psychobiology who also are not fully aware of the applicability of this field to their own work. We say this knowing, of course,

---

[1] The neuroembryologist Viktor Hamburger, who was at the University of Chicago with Lashley for several years in the 1930s, has related to me (R.W.O.) that he recalls having had many long conversations with Lashley about embryology, behavior, and neural development.

TABLE 1. CLASSES OF FACTORS IN BEHAVORIAL DEVELOPMENT[a]

No.	Class	Source, mode of action, etc.
I	Genetic	Physiological properties of the fertilized ovum
II	Chemical, prenatal	Nutritive or toxic influence in the uterine environment
III	Chemical, postnatal	Nutritive or toxic influence; food, water, oxygen, drugs, etc.
IV	Sensory, constant	Pre- and postnatal experience normally inevitable for all members of the species
V	Sensory, variable	Experience that varies from one member of the species to another
VI	Traumatic	Physical events tending to destroy cells: an "abnormal" class of events to which an animal might conceivably never be exposed, unlike Factors I to V

[a]From Hebb, 1972, by permission.

that it is becoming increasingly fashionable to draw on certain embryological ideas for explicating neurobehavioral development. The writings of the late embryologist C. H. Waddington, for instance, have had some influence on a number of contemporary developmental psychobiologists (see, e.g., Scarr, 1976; Bateson, 1978), particularly his related concepts of the *epigenetic landscape* and *canalization,* which in principle at least can be easily utilized in conceptions of behavioral development.

Nonetheless, we are not convinced that such ideas have penetrated very deeply into the thinking of most behaviorists interested in ontogeny. Again, without wishing to draw undue attention to any particular individual, we note that in a recent chapter on "Genetics and Behavior," which is, in many ways, exemplary, the behavioral biologist G. Barlow has commented that, "most of us, with apologies to the embryologists, are concerned with the role of the environment that is *external to the animal.* . . . experience received via sensory systems, since sensory competence develops late in the embryo's life . . . *we are mainly interested in experience after hatching or birth*" (Barlow, 1981, p. 237, italics added). This, of course, is basically the same position taken many years ago by the ethologist K. Lorenz. In rebutting one of the points made by Lehrman (1953) in his classic critique of ethology, Lorenz argues that, "not being experimental embryologists but students of behavior, we begin our query not at the beginning of growth, but at the beginning of . . . innate mechanisms" (Lorenz, 1965, pp. 43–44). In addition to relegating embryology to a secondary position for the understanding of neurobehavioral development, such views also tend to assume that there are few, if any, significant contributions of experience (or of the external environment) to behavior prior to birth or hatching. In fact, there is now considerable evidence indicating that this is probably not a valid assumption (Gottlieb, 1976; Mistretta and Bradley, 1978; Oppenheim, 1982a). In apparent contrast to these views, we believe that it is impossible to formulate a valid theory of developmental psychobiology without including concepts, principles, and findings from neuroembryology and without considering the possible roles of experience, practice, or use during the embryonic and prenatal periods.

II. Some Embryological Findings

Although it seems entirely plausible to us to consider developmental psychology and psychobiology as branches of a single parent discipline, embryology (or developmental biology), we realize that this is not necessarily a widespread view. It is still somewhat surprising, however, that this is not the case. For it was experiments in embryology, starting even before the turn of the present century, that led to the establishment of the modern theory of epigenesis, the conceptual touchstone for all valid theories of ontogeny, whether behavioral, biological, or molecular.

In the late 1880s and early 1890s a major controversy was being debated that involved the proponents of two opposing, and seemingly irreconcilable, views of biological development: *neoepigenesis versus neopreformationism* (Oppenheim, 1982b). According to the supporters of neopreformationism, the full complement of hereditary material that was present in the zygote following conception was believed to be progressively reduced during development such that each cell type ended up with only a specific subset of genes. A nerve cell, for example, would retain only those genes necessary for carrying out neural functions (e.g., synaptic transmission) and those needed for other *general* metabolic functions common to all living cells. In other words, differentiation was thought to be mainly due to the actual loss of all specific genetic material not directly involved in the phenotypic expression of each cell type; thus each cell type in the mature animal was thought to retain only a small fraction of the genetic material (genes, chromosomes) originally present in the nucleus of the fertilized egg. In its most extreme form, neopreformationism held that by this means, differentiation occurred autonomously with little, if any, contribution from the environment.

By contrast, the advocates of neoepigenesis argued that all cells, at all stages of development, contained identical, complete complements of hereditary material. Thus differentiation was thought to be controlled not by the autonomous loss of specific genes or chromosomes from cells, but rather by a progressive, precisely timed sequence of intra- and extracellular events that selectively activated or suppressed the *expression* of particular genes. More recent experiments involving nuclear transplantation from differentiating or differentiated cells to unfertilized eggs, have shown conclusively that this viewpoint is substantially correct; even after differentiation is completed, sufficient genetic information exists in the nucleus of individual cells to support the development of a complete embryo (Gurdon, 1974) (Figure 1). Thus we now know that genes are suppressed or inhibited, and not acutally lost, during ontogeny. In its most extreme form, however, the theory of neoepigenesis implied that all the genetic material in the fertilized egg was homogeneous and that the subsequent diversity of cells and functions was therefore due *entirely* to environmental influences.

In 1888 the pioneer embryologist W. Roux published the results of an experiment that appeared to refute the theory of neoepigenesis. Roux argued that if the neopreformation theory was correct, then when the fertilized egg divides for the first time, each cell (blastomere) should only contain the appropriate genetic material necessary to form one-half of the embryo. By contrast, if the neoepigenesis theory was true, then each blastomere should be capable of producing a complete, albeit quantitatively small, embryo. The outcome of Roux's experiment is illustrated in Figure 2. Since the destruction of one blastomere resulted in the development of a half embryo, the experiment seemed to settle unequivocally the dispute in favor of neopreformationism. A few years later, however, similar experiments, done independently by H. Driesch, T. H. Morgan, and H. Spemann, produced different results: a single blastomere was, in fact, shown to be capable of forming a complete, small embryo (see Spemann, 1938). The discrepancy apparently resulted from an artifact of Roux's experimental procedures; by leaving the damaged or destroyed blastomere in contact with the intact cell, normal development of the intact blastomere was impaired. If one blastomere was completely isolated by a different procedure, then it was found that these cells formed a complete embryo (Figure 3).

Although these later experiments were interpreted as supporting neoepigenesis, additional experiments by Spemann and others showed that the ability of a single, isolated blastomere to form a complete embryo was restricted: after the four- to six-cell stage, only partial embryos were formed from single blastomeres. Collectively, such findings led to the formulation of the concepts of *determination*

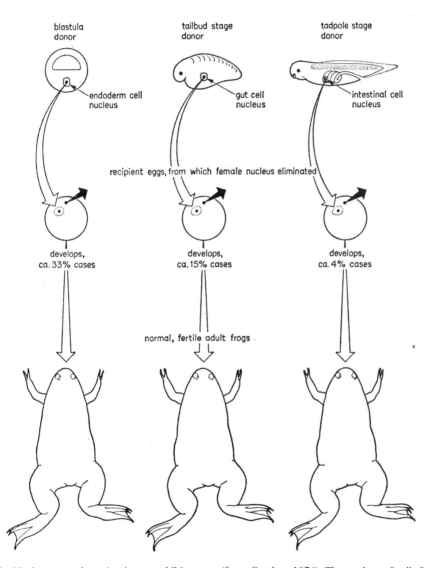

Fig. 1. Nuclear transplantation into amphibian eggs (from Gurdon, 1974). The nucleus of cells from embryos and tadpoles at different stages of development when placed into egg cells support the development of normal adult frogs.

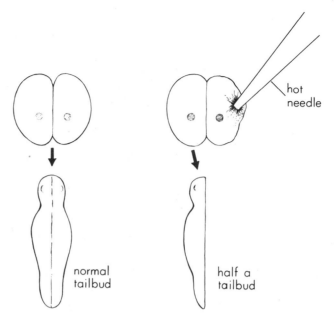

Fig. 2. Roux's experiment. One blastomere of a two-cell stage of a frog egg was killed with a hot needle. The surviving cell gave rise to a half embryo. (From Hamburg, 1971). As explained in the text, this result was obtained because of an artifact of the technique employed to isolate the blastomere. See Figure 3 for a different result when the artifact is excluded.

and *critical* or *sensitive periods*. Over the next 50 years, repeated demonstrations in a variety of tissues using diverse experimental designs showed that during development, cells gradually lose the potential to regulate or alter their fate when perturbed. In other words, the phenotypic expression of each cell becomes gradually restricted (determined) according to a rather rigidly controlled timetable (critical period). A particularly striking experiment (see Spemann, 1938) that clearly exemplified these phenomena is summarized in Figure 4.

Pieces of ectoderm were exchanged between two frog embryos that were in the beginning of the gastrula phase of early development. The dark tissue is derived from the future brain region of one frog species and the light piece from the future belly-skin region of another species. (In the absence of modern cell marking techniques, the embryologist ingeniously used the cells from two differently pigmented species as a naturally occurring marker for tracing the fate of the transplanted tissue). Following the transplantation of brain cells to the belly region or vice versa, the "foreign" tissue healed completely and continued to develop. The transplants adapted remarkably well to their new surroundings, and except for the different pigmentation, differentiated normally and fulfilled in every respect the demands of their new location. The transplanted skin cells became neurons and participated in the formation of the brain region to which they were transplanted. It would obviously be of considerable interest to know whether the transformed skin cells also *function* as nerve cells and participate in the mediation of behavior patterns appropriate to the brain region in which they are located. In any event, it can be concluded from this experiment that at the onset of the gastrula stage, brain and skin cells are not irrevocably determined. Rather, they are still able to alter their ultimate fate (phenotype) if provided with the appropriate environmental influences.

By contrast, if the very same experimental manipulations are done at the *end*, rather than the beginning, of the gastrula stage—a difference of only a few hours—the outcome is strikingly different. Although the tissues at this stage, when examined in histological sections, are virtually indistinguishable from those used in the early gastrula transplants, they respond quite differently to the same perturbation. Despite being placed in the same region of the future brain, the prospective skin now continues to develop into skin and persists in the brain as a disruptive foreign tissue. Similarly, the piece of transplatned brain tissue separates from the surrounding skin in the belly region and self-differentiates at this location into whatever part of the brain it would have formed in its original location; in the

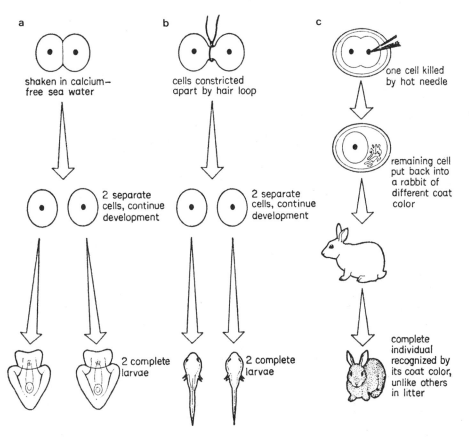

Fig. 3. Examples of the adaptability of embryonic cells at the two-cell stage. Results of separating the first two blastomeras in (a) Echinoderms; (b) amphibians; (c) the rabbit. (From Deuchar, 1975.)

specific case shown here, the prospective neural tissue develops into an eye and adjacent diencephalon.

Obviously, something decisive has happened to the ectoderm cells between the times of early versus late gastrulation. The *early* gastrula cells still have the potential to become either skin or brain (or perhaps a variety of other tisuses, depending upon the site to which they are transplanted); that is, they have not yet been determined to form any specific cell type. By the *end* of gastrulation, however, those specific molecular and biochemical events that mediate the process of determination in these tissues (which are largely unknown) are completed. Even if the cells from a *late* gastrula are transplanted to a heterotopic position in an early gastrula embryo—where one might imagine conditions would be more favorable for transformation—they are still unable to alter their fate (e.g., "late" brain tissue in the

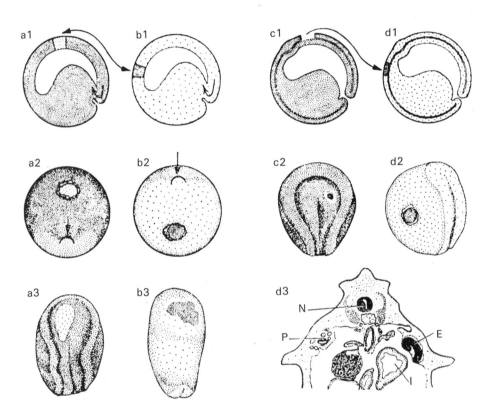

Fig. 4. H. Spemann's classical exchange transplant experiments on newt embryos, performed either at the beginning (a and b) or after completion of gastrulation (c and d). (a1 and b1) Diagram of the operation whereby neural ectoderm from a1 (dark) is exchanged with skin ectoderm from b1 (light). The arrow within the dorsal blastopore lip indicates the direction of invagination. (a2 and b2) Location of the newly implanted pieces in relation to the blastopore (arrow). (a3) Implant is located in the brain part of the neural plate. (b3) The other implant is located in the ectoderm on the underside of the head. (c1 and c2) Donor of neural material (gap). (d1 and d2) Host neurula with implant taken from c in its flank. (d3) Further development of the larva (section). Implant develops into an eye (E). Host organs: (N) neural tube, (P) pronephros, (I) intestine. (From Hadorn, 1974.)

belly region develops into brain). These experiments illustrate that whatever has transpired between the two stages involves changes inherent to the cells themselves and is not simply the result of a failure of the surrounding tissue environment of the late gastrula to provide a signal. Once the cells have been determined, their subsequent course of development is virtually inevitable as well as irreversible. It is worth noting that experiments such as this provide an especially compelling illustration of the principle of sensitive or critical periods in development (see Erzurumlu and Killackey, 1982).

As already mentioned, the rationale for making transplants between differently pigmented species (heterospecific transplants) in these experiments was to provide a convenient cell marker. However, similar heterospecific transplants have often been used in embryology to reveal a number of important developmental principles. One example is shown in Figure 5. This experiment capitalizes on the fact that salamander and frog tadpoles have different kinds of mouth parts that they use to stabilize themselves in the water; salamanders have balancers, whereas frogs have adhesive suckers.

It was known from previous experiments that when the oral ectoderm of a frog embryo is removed and then replaced by a piece of ectoderm from its own belly region, the belly ectoderm responds to signals from surrounding cells and develops into adhesive suckers rather than belly skin. This result is conceptually similar to the "brain-skin" transplants described earlier. What would happen, however, if the belly skin of a frog embryo were transplanted to the oral region of a

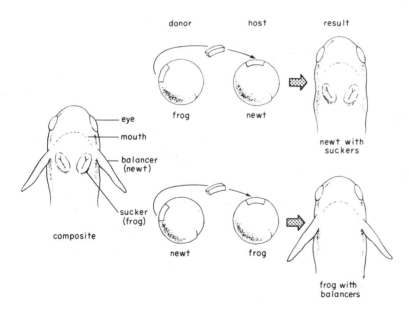

Fig. 5. Schematic representation of Spemann and Schotte's experiment. Note prospective oral epithelium of a frog transplanted to a newt gave rise to oral suckers, typical of the mouth region of a frog. Conversely, oral epithelium from a newt transplanted to a host frog gave rise to balancers, typical mouth structures found in newts. (From Hamburg, 1971.)

salamander (newt) embryo or vice versa? The results of this experiment, which was first done by O. Schotte (see Spemann, 1938), are shown in Figure 5.

As expected, the frog belly ectoderm when transplanted to the oral region of the newt embryo develops mouth parts rather than skin. Most interestingly, however, rather than developing mouth parts characteristic of the newt (i.e., balancers), the frog belly ectoderm now develops into typical frog suckers. The same thing occurs in the reciprocal newt → frog transplants (i.e., newt belly ectoderm placed in the oral region of a frog develops into balancers).

The results of this rather remarkable experiment reveal an important developmental principle, namely, that the cells comprising a given piece of tissue (e.g., *frog* belly skin) can respond to signals from a heterospecific source (*newt* oral tissue) only within the limits of its own species-specific properties. Despite their ability to alter their fate when perturbed, the belly ectoderm cells can only express the specific properties programmed in their genome.

These kinds of experiments are entirely representative of a large literature in embryology that has repeatedly illustrated the inexorable interrelationship that exists between the genome and cytoplasm and between cells, tissues, and their internal and external environments. By the turn of the century, experiments such as these already had convinced most biologists that the extreme forms of both neoepigenesis and neopreformationism were invalid conceptualizations and that some sort of rapprochement was needed. E. B. Wilson, a leading biologist of the period, expressed what was rapidly to become a consensus on this issue. Writing in 1900, in the second edition of his popular book *The Cell in Development and Heredity,* Wilson stated:

> Every living organism at every stage of its existence reacts to its environment by physiological and morphological changes. The developing embryo is a moving equilibrium—a product of the response of the inherited organization to the external stimuli working upon it. If these stimuli be altered, development is altered. However, we cannot regard specific forms of development as directly caused by these external conditions, for the character of the response is determined not by the stimulus but by the inherited organization [p. 428].

Writing at about the same time, the embryologist (and later animal behaviorist) C. O. Whitman (1894) expressed the same point even more succinctly: "Therefore, the intra and extra do not exclude each other but coexist and cooperate from the beginning to the end of development" (p. 222).

It is no coincidence that the views expressed here by both Wilson and Whitman have a remarkably modern ring to them. Despite their ignorance of most of the details of ontogeny, by the turn of the century embryologists were beginning to develop a conceptual framework (the modern theory of epigenesis) that has continued to guide the entire field right up to the present time. Although it does not take a gret deal of imagination to recognize the significance of this framework for the study of neurobehavioral development, it has nonetheless taken a long time for most students of behavioral development to begin to appreciate this fact. For most of this century, behavioral scientists have often been involved in disputes over ideas and issues (e.g., nature versus nuture) that, conceptually at least, had been resolved by most biologists over three-quarters of a century ago. The reasons behind this constitute a fascinating episode in the history of science. However, since these matters have been recently discussed in detail by one of us (see Oppenheim, 1982b) and because it constitutes somewhat of an aside to the major focus of the present essay, we shall not attempt to repeat that material here.

III. Early Neural Development

The embryological studies reviewed previously help to reveal important principles of development. Although someone strictly concerned only with neurobehavioral development may find them interesting, they will probably also consider them rather remote from, or tangential to, behavioral problems. After all, it is still widely thought that those aspects of neural development most germane to behavior must involve interactions between the entire organism and the external (i.e., sensory) environment. While there is clearly some validity to this point, we would argue that the factors involved in behavioral development represent a continuum ranging from the kinds of cell and tissue interactions discussed previously all the way to the devlopmental role of complex social interactions. Consequently, what we wish to explore in this section are problems and issues that were motivated largely by embryological concerns but that also have important and, we believe, more clearly recognizable relevance to neurobehavioral development.

The first example in this section involves a population of transient embryonic cells known as the *neural crest*. The neural crest is composed of ectoderm cells that become segregated from the surrounding ectoderm during closure of the neural tube (Figure 6). Neural crest cells migrate throughout the head and trunk of the embryo and eventually give rise to a variety of cellular phenotypes, including peripheral neurons and glia, cartilage, endocrine cells, and pigment cells. For the specific experiments we shall discuss, it is important to point out that in the trunk region, neural crest cells are the source of all the *adrenergic* neruons in the sympathetic paravertebral ganglia, as well as the source of all the *cholinergic* neurons in the parasympathetic ganglia. Adrenergic sympathetic neurons use norepineprine (NE) as a neurotransmitter, whereas the cholinergic parasympathetic neurons use acetylcholine (ACh).

On the basis of a large number of experiments too numerous to review here (see Le Douarin, 1980; Bunge *et al.*, 1978; Weston, 1982, for reviews), it now seems likely that most individual neural crest cells are not initially programmed to develop a specific phenotype. Rather, signals provided by other cells encountered during (or following) the migration of the neural crest are essential in determining, for instance, whether a crest cell will express adrenergic or cholinergic properties. As indicated in Figure 7, differentiated adrenergic and cholinergic ganglionic neurons can be reliably distinguished by cytological criteria, as well as by their biochemical, histochemical, and physiological properties.

Experiments by Paul Patterson and his colleagues (Walicke *et al.*, 1977; Patterson, 1978) have suggested that one of the extrinsic factors involved in the determination of phenotypic expression of neural crest cells is synaptic activity. They have shown that embryonic neural crest cells, placed in tissue culture and exposed to ionic changes that mimic electrical (synaptic) activity, develop primarily into adrenergic neurons, whereas the same cells when left untreated develop cholinergic properties. Although it is believed that the time period during which these same events occur *in vivo* coincides with the onset of innervation of the ganglion neurons by preganglionic cells in the spinal cord, it has not yet been shown that ganglionic blockade does, in fact, result in altered neural crest phenotypes. If this assumption is correct, however, then the results of the *in vitro* experiments imply that synaptic transmission (or its absence) may be important for regulating the initial expression of specific neurotransmitter properties of embryonic neural crest cells. A variety of other experiments have also implicated both pre- and postganglionic synaptic transmission in the regulation of other properties of adrenergic ganglion cells, albeit at somewhat later stages of development (Black, 1982).

Another example in which nerve activity plays an important role in development involves the relationship between motoneurons and the differentiation of dif-

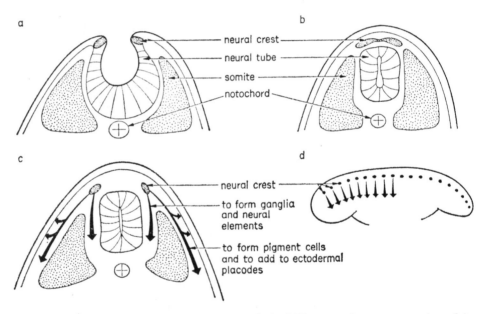

Fig. 6. The origins and interactions of neural crest cells. (a–c) Diagrammatic transverse sections of the axial tissues of a vertebrate embryo, during and after the end of neurulation. Neural crest cells (shaded) become detached from the edges of the neural tube, then form dorsolateral clusters from which individual cells migrate in the directions shown by arrows in (c) and (d), to become pigment cells, nerve ganglion cells, or other cell types. In (d) a lateral view of the embryo is shown, with the neural crest clusters segmentally arranged. (From Deuchar, 1975).

ferent types of skeletal muscle. Most adult vertebrates have two basic types of skeletal muscle, fast-twitch and slow-tonic, which differ in their morphology, biochemistry, innervation pattern, and contraction properties (Goldspink, 1980). The fast-twitch or phasic muscles are those used primarily during short bursts of rapid movements, such as when a leopard is chasing prey, whereas the slow-tonic muscles are used primarily for the maintenance of posture, as occurs during

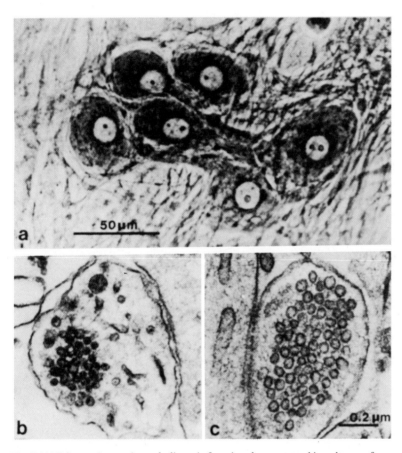

Fig. 7. The "shift" from adrenergic to cholinergic function demonstrated in cultures of neonatal autonomic neurons. The pictures are of neurons from the superior cervical ganglion of newborn rats in dissociated cell cultures with supporting cells. (a) Light micrograph of a cluster of six neurons after 1 month in culture. The neurons are enmeshed in a network of neuronal processes, Schwann cells, and fibroblasts. (b) Synaptic vesicles after 1 week in culture [as in (a)]. This electron micrograph shows an axonal terminal on a neuronal soma after incubation in 10^{-5} M norepinephrine and fixation in potassium permanganate. As is characteristic of synaptic vesicle clusters in dissociated perinatal neurons after 1 week in culture, most vesicles contain dense cores, a cytochemical index of the norepinephrine content. Magnification as in (c). (c) Synaptic vesicles after 4 weeks in cultures. Preparation as in (b). Characteristically, the synaptic vesicles now show few dense cores. Correlating with this change in vesicle morphology, the cultures show increased levels of choline acetyltransferase and an increasing incidence of cholinergic synaptic interacitons between the cultured neurons. (From Bunge *et al.*, 1978.)

amplexus of frogs or in the ability of birds to keep the wings from drooping when at rest.

There is considerable evidence that in both birds and mammals the nerves that innervate an embryonic muscle are involved in the developmental regulation of specific fast and slow muscle properties (Vrbová *et al.*, 1978). For instance, cross innervation of fast muscles with nerves that normally innervate slow muscles, when carried out early in development, alters subsequent differentiation, such that the prospective fast muscles take on many properties of slow muscles and vice versa. This influence appears to involve synaptic activity at the neuromusuclar junction, and/or muscle contraction, since modifications of such activity, even in the absence of cross-innervation, result in similar changes in muscle properties (Figure 8).

Not only does neuromuscular activity regulate the differentiation of muscle, but recently it has been reported that the survival and maintenance of the moto-neurons innervating skeletal muscle may also be regulated by such activity. During the embryonic development of vertebrates, many thousands of nerve cells begin to develop, only to die subsequently (Oppenheim, 1981; Hamburger and Oppen-heim, 1982). This naturally occurring cell loss is especially clear in the case of chick spinal motoneurons, in which 40–60% of the cells in a given population die during the early stages of embryogenesis (Figure 9). The number of motoneurons that die can be altered by increasing or decreasing the size of the peripheral targets (skel-etal muscle). Removal of a limb-bud early in development results in the death of virtually all the motoneurons in the limb-innervating population, whereas the addi-

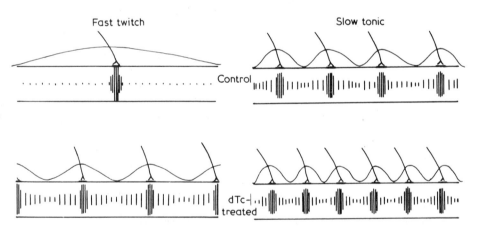

Fig. 8. A schematic representation of the distribution of ACh sensitivity of skeletal muscle fibers in relation to their innervation is shown. The density and size of the vertical bars indicate the degree of chemosensitivity. The curve above each "muscle fiber" indicates the area along which the depolarization produced on nerve activity may spread and reduce chemosensitivity; on the fast twitch muscle fiber this spread would cover most of the muscle fiber surface, and on the slow tonic muscle fiber each nerve ending produces a depolarization that spreads only over a small distance. After curare, since the end-plate potential is reduced, the area of spread of the depolarization is reduced, and chemosensitivity will decrease over a smaller area. The fast twitch muscle fibers will become multiply innervated and the distance between successive endplates on the slow tonic muscle will become reduced. (From Vrbová, *et al.*, 1978.)

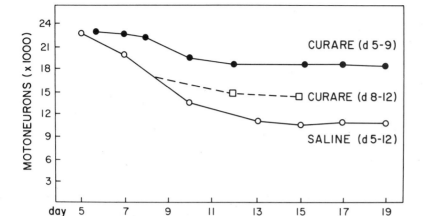

Fig. 9. Cell number in the lateral motor column of the embryonic chick lumbar spinal cord. Two mg of curare in 0.2 ml saline were administered daily from either day 5 to day 9 or day 8 to day 12. Saline (0.2 ml) was used as a control. (Data from Pittman and Oppenheim, 1979; Oppenheim, unpublished data.) The E19 value for the curare d 5–9 group is retained for at least 4 days after hatching (Oppenheim, 1982c). (From Hamburger and Oppenheim, 1982.)

tion of a supernumerary limb-bud rescues many of the motoneurons that normally would have died. These results have strongly implicated neuromuscular interactions in the regulation of motoneuron survival.

Recently, it has been shown that synaptic activity is also somehow involved in this phenomenon (Pittman and Oppenheim, 1978, 1979). Blocking neuromuscular activity with drugs such as curare during the entire period when motoneurons are normally dying results in the prevention of virtually all such cell loss (Figure 9). This result can be accomplished by either pre- or postsynaptic blockade of the neuromuscular junction. If the pharmacological blockade is stopped during embryonic development and neuromuscular activity is allowed to recover, the excess (rescued) motoneurons then degenerate. Furthermore, electrical stimulation of the nerves and limb muscles during the period of naturally occurring motoneuron death enhances or accelerates the normal cell loss (Oppenheim and Nunez, 1982). Although it remains to be seen whether other kinds of neurons respond in the same way (but see Oppenheim *et al.,* 1982), collectively, these findings indicate that, at least for motoneurons, even such a basic aspect of neurogenesis as the control of final cell number in a population may involve neural activity.

The experiments cited previously illustrate some of the ways by which neural activity, use, or experience can influence the normal development of the nervous system during the early embryonic or prenatal period. We do not wish to give the impression, however, that we believe such factors to be involved in all stages or aspects of neurobehavioral development. While it is true that the roles of neural activity are receiving increasing attention in neuroembryology (Harris, 1981), there are also many interesting cases where such activity seems to play little, if any, role. Because we shall reveiw some of this literature in detail, in the present context we only wish to mention the well-known studies of Sperry (1951), Weiss (1941), Detwiler (1936), Coghill (1929), and others in which embryological procedures were used to demonstrate the capacity of portions of the early nervous system to develop autonomously, independent of environmental influences, and despite the maladaptive nature of the resulting behavior.

An interesting example illustrating this latter point involves the morphological and functional development of spinal cord regions in the chick embryo. The histological and morphological appearance of the various levels of the adult spinal cord (cervical, thoracic, lumbar, etc.) exhibit characteristic differences. Limb-innervating regions, for instance, have large motor columns and lack autonomic preganglionic neuronal groups (Figure 10). By transplanting portions of prospective spinal cord early in development from limb to nonlimb regions (and vice versa), Wenger (1951) showed that by the time of neural tube closure, the regional-specific spinal cord differences were already determined. That is, even though the limb and nonlimb regions appear indistinguishable at the time of neural tube closure, they each exhibit characteristic developmental differences when placed in the same environment. For instance, the neural tube from the brachial region develops normally even when placed in the cervical or thoracic regions. In the reverse experiment, Wenger noted that wings innervated by transplanted cervical or thoracic cord showed spontaneous neuromuscular activity. However, his study did not attempt to analyze whether such movements were characteristic of, or different from, wings innervated by normal brachial spinal cord.

Two more recent experiments have, however, carried the functional analysis of these preparations a step further. Independently, Straznicky (1967) and Narayanan and Hambruger (1971) have shown that, in the chick, replacing brachial (wing-innervating) spinal cord with lumbo-sacral (leg) segments, or vice versa, leads to characteristic abnormalities in limb function. After hatching, wings innervated by lumbo-sacral cord behave like legs in that they perform alternating movements that are in synchrony with the alternating leg movements seen during normal walking. Similarly, legs innervated by brachial (wing) spinal cord do not exhibit alternating movements but rather move together in synchrony with the simultaneous movements (wings–flaps) of the wings.

Although there obviously are a great number of epigenetic events intervening between the time of neural tube closure and hatching, these findings demonstrate that the gradual structural and functional differentiation of these spinal cord segments are, nonetheless, irreversibly determined by 48 hr of incubation. While there appears to be little, if any, capacity for regulation of this basic structure–function pattern following translocation of gross spinal cord regions after this time, it remains to be seen whether more detailed analyses of the microanatomy, physiology, and biochemistry of the transplanted tissue will support the implications of the behavioral results. Even if this should prove to be the case, it will remain to be shown that such results are applicable to nervous tissue that subserves more complex behavioral processes (e.g., cerebral cortex) or whether the same phenomena can be demonstrated in mammals (but see Lund and Hauschka, 1976; Perlow,

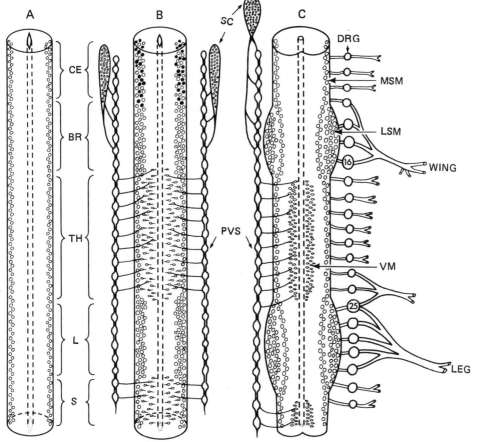

Fig. 10. Origin and migration of motoneurons in the spinal cord of the chick embryo at 3 days (A), 4.5 days (B), and 8 days (C) of incubation. The motoneurons originate uniformly at all levels from the germinal zone surrounding the central canal. Visceromotor neurons (spindle-shaped) migrate toward the central canal to form the visceromotor column (of Terni) at thoracic levels. Degenerating motoneurons at cervical levels are shown in black. The somatic motoneurons form into a medial column that supplies axial muscles and into a lateral motor column that supplies limb muscles. In (C) the spinal nerves and ganglia have been omitted on one side and the sympathetic ganglia on the other. (BR, Brachial; CE, cervical; DRG, dorsal root ganglia; L, lumbar; LSM, lateral somatic motor column; MSM, medial somatic motor column; S, sacral; PVS, sympathetic ganglia; SC, superior cervical ganglion; TH, thoracic, VM, visceromotor.) (From Jacobson, 1978.)

1980; Roberts, 1983). Nonetheless, the results are encouraging and indicate that an embryological perspective on these issues may be a valuable addition to the range of experimental approaches used by developmental psychobiologists.

IV. NEUROBEHAVIORAL DEVELOPMENT: THE ROLE OF NEURAL ACTIVITY

The examples presented in the previous section have demonstrated some of the means by which the environment external to the nervous system can influence neural development. They also indicate that functional activity *within* the nervous system may influence the ontogeny of that system. Growing neurons exhibit membrane depolarizations, propagated action potentials, synaptic transmitter release, and many other functional properties of mature neurons. One reflection of this activity within the developing nervous system is spontaneous embryonic motility—a phenomenon observed in every species in which it has been studied (Corner *et al.*, 1979; Bekoff, 1981; Hamburger, 1972; Oppenheim, 1974). In those systems that are amenable to manipulation, these embryonic movements have been shown to be, at least initially, entirely independent of sensory control (Hamburger *et al.*, 1966). It is not surprising, therefore, that this phylogenetically stable phenomenon has been conscripted frequently into the repertoire of interactions that guide normal development (see Jacob, 1977).

As noted in the previous section, however, functional activity within the developing nervous systm is not involved in all aspects of neural and behavioral development. In the experiments discussed earlier in which a pharmacological blockade of neuromuscular transmission prevented naturally occurring motoneuron death, it was shown that recovery from the blocking agent during embryonic life leads to a gradual return of normal levels of motility (and belated degeneration of rescued cells). At least some of these embryos later hatch and appear to behave relatively normally after hatching (Oppenheim *et al.*, 1978). Indeed, it has been possible to bring a few animals through hatching (at 21 days of incubation) that have been exposed continually to curare since the normal time of cell death (5–10 days of incubation). For reasons not yet understood, the *posthatch* return of movement that occurs in these animals does not result in the delayed degeneration of excess motoneurons that had been rescued from naturally occurring cell death during embryonic life. Even in the presence of excessive numbers of motoneurons, however, observations during the first 3–4 days after hatching show these chicks to perform many normal-appearing behaviors vigorously (although detailed analyses could not be carried out because of the effects of immobilization on joint and muscle development; Oppenheim, 1982c). Thus the absence of motor activity during substantial portions of embryonic development does not appear to have any significant effects on at least some aspects of subsequent behavior.

These observations in the chick are reminiscent of the results of the classic deprivation experiments performed by Harrison (1904), Carmichael (1926, 1927), and Matthews and Detwiler (1927). In each of these studies, amphibian embryos were immersed in the immobilizing agent chloretone prior to the onset of embryonic movements. When they were removed from the drug solution to pond water a number of days later, the swimming of the experimental embryos soon appeared equivalent to that of normally reared animals. The conclusion thus drawn from these frequently cited works was that function or practice was not necessary for the normal development of amphibian swimming behaivor. These classic studies have been influential in the formulation of a basic tenet of develomental neuroembryology (i.e., that during the *earliest* stages of neurobehavioral ontogeny, the nervous system develops in "forward reference" to, and without benefit from, functional activity). For a number of reasons, however (most notably the purely observational nature of behavioral scoring, with no application of quantitative measures, as well as the lack of direct studies of neuronal development), the adequacy of these works as bases for such a conclusion is questionable.

A replication of these early works has recently been performed in which the basic paradigm of the original studies was extended and many of the procedural and interpretational ambiguities of the classic studies addressed (Haverkamp, 1983). Embryos of the amphibian *Xenopus laevis* were immersed very early in development (late neural fold state) in solutions of either chloretone or lidocaine (which blocks conduction of neural impulses) or were injected with alpha-bungarotoxin (which blocks neuromuscular transmission). Approximately 36 hr later, when stage-matched, normally reared control embryos from the same spawning were swimming well, the experimental animals were removed, or had recovered, from the immobilizing drugs. Two hours after regaining motility, many of the previously immobilized embryos *appeared* to behave quite normally. However, detailed quantitative analyses performed at this time revealed significant deficits in their swimming. When behavioral quantifications were repeated approximately 24, 48, and 168 hr after recovery from the drug, however, the previously immobilized animals' swimming was indistinguishable from that of control embryos (Figure 11). Therefore preventing all neural activity during that period when the motor system is forming had no *permanent* effects upon later behavioral performance. The primary conclusions of Harrison's (1904) and the later studies were thus supported in a different species, using a number of drugs to induce immobilization, and with the application of sensitive quantitative measures.

These observations that the initial development of motor capacities can proceed normally in the absence of neural activity imply that the growth, differentiation, and connectivity of the neurons underlying these behaviors also proceeds normally when impulse activity is blocked. Nonetheless, the many demonstrations of seemingly normal behavior in the presence of distinctly abnormal neuroanatomy (e.g., Brodal, 1981; Caviness and Rakic, 1978) preclude the ready acceptance of such an assumption. Also, in the experiment just described, there was a period of time, immediately after removal of experimental embryos from the drug solutions, during which deficits in quantitative measures of behavior could be demonstrated. While the source of these transient effects may simply have been the effects of a residual drug, it might also be argued that this period of behavioral deficit was a time during which developmental events dependent upon function belatedly occurred (i.e., a period of rapid behavioral "practice").

Further experiments upon *Xenopus* embryos treated under the immobilization paradigm just outlined indicate, however, that at least certain essential components of the motor system had differentiated normally in the absence of impulse activity. Physiological recordings of activity in the motor nerves of previously immobilized animals revealed no effects of the drug treatments (Figure 12). When these recordings were quantified, no significant differences between experimental and control animals were seen, either in the period of time between successive motor nerve bursts during fictive swimming episodes or in the phase relationship of bursts on either side of the animal. Experimental and control animals did not differ in these measures even during that time after drug release when behavioral effects of immobilization could be demonstrated. In addition, anatomical studies of the dendritic arborization of the primary motoneurons of *Xenopus* embryos that had *never* been allowed to recover from the anesthetizing agents, revealed no significant effects of neural blockade on this measure of differentiation (Figure 13).

Conceptually related studies done in tissue culture by Crain and his co-workers (Crain *et al.*, 1968; Model *et al.*, 1971) also support the premise that during the earliest stages of neural development, differentiation and connectivity take place normally in the absence of impulse activity. When small pieces of embryonic neural tissue are maintained in solutions composed of certain salts and sera, cells will grow and differentiate in an apparently normal manner. Synapses will form within such

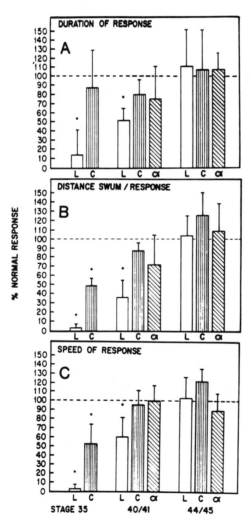

Fig. 11. Quantitative measures of swimming in response to tactile stimulation by *Xenopus laevis* embryos, immobilized by immersion in chloretone (C) or lidocaine (L) solutions from st. 17–35, or through injection of alpha-bungarotoxin (α) at st. 23. The bars (±S.D.) represent group means of measures obtained 2 hr after removal from the drug solutions (st. 35), approximately 24 hr later (st. 40/41), and approximately 48 hr after recovery from the drug (st. 44/45). Stars denote values that significantly differ from those of normally reared embryos.

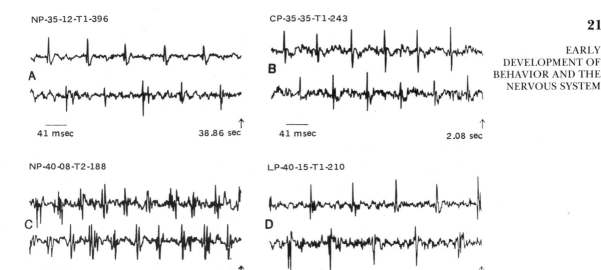

Fig. 12. Representative traces of ventral root activities recorded from *Xenopus* embryos. Each pair of traces are those recorded simultaneously from the left and right sides of an immobilized embryo. Traces (A) and (B) are from st. 35 embryos (approx. 50 hr postfertilization), traces (C) and (D) from st. 40 embryos (approx. 65 hr postfertilization). Traces (A) and (C) were obtained from normally reared embryos, while (B) and (D) were obtained from embryos reared in chloretone and lidocaine, respectively, from st. 17–35. The animal from which the traces in (B) were obtained had been removed from the anesthetic solution just 2 hr prior to recording. In each of the tracings, note the regular, alternating bursting pattern of ventral root activity on either side.

explants and, after a period of time, the tissue will evince complex electrical activities in response to local stimulation. When Crain's group maintained explants of embryonic rat cortex in solutions that contained the neural conduction blocker lidocaine, they observed that the treated tissue formed synapses that were ultrastructurally indistinguishable from those found in normally reared explants. Furthermore, these explanted tissues produced normal-appearing electrically evoked responses immediately upon withdrawal of the drug from the tissue environment.

This ability of neurons to develop structurally and functionally effective connectivity in the absence of activity has been shown in a number of different systems, using a variety of means to block neural conduction or synaptic transmissioin (e.g., Bird, 1980; Obata, 1977; Romijn *et al.*, 1981). It should be noted, however, that number of these works have demonstrated certain *quantitative* effects of such blockade (e.g., decreased numbers of synapses formed; Janka and Jones, 1982). In addition, one set of studies (Bergey *et al.*, 1981; Christian *et al.*, 1980) has shown that blocking neural activity in *dissociated* cell cultures with the potent neurotoxin, tetrodotoxin (TTX), results in the death of one of the interacting cell types contained in the culture.

A particularly elegant demonstration in the intact animal of normal differentiation and connectivity in the absence of neural activity has been performed by Harris (1980). It was first demonstrated by the embryologist Twitty (1937) that

TTX, which blocks all neural impulse propagation, is a normal constituent of the body fluids of the California newt. Though this newt's tissues are, of course, insensitive to the effects of TTX, neural tissue from other salamander species, transplanted into the California newt, is incapable of generating action potentials. Harris exploited this system by transplanting the eye primordia of the Mexican axolotl into the head region of newt embryos (Figure 14). After the animals matured, Harris found that not only had the axolotl eyes correctly projected to the optic tectum, but the spatial distribution of these projections on the tectum was also normal and the synapses formed were indistinguishable from the tectal synapses formed by the host retina. Thus in the absence of not only light-induced activity but also spontaneous discharge, retinal ganglion cells had apparently differentiated normally and projected to form connections with that portion of their target appropriate to the position of the ganglion cell within the eye. This finding is all the more remarkable considering that the transplanted eyes were in direct competition at the target site with the normal, functionally intact, host eye.

Results of a recent study by Meyer (1982) are similar to those of Harris' work yet demonstrate that while functional activity may play no role in one aspect of a system's development, it may be integral to another step in the same system's ontogeny. When the two optic nerves of a goldfish are transected and forced to reinnervate a single optic tectum, the projections from either eye initially show

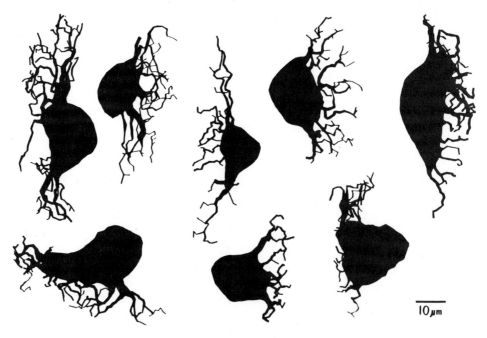

Fig. 13. Computer-generated reconstructions of the dendritic arborizations of primary motoneurons of st. 35 *Xenopus* embryos. The cells were visualized through reaction of peripherally applied horseradish peroxidase, and morphometric quantifications made by computer-assisted digitization of their processes. Depicted here are representative neurons from both normally reared embryos, as well as from embryos continuously exposed to either chloretone or lidocaine from st. 17.

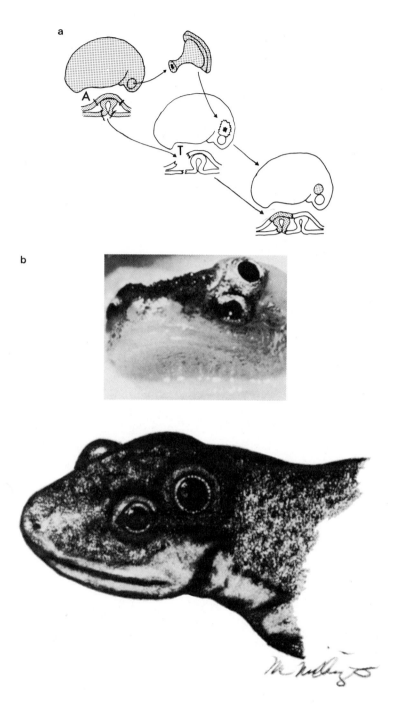

Fig. 14. (a) Method of transplanting axolotl (A) eyes to newts (T). Top: The eye cup is excised along with the entire optic stalk. Middle: Wounds are made in the host's epidermis and neuroepithelium, dorsal and posterior to the eye on the same side as the one being transplanted. Bottom: The transplant is put into place. Underneath each embryo is a schematic cross section of the operation. (b) Photograph and drawing of metamorphic newt with a third eye from a axolotl. (From Harris, 1980.)

extensive overlap at the target site. After a few weeks of development, however, the projections from the left and right eyes segregate, forming multiple bands on the tectum, each of which contain terminals from only one optic nerve. Meyer (1982) injected TTX into the eyes of goldfish treated in this manner and found that the eyes nevertheless projected to, and distributed over the surface of the tectum normally (as Harris found for the supernumerary axolotl eyes). The subsequent segregation of retinal afferent terminals on the tectal surface did not occur, however, with the TTX-induced block of neural activity in both eyes. Interestingly, when TTX was injected into only one eye, the segregation seen in control animals occurred normally.

This surgically induced column formation on the goldfish optic tectum is analogous to the formation of ocular dominance columns in the mammalian visual cortex. In this latter system there is a similar initial extensive overlap of afferent terminals driven by either eye. This overlap is later modified so that bands are formed in the cortex that show primary responses to stimulation of one or the other retina. This segregation of ocular dominance columns can be prevented, however, as in the goldfish, by bilateral intraocular injections of TTX into neonatal kittens (Stryker, 1981).

In addition to pointing out the selective effects of neural activity on neural development, these experiments on ocular dominance column formation further illustrate the nonequivalence of sensory deprivation and total blockade of neural activity. While binocular deprivation (lid suturing, dark rearing) may retard the development of the ocular dominance columns in the cortex, such treatment is not effective in totally preventing their formation in the manner that TTX injections are. To the same point, while monocular visual deprivation has distinct effects on retinal projections to the lateral geniculate nucleus, Archer *et al.* (1982) have shown that monocular TTX injections alter retino-geniculate projections in a manner totally different from that which results from sensory deprivation.

V. BEHAVIORAL EMBRYOLOGY: PRACTICE AND PERCEPTUAL EXPERIENCE

In the preceding examples the overlap of concern of developmental psychology and of developmental biology is apparent. The role that environment plays in behavioral development is readily seen as one aspect of the study of the modes and effects by which cellular and tissue interactions bring about development of the embryonic nervous system. Equally obvious is the fact that an understanding of behavioral and biological development cannot be gained by confining our investigations to a period after an often arbitrary (albeit significant) developmental landmark such as birth or hatching (or metamorphosis in invertebrates and amphibians).

Locomotor development in a number of species further illustrates this point. The ability of butterflies and moths to fly soon after emergence from their pupal cases is a striking event, so much so that it has been used as a paradigm case of the lack of necessity for practice in the performance of complex motor patterns (see, e.g., Hinde, 1970). Recordings from the motor nerves of moth pupae reveal, however, that a primitive form of the flight motor pattern is apparent long before emergence, and it is gradually refined to adult form while the animal is still encased in the pupal sheath (Kammer and Kinnamon, 1979). The formation and refinement of intra- and interlimb coordination are also apparent in the chick embryo during early embryonic development (Bekoff, 1981). In the frog, patterned activity in the ventral roots that innervate the hindlimbs is first apparent during those stages of development during which the limbs are merely buds of undifferentiated mesenchyme (Stehouwer and Farel, 1983). Thus even such a basic behavior as locomotion has an embryonic history that must be taken into account for any consideration of behavioral development. The question of what role this precocious function plays in development of mature movements is unanswered, however. The demonstrations of normal development of swimming behavior in immobilized tadpoles, discussed earlier, while compelling, do not necessarily imply that more complex forms of motor activity are similarly independent of function during ontogeny (but see Oppenheim, 1982a).

A particularly striking example of prenatal experience that clearly *does* affect later behavior has been elucidated by G. Gottlieb (1980a,b, 1982, 1983). Gottlieb has demonstrated that the embryo of the domestic mallard duck is specifically responsive to the mallard maternal call prior to any auditory experience. In normal ducklings this responsiveness is largely directed toward the repetition rate of the call, and the specificity of the response is maintained through hatching. If, however, duckling embryos are devocalized and incubated in isolation, this selective responsiveness is lost and the ducklings will respond equally well to calls of a number of other species (performed at different species-specific rates). The experiental variable missing from the environment of these experimental animals appears to be the perception of their own, or sib's, volcalizations while *in ovo*. If the devocalized embryos are exposed to recordings of their own "contact-contentment" calls, selective responsiveness to species-specific maternal vocalizations remains intact in the posthatch ducklings.

Surprisingly, when Gottlieb performed a finer analysis of the perceptual development of these embryos reared in auditory isolation, he found that a 48-hour "consolidation" period was necessary for the ducklings to exhibit normal responsiveness to recordings of maternal vocalizations. When tested 24 hr after exposure to contact calls, the experimental animals did not show the selectivity of response that could be demonstrated 24 hr later. Gottlieb noted that in the natural environment, the ducklings and embryos hear not only contact-contentment calls (normally produced at a repetition rate of 4.0 notes/sec), but a wide range of other calls from sibs and self as well, which vary in repetition rates from 2.1 to 5.8 notes/sec. A group of devocalized, isolated embryos was therefore exposed to the contact call replayed at three repetition rates: of 2.1, 4.0, and 5.8 notes/sec. Unlike experimental embryos exposed only to the contact call replayed at its normal rate, embryos exposed to all three rates of the call did not require a 48-hr consolidation period, but would respond to the correct call in choice tests given 24 hr after stimulation.

While exposure to a single type of duckling call is a sufficient stimulus, therefore, for the young birds to later exhibit a preference for the maternal call, exposure to the normally encountered range of repetition rates at which different duckling vocalizations are normally produced is necessary for this preference to be continuously maintained. In further experiments, Gottlieb has shown that the frequency modulation of the contact call is also required for later preference for the repetition rate of the maternal call to be exhibited; exposure of the ducklings to white noise pulsed at the correct rate is an ineffective stimulus.

As Gottlieb (1982) points out, these often subtle and "nonobvious" requirements for normal perceptual development in embryonic ducklings are quite reminiscent of the requirements in cellular differentiation for interaction with the normally occurring *tissue* environment. In the same manner that the particular stimulus and specific response may differ between species in this type of cellular interaction with the environment, behavioral, and perceptual developmental interactions with the environment also show species variabilities.

A variation on this theme of prenatal auditory experience affecting postnatal perception as demonstrated in the mallard duck is seen in a related species, the wood duck. Embryos of this species also develop a selective responsiveness to a component of the maternal call, in this case a descending frequency modulation of the notes. As with the mallard duckling, specific auditory experience is required to maintain this preference. In the wood duck, however, this requirement is for exposure to the duckling's alarm call (rather than the contact call). A further difference between the two species is that while self-stimulation is sufficient to maintain perceptual preference in the mallard duckling, the young of the wood duck must be exposed to the calls of siblings to maintain a preference.

These demonstrations of the role of sensory experience in the development of ducklings bring to mind the many investigations performed on the development of adult song in various other avians. These works have shown song development in passarines to be dependent upon the complex interactions of genetic, hormonal, and experiential factors. The degrees to which vocal development can be regulated by the manipulation of each of these factors likely describes a continuum of control, but may be grouped into three broad categories: (1) no auditory feedback is required for normal song development, (2) auditory feedback (self-stimulation) is necessary to normal song development but experience in hearing conspecifics is not, and (3) exposure to conspecific song is necessary to normal development of adult song (Nottebohm, 1968).

As the anatomical, physiological, and pharmacological bases of these differences are further revealed, this system may provide valuable insights into the neural mechanisms underlying a variety of types of behavioral polymorphism and developmental strategies (see Chapter 4). The recent extension of neurobiological methods to the study of these systems (see, e.g., Nottebohm, 1980) does not imply, of course, that further behavioral analyses will not continue to be of great value.

While behavioral investigations will continue to suggest courses of study at the suborganismal level of analysis, so will these latter works undoubtedly contribute to defining routes for further behavioral studies.

This impetus to behavioral investigation and interpretation may be either indirect (i.e., by analogy to interactions at the cellular level) or direct (i.e., a testing or proposal of behavioral mechanisms from the demonstrated function of neurons and their assemblies). A number of examples of both types of crossover have been noted in the previous papers. A more recent example, relating to a system just discussed, is the epigenetic narrowing of the vocal repertoire of the male swamp sparrow. Marler and Peters (1982) have demonstrated that, prior to these birds' adult performance of a stabilized song, they exhibit a phase during which their vocalization contains up to five times as many song types as in the mature state. During the intervening period the components of adult song are derived from this enlarged repertoire by an actively selective process. Marler and Peters (1928) draw the apt analogy of this developmental restriction of song types to the phenomenon of neuronal cell death and related regressive phenomena during development, citing the initial overproliferation of units in each, followed by their subsequent attrition.

Another, more detailed, analysis that relates behavioral to physiological data has been published recently by Bischof (1983). Two of the best studied phenomena in behavioral development and developmental neurobiology are, respectively, imprinting and the development of the mammalian visual cortex. Athough the similarities in the two systems are too extensive to relate here (see, e.g., Figure 15), it is for this very reason (i.e., extensive descriptive coincidence) that the two phenomena have often been linked, and it has even been suggested that the same sort of neural mechanisms are operating in each (Rose, 1981). While we do not believe

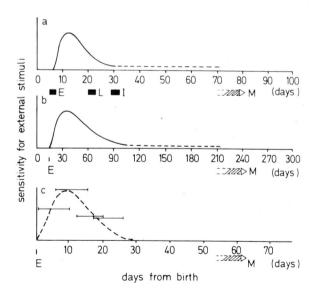

Fig. 15. Time course of the efficiency of external stimulation in three different paradigms: (a) Sexual imprinting in the zebra finch (Immelmann, 1972), (b) Ocular dominance in the cat (Blakemore, 1980), and (c) Sexual imprinting in the Japanese quail (Gallagher, 1967). Ordinate scale is arbitrary. (E, eye opening, L, leaving the nest, I, independent, M, sexual maturity.) (From Bischof, 1983.)

that such linkage is necessarily indicated by the data from each field, this type of exercise in contrast and comparison is exactly the sort of insight into the methods and findings of separate fields that is too often lacking in many research approaches. Not only can the works of developmental psychologists and psychobiologists be extended through such comparisons of data, but that of neurobiologists and embryologists can profit as well. Through his comparative analysis of imprinting and visual cortical plasticity, Bischof (1983) concludes that a similarity of mechanisms underlying the two systems is unlikely but that each may reflect the expression of a common developmental process. He goes on to propose some characterizations of this process (i.e., its self-limiting nature and its basis in the alteration of neural development). Whether such characterizations are justified or will hold up to experimentation is irrelevant to this illustration of the new insight and predictive value that such comparative consideration of data from different levels of analysis imparts to each of the fields.

These last few examples have extended the range of discussion from the prenatal to the postnatal period, yet continue to emphasize, it is hoped, the value of relating the findings of different levels of analyses. Investigations of bird song development, mentioned earlier, are revealing a system that illustrates the interacting contributions of the genome (sexual genotype), hormones (presence of testosterone), and experience (the role of practice and learning) in neural and behavioral development.

Another fascinating example of the interplay of sexual genotype and hormonal influence is to be found in mammalian development. Hauser and Gandelman (1983) have each presented evidence that during intrauterine development, the sex of adjacent embryos markedly affects adult behavior of rats. To cite one portion of these data, male pups that develop between two female litter mates were significantly more sexually active and less aggressive as adults than were male pups that gestated between two other males. This effect is presumably due to increased levels of estrogens available to the female-contiguous male pups. In the same manner, prenatal androgen levels are held responsible for the altered avoidance responding of adult female rats that have developed between two males *in utero*.

As a final example of such prenatal influences on later behavior, we shall mention a number of studies performed on nipple attachment and nursing in newborn mammals. Two of the sensory systems that are most saliently associated with this behavior (olfaction and taste) become functional during prenatal life. A response to substances injected into the amniotic fluid that presumably stimulate receptors of both types can be perceived prior to birth (Bradley and Mistretta, 1973). When such substances are paired with negative simuli *in utero*, postnatal exposure to that odor results in an avoidance of the odiferous source postnatally (Smotherman, 1982). When exposure to such substances is not paired with negative stimuli *in utero*, and exposure is extended to the period immediately following birth, the infant pup will perferentially attach to a nipple that is coated with the same substance (Figure 16). The significance in the natural state of this fetal experiential ability and its postnatal retention is hinted at by the observation that the mother rat, at birth, may have significant oral contact with amniotic fluid and her own nipples. When the nipple is washed, her pups fail to attach to the nipple and do not nurse (Pedersen and Blass, 1982).

These examples have clearly shown, then, the fallacy in assuming that a prior behavior is innate and independent of experiential effect simply because initial perceptual responses or behavioral performances are performed soon after birth. It should be obvious that investigations of behavioral development cannot, therefore, be confined to the influences of obvious postnatal experience and easily recognizable performance. The examples given earlier have pointed out not only the necessity of extending manipulation and analysis of behavioral development to the prenatal period, but also the advantages of viewing neural and behavioral development from an embryological perspective.

Modifications of the nervous system, which are, of course, the bases of behavioral change, may be either transient or resistant to further alterations. Whatever the time course or persistence of these changes, they are always the result of activity within cells and due to the interactions of the neurons with both the intra- and extraorganismal environment. Decades of study in developmental biology have led to an increased understanding of many of the epigenetic interactions that influence ontogeny, including the one class of those interactions reviewed here, the roles of neural activity, use, and sensory input on behavioral development. We hope, therefore, that through reflection on the experimental examples outlined previously, the reader may come to share our belief that the development of behavior is an inextricable part of developmental biology and may be fully understood only when viewed within that framework. To attempt a complete understanding of a system's

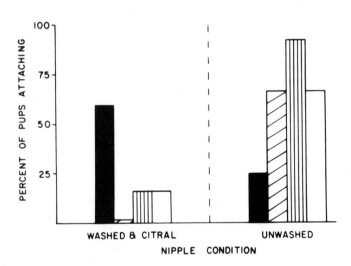

Fig. 16. Percent of pups experiencing citral pre- and postnatally (dark bar), citral prenatally (hatched), or citral postnatally (striped), or no citral (open bar) and attaching to nipples washed and scented with citral (left) or unwashed (right). (From Pedersen and Blass, 1982.)

functions, it is indeed necessary at times to "begin our query at the beginning of growth."

References

Archer, S. M., Dubin, M. W., and Stark, L. A. Abnormal development of kitten retino-geniculate connectivity in the absence of action potentials. *Science,* 1982, *217,* 743–745.

Bateson, P. P. G. How does behavior develop? In P. P. G. Bateson and P. Klopfer (Eds.), *Perspectives in ethology.* vol. 3. Cambridge: Cambridge University Press, 1978.

Barlow, G. Genetics and development of behavior, with special reference to patterned motor output. In K. Immelmann, G. W. Barlow, L. Petrinovich, and M. Main (Eds.), *Behavioral development.* Cambridge: University of Cambridge Press, 1981.

Bekoff, A. Embryonic development of the neural circuitry underlying motor coordination. In W. M. Cowan (Ed.), *Studies in developmental neurobiology: Essays in honor of Viktor Hamburger.* New York: Oxford University Press, 1981.

Bergey, G. K., Fitzgerald, S. C., Schrier, B. K., and Nelson, P. G. Neuronal maturation in mammalian cell culture is dependent on spontaneous electric activity. *Brain Research,* 1981, *207,* 49–58.

Bird, M. M. The morphology of synaptic profiles in explants of foetal and neonatal mouse cerebral cortex maintained in a magnesium-enriched environment. *Cell and Tissue Research,* 1980, *206,* 115–122.

Bischof, H.-J. Imprinting and cortical plasticity: A comparative review. *Neuroscience and Biobehavioral: Reviews,* 1983, *7,* 213–225.

Black, I. Stages of neurotransmitter development in autonomic neurons. *Science,* 1982, *215,* 1198–1204.

Bradley, R. M., and Mistretta, C. M. Fetal sensory receptors. *Physiological Review,* 1973, *55,* 352–381.

Brodal. A. *Neurological anatomy in relation to clinical medicine,* 3rd ed. New York: Oxford University Press, 1981.

Bunge, R., Johnson, M., and Ross, C. D. Nature and nurture in development of the autonomic neuron. *Science,* 1978, *199,* 1409–1416.

Carmichael, L. The development of behavior in vertebrates experimentally removed from the influence of external stimulation. *Psychological Review,* 1926, *33,* 51–58.

Carmichael, L. A further study of the development of behavior in vertebrates experimentally removed from the influence of external stimulation. *Psychological Review,* 1927, *34,* 34–47.

Carmichael, L. The onset and early development of behavior. In L. Carmichael (Ed.), *Manual of child psychology.* New York: Wiley, 1946.

Caviness, V. S., Jr., and Rakic, P. Mechanisms of cortical development: A view from mutations in mice. *Annual Review of Neuroscience,* 1978, *1,* 297–326.

Christian, C. N., Bergey, G. K., Daniels, M. P., and Nelson, P. G. Cell interactions in nerve and muscle cell cultures. *Journal of Experimental Biology,* 1980, *89,* 85–101.

Coghill, G. E. *Anatomy and the problem of behavior.* Cambridge: Cambridge University Press, 1929.

Corner, M. A., Bour, H. L., and Mirmiran, M. Development of spontaneous motility and its physiological interpretation in the rat, chick, and frog. In E. Meisami and M. A. B. Brazier (Eds.), *Neural growth and differentiation.* New York: Raven Press, 1979.

Crain, S. M., Bornstein, M. B., and Peterson, E. R. Maturation of cultured embryonic CNS tissues during chronic exposure to agents which prevent bioelectric activity. *Brain Research,* 1968, *8,* 363–372.

Detwiler, S. R. *Neuroembryology: An experimental study.* New York: Macmillan, 1936.

Deucher, E. *Cellular interactions in animal development.* London: Chapman and Hall, 1975.

Erzurumlu, R. S., and Killackey, H. P. Critical and sensitive periods in neurobiology. *Current Topics in Developmental Biology,* 1982, *17,* 207–240.

Gesell, A. *The embryology of behavior.* New York: Harper, 1945.

Goldspink, D. F. *Development and specialization of skeletal muscle.* Cambridge: Cambridge University Press, 1980.

Gottlieb, G. The role of experience in the development of behavior and the nervous system. In G. Gottlieb (Ed.), *Neural and behavioral specificity.* New York: Academic Press, 1976.

Gottlieb, G. Development of species identification in ducklings: VI. Specific embryonic experience required to maintain species-typical perception in peking ducklings. *Journal of Comparative and Physiological Psychology*, 1980a, *94*, 579–587.

Gottlieb, G. Development of species identification in ducklings: VII. Highly specific early experience fosters species-specific perception in wood ducklings. *Journal of Comparative Physiological Psychology*, 1980b, *94*, 1019–1027.

Gottlieb, G. Development of species identification in ducklings: IX. The necessity of experiencing normal variations in embryonic auditory stimulation. *Developmental Psychobiology*, 1982, *15*, 507–517.

Gottlieb, B. Development of species identification in ducklings: X. Perceptual specificity in the wood duck embryo requires sib stimulation for maintenance. *Developmental Psychobiology*, 1983, *16*, 323–334.

Gurdon, J. B. *The control of gene expression in animal development*. Cambridge, Mass.: Harvard University Press, 1974.

Hadorn, E. *Experimental studies of amphibian development*. New York: Springer Verlag, 1974.

Hamburg, M. *Theories of Differentiation*. New York: American Elsvier, 1971.

Hamburger, V., and Oppenheim, R. W. Naturally-occurring neuronal death in vertebrates. *Neuroscience Commentaries*, 1982, *1*, 39–55.

Hamburger, V. Anatomical and physiological basis of embryonic motility in birds and mammals. In G. Gottlieb (Ed.), *Studies in the development of behavior and the nervous system*. Vol. 1. *Behavioral embryology*, New York: Academic Press, 1973.

Hamburger, V., Wenger, E., and Oppenheim, R. Motility in the chick embryo in the absence of sensory input. *Journal of Experimental Zoology*, 1966, *162*, 133–160.

Harris, W. A. Neural activity and development. *Annual Review of Physiology*, 1981, *43*, 689–710.

Harris, W. A. The effects of eliminating impulse activity on the development of retinotectal projections in salamanders. *Journal of Comparative Neurology*, 1980, *194*, 303–317.

Harrison, R. G. An experimental study of the relation of the nervous system to the developing musculature in the embryo of the frog. *American Journal of Anatomy*, 1904, *3*, 197–220.

Hauser, H., and Gandelman, R. Contiguity to males *in utero* affects avoidance responding in adult female mice. *Science*, 1983, *220*, 437–438.

Haverkamp, L. J. Neurobehavioral development with blockade of neural function in embryos of *Xenopus laevis*, Ph.D. dissertation, University of North Carolina, Chapel Hill, 1983.

Hebb, D. O. *Organization of behavior*. New York: Wiley, 1949.

Hebb, D. O. *Essay on mind*. Hillsdale, N.J.: Erlbaum, 1980.

Hinde, R. A. *Animal behavior: A synthesis of ethology and comparative psychology*. New York: McGraw-Hill, 1970.

Jacob, F. Evolution and tinkering, *Science*, 1977, *196*, 1161–1166.

Jacobson, M. *Developmental Neurobiology*. New York: Plenum Press, 1978.

Janka, Z., and Jones, D. G. Junctions in rat neocortical explants cultured in TTX-, GABA-, and MG^{++}-environments. *Brain Research Bulletin*, 1982, *8*, 273–278.

Kammer, A. E., and Kinnamon, S. C. Maturation of the flight motor pattern without movement in *Manduca sexta*. *Journal of Comparative Physiology*, 1979, *130*, 29–37.

Lashley, K. Experimental analysis of instinctive behavior. *Psychological Review*, 1938, *45*, 445–471.

Le Douarin, N. Migration and differentiation of neural crest cells. *Current Topics in Developmental Biology*, 1980, *16*, 32–85.

Lehrman, D. S. A critique of Konrad Lorenz's theory of instinctive behavior. *Quarterly Review of Biology*, 1953, *28*, 337–363.

Lorenz, K. *Evolution and modification of behavior*. Chicago: University of Chicago Press, 1965.

Lund, R. D., and Hauschka, S. D. Transplanted neural tissue develops connections with host strain. *Science*, 1976, *193*, 582–584.

Marler, P., and Peters, S. Developmental overproduction and selective attrition: New processes in the epigenesis of birdsong. *Developmental Psychobiology*, 1982, *15*, 369–378.

Matthews, S. A., and Detwiler, S. R. The reaction of *Amblystoma* embryos following prolonged treatment with chloretone. *Journal of Experimental Zoology*, 1926, *45*, 279–292.

McGraw, M. *Growth: A study of Johnny and Jimmy*. New York: Appleton, 1935.

Meyer, R. L. Tetrodotoxin blocks the formation of ocular dominance columns in goldfish. *Science*, 1982, *218*, 589–591.

Mistretta, C. M., and Bradley, R. M. Effects of early sensory experience on brain and behavioral development. In G. Gottlieb (Ed.), *Early influences*. New York: Academic Press, 1978.

Model, P. G., Bornstein, M. B., Crain, S. M., and Pappas, G. D. An electron microscopic study of the development of synapses in cultured fetal mouse cerebrum continuously exposed to xylocaine. *Journal of Cell Biology*, 1971, *49*, 362–371.

Narayanan, C. H., and Hambruger, V. Motility in chick embryos with substitution of lumbosacral by brachial and brachial by lumbosacral spinal cord segments. *Journal of Experimental Zoology*, 1971, *178*, 415–432.

Nottebohm, F. Auditory experience and song development in the chaffinch, *Fringilla coelebs. Ibis*, 1968, *110*, 549–568.

Nottebohm, F. Brain pathways for vocal learning in birds: A review of the first 10 years. *Progress in Psychobiology and Physiological Psychology*, 1980, *9*, 85–124.

Obata, K. Development of neuromuscular transmission in culture with a variety of neurons and in the presence of cholinergic substances and tetrodotoxin. *Brain Research*, 1977, *119*, 141–153.

Oppenheim, R. W. The ontogeny of behavior in the chick embryo. In D. S. Lehrman, R. A. Hinde, E. Shaw, and J. Rosenblatt (Eds.), *Advances in the study of behavior*. Vol. 5. New York: Academic Press, 1974.

Oppenheim, R. W. Neuronal cell death and some related regressive phenomena during neurogenesis: A selective historical review and progress report. In W. M. Cowan (Ed.), *Studies in developmental neurobiology: Essays in honor of Viktor Hamburger*. New York: Oxford University Press, 1981.

Oppenheim, R. W. The neuroembryology of behavior: Progress, problems, perspectives. *Current Topics in Developmental Biology*, 1982a, *17*, 257–309.

Oppenheim, R. W. Preformation and epigenesis in the origins of the nervous system and behavior. In P. P. G. Bateson and P. Klopfer. *Perspectives in ethology*. vol. 5. New York: Plenum Press, 1982b.

Oppenheim, R. W. Cell death of motoneurons in the chick embryo spinal cord: VIII. Motoneurons prevented from dying in the embryo persist after hatching. *Dev. Biol.*, 1982c, *101*, 35–39.

Oppenheim, R. W., Maderdrut, J. L., and Wells, D. Reduction of naturally-occurring cell death in the thoraco-lumbar preganglionic cell column of the chick embryo by nerve growth factor and hemicholinium-3. *Developmental Brain Research*, 1982, *3*, 134–139.

Oppenheim, R. W., and Nunez, R. Electrical stimulation of hindlimb increases neuronal cell death in chick embryo. *Nature*, 1982, *295*, 57–59.

Oppenheim, R. W., Pittman, R., Gray, M., and Maderdrut, J. L. Embryonic behavior, hatching and neuromuscular development in the chick following a transient reduction of spontaneous motility and sensory input by neuromusucular blocking agents. *Journal of Comparative Neurology*, 1978, *179*, 619–640.

Patterson, P. H. Environmental determination of autonomic neurotransmitter functions. *Annual Review of Neuroscience*, 1978, *1*, 1–17.

Perlow, M. J. Functional brain transplants. *Peptides*, 1980, *1*, 101–110.

Pedersen, P. E., and Blass, E. M. Prenatal and postnatal determinants of the 1st suckling episode in albino rats. *Developmental Psychobiology*, 1982, *15*, 349–355.

Piaget, J. and Inhelder, B. *The psychology of the child*. London: Routledge and Kegan, 1969.

Pittman, R., and Oppenheim, R. W. Neuromuscular blockade increases motoneuron survival during normal cell death in the chick embryo. *Nature*, 1978, *271*, 364–366.

Pittman, R., and Oppenheim, R. W. Cell death of motoneurons in chick embryo spinal cord: IV. Evidence that a functional neuromuscular interaction is involved in the regulation of naturally-occurring cell death and the stabilization of synapses. *Journal of Comparative Neurology*, 1979, *187*, 425–446.

Roberts, L. Brain grafting: Surgery reduces neurological damage. *Bioscience*, 1983, *33*, 80–83.

Romijn, H. J., Mud, M. T., Habets, A. M. M. C., and Wolters, P. S. A quantitative electron microscopic study of synapse formation in dissociated fetal rat cerebral cortex *in vitro. Developmental Brain Research* 1981, *1*, 59–605.

Rose. S. P. R. From causation to translations: What biochemists can contribute to the study of behaviour. In P. O. G. Bateson and P. H. Klopfer (Eds.), *Perspectives in ethology. IV. Advantages of diversity*, New York: Plenum Press, 1981.

Roux, W. Contributions to the developmental mechanics of the embryo (1888). In B. H. Willier and J. Oppenheimer (Eds.), *Foundations of experimental embryology*. Englewood Cliffs, N.J.: Prentice-Hall, 1967.

Scarr-Salapatek, S. An evolutionary perspective on infant intelligence: Species patterns and individual variations. In M. Lewis (Ed.), *Origins of intelligence*. New York: Plenum Press, 1976.

Smotherman, W. P. Odor aversion learning by the rat fetus. *Physiology Behavior*, 1982, *29*, 769–771.

Spemann, H. *Embryonic development and induction*. New Haven: Yale University Press, 1938.

Sperry, R. W. Mechanisms of neural maturation. In S. S. Stevens (Ed.), *Handbook of experimental psychology.* New York: Wiley, 1951.

Stehouwer, D. J., and Farel, P. B. Development of hindlimb locomotor activity in the bullfrog *(Rana catesbeiana)* studied *in vitro. Science,* 1983, *219,* 516–518.

Straznicky, K. Function of heterotopic spinal cord segments investigated in the chick. *Acta Biologica Hungarium,* 1967, *14,* 145–155.

Stryker, M. P. Late segregation of geniculate afferents to the cat's visual cortex after recovery from binocular impulse blockade. *Society for Neuroscience Abstracts,* 1981, *7,* 842.

Twitty, V. C. Experiments on the phenomenon of paralysis produced by a toxin occurring in *Triturus* embryos. *Journal of Experimental Zoology,* 1937, *76,* 67–104.

von Saal, F. S., Grant, W. M., McMullen, C. W., and Laves, K. S. High fetal estrogen concentrations: Correlation with increased adult sexual activity and decreased aggression in male mice. *Science,* 1983, *220,* 1306–1309.

Vrbová G., Gordon, T., and Jones, R. *Nerve–muscle interaction.* Chapman and Hall: London, 1978.

Walicke, P. A., Campenot, R. B., and Patterson, P. H. Determination of neurotransmitter function by neuronal activity. *Proceedings National Academy of Sciences USA,* 1977, *74,* 5767–5771.

Weiss, P. Self-differentiation of the basic patterns of coordination. *Comparative Psychology Monographs,* 1941, *17,* 1–96.

Wenger, B. S. Determination of structural patterns in the spinal cord of the chick embryo studied by transplantation between brachial and adjacent levels. *Journals of Experimental Zoology,* 1951, *116,* 123–146.

Weston, J. A. Neural crest cell development. In M. Burger and R. Weber (Eds.), *Embryonic development. Part B, cellular aspects.* New York: Alan Liss, 1982.

Whitman, C. O. Evolution and epigenesis. *Woods Hole Biological Lectures,* 1894, No. 10, 203–224.

Wilson, E. B. *The cell in development and heredity,* 2nd ed. New York: Macmillan, 1900.

Motor Patterns in Development

JOHN C. FENTRESS AND PETER J. MCLEOD

I. INTRODUCTION

It is through the production of integrated sequences of movement that animals express rules by which they interact with, and adapt to, their physical, biological, and social environments. The diversity of these motor patterns in the behavior of animals can provide a valuable assay of processes of developmental organization. Our aim in this chapter is to examine motor patterns in development at different levels of organization and from several complementary perspectives.

A. PRELIMINARY ISSUES

The first issue we address concerns categorizations of motor patterns. To create meaningful taxonomies it is necessary to fractionate processes of movement into basic actions and determine how these actions and their underlying dimensions are combined. Within a developmental context it is important to formulate these taxonomies from explicitly defined criteria that capture both the dynamic and relational properties of observed movement sequences. To illustrate this we have placed special emphasis upon the theme of pattern formation in integrated movement.

With appropriate descriptive taxonomies a second issue fundamental to motor patterns in development, the balance between eliciting factors and tendencies toward spontaneous expression, can be examined. This issue has been one of con-

JOHN C. FENTRESS   Departments of Psychology and Biology, Dalhousie University, Halifax, Nova Scotia, Canada B3H 4J1.   PETER J. MCLEOD   Department of Psychology, Dalhousie University, Halifax, Nova Scotia, Canada B3H 4J1.

siderable historical debate in ethology, psychology, and neuroscience (see e.g., Bekoff, 1981; Carmichael, 1970; Hamburger, 1977; Hoyle, 1984; Lorenz, 1981; Oppenheim, 1981). As analyses become refined, a third issue arises concerning interactions among different classes of action, both in terms of moment-to-moment expression and over a developmental time frame (see e.g., Bateson, 1981; Fentress, 1984; Hinde, 1982; Moran *et al.*, 1983).

We examine these issues primarily from an ethological perspective. This perspective emphasizes the importance of accurate descriptions in attempts to understand the causal antecedents and functional consequences of behavior (see, e.g., Tinbergen, 1963). Ethologists have also been concerned with problems of the hierarchical *expression* of behavior (Dawkins, 1976; Fentress, 1983a,b; Gallistel, 1980). In keeping with this perspective, we seek themes of motor patterns in development at several levels of organization. We also hope to provide the reader with some sense of the diversity of motor patterns in animal behavior, as only in this way can meaningful generalizations, and their limitations, be brought into clear focus.

B. Approach Taken

In addition to the preceding considerations, we have found it useful to form our review of the literature around two more abstract polarities of motor organization: (1) the relative stability versus change within underlying dimensions and (2) the degree of separation (discontinuity) versus cohesion (continuity) among these dimensions. Each of these conceptual polarities can be applied to different time frames and levels of organization (Fig. 2-1). By placing appropriate emphasis upon the problem of dynamic relations among movement dimensions, rigid "unit" concepts of motor patterns in development can be avoided, thereby clarifying the operation of underlying processes (Fentress, 1984).

Following a review of these issues (Section 2), we summarize relevant data from our own work (Section 3). This is, in turn, followed by a more in-depth review of developmental processes in motor behavior (Section 4).

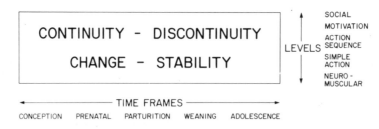

Fig. 1. Summary of basic organizational polarities relevant to motor patterns in development. Continuity–discontinuity refers to the degree of separation among behavioral phenomena and their underlying processes (mechanisms). Change–stability refers to the extent that properties and processes of behavioral organization remain fixed in time. Each of these two polarities can be examined over a number of levels of organization as well as across complementary time frames. (Adapted from Fentress, 1984.)

A. Basic Dimensions of Movement

1. Within and Between Limb Coordination. Based upon her observations of chick motor development and a review of related studies, Bekoff (1976, 1981) suggests as a general principle that intrajoint coordination tends to precede interjoint coordination, which, in turn, precedes interlimb coordination (with homologous limbs preceding homolateral limbs). Further, there is evidence for cephalocaudal and proximodistal progressions in movement. While there may be important species differences, and differences among functionally defined classes of behavior within species (see, e.g., Anokhin, 1964; Carmichael, 1970), similar principles of organization appear to be widespread in the animal kingdom. To cite an invertebrate correlate to Bekoff's chick studies, Bentley and Hoy (1970) demonstrated that interlimb coordination between forewings and hindwings in crickets develops after coordination of muscles within each wing (cf. Altman, 1975; Kutsch, 1971).

In an early study of motor development in fetal rats, Angulo y González (1932) noted that forelimb movements and trunk movements tended to occur together, as did tongue protrusions and movements of the head. Bekoff *et al.* (1975) found that in 7-day chick embryos alternation of muscle antagonists is imprecise, with partial overlap in some phases. In a study of swimming behavior in postnatal rats, Bekoff and Trainer (1979) report that early independence of limb movements is followed by a progressive coupling within limb pairs. Once the hindlimbs become well coordinated, the forelegs cease paddling movements (cf. Fentress, 1972; 1978).

Bernstein (1967) argued that in the development of human motor patterns there is a trend toward a reduction in the degrees of freedom among movement dimensions. The overall movement endpoints remain more or less constant through increased compensatory responses. A recent study of human marksmanship by Arutyunyun *et al.* (1981) suggests that during learning of this skill, movements of the wrist and shoulder joints become mutually constrained (cf. Kelso and Tuller, 1983). As shown by Johnston (1980) even young pigtail macaques *(Macaca nemestrina)* are able to compensate for variations in the rigidity of the environmental substrate by varying joint angles. In this way they can maintain an invariance in jumping trajectory.

These references highlight two important issues. First, certain properties of movement may become more finely separated from one another during development, while other movement parameters become mutually constrained. Second, once cohesions among movement dimensions are established, compensatory variation among these dimensions may preserve the functional integrity of the movement as a whole.

2. Skilled Actions and Locomotion. Bruner (1974) has argued that for human motor development, decompositions and recompositions of individually defined actions are commonplace (cf. Bower, 1979). Collard and Povel (1982) pro-

pose that in the development of a variety of skilled actions, rules for overall spatiotemporal structure preserve the integrity of movement when perturbations are applied (cf. Weismer and Fennell, 1983). A study by von Hofsten (1979) on visually guided reaching in human infants provides evidence that the development of postural compensatory mechanisms is an important part of skill acquisition. Such data suggest hierarchically organized motor performance (Connolly, 1977). Moss and Hogg (1983) found that for infants between 12 and 18 months, *increased* variability in movement subunits accompanied improvement in the proficiency of an acquired motor task. This places restrictions on any "brick-by-brick" notions of motor development. As Evarts (1975) and others have noted, relations among abstracted parts of movement provide the important keys for understanding motor patterning.

Such studies in normal developmental contexts can also provide information relevant to a variety of motor disorders. For example, Sherwood (unpublished manuscript) has found that clumsy children tend to recruit fewer motor components in response to experimentally applied perturbations than do normal children of the same age. It is interesting that adults employ an intermediate number of motor components in compensating for the same applied perturbations. This illustrates that it is not just individual components (or dimensions) of movement that we need to examine, but also the rules by which coherent groupings of action are produced.

An increasing number of investigators have looked for developmental changes in movement properties (dimensions) that may transcend motor "acts." Golani (1981) speaks of "attractors" in development to emphasize the changing nodes of relative invariance within which other properties of motor organization vary. In this way both change and stability can be examined together in relative terms.

In her classic study of motor development in macaques, Hines (1942) emphasized that "regression" along certain movement dimensions is prerequisite to "progression" along others. Early fetal reflexes were characterized by briskness and irradiation that involved many muscle groups not activated at later ages. Most of these early movements were "initiated in the proximal muscles and [are] incapable of any variation or rearrangement as it proceeds down the extremity" (p. 157). Independent use of the index finger and thumb was not fully attained until the animals were approximately 1 year of age. Lawrence and Hopkins (1976) have shown that cortical connections to motoneuronal cell groups do not develop until about 6 months in rhesus monkeys. Prior to this time independent finger movements are rarely, if ever, observed (cf. Kuypers, 1981). McGraw (1940) has argued that for human infants, cortical development can lead both to the coordination of previously fragmented actions and to the suppression of previously active pathways. The simple dichotomy of differentiation versus integration cannot be applied to such complex alterations of motor organization (cf. review in Hofer, 1981).

The development of locomotor patterns has been studied thoroughly. Forssberg and Wallberg (1981) examined locomotor patterns in human infants between 5 hr and 6 weeks of age. These infants were held under the arms for support, with their feet on a treadmill. Five infrared diodes were used to feed leg movement data into a computer for analysis. EMG records were also obtained from the tibialis anterior, lateral head of the gastrocnemius, medial hamstring, and quadriceps. All infants in the first 6 weeks could perform at least part of a single step, and many performed consecutive leg swings. In comparison with older children, the infant walking patterns were highly irregular, with components that seemed "extreme." In particular, infants differed from older children in the following ways: the swing phase was produced by the hip only, with the result that the foot was lifted to an extreme position; early excessive extensor activity led to a digitigrade rather than the more mature plantigrade walking pattern; EMG records were irregular with high degrees of coactivation and the two limbs often became entangled with one another, demonstrating an imperfect interlimb coordination. Thelen *et al.* (1983) have shown that interlimb coordination tends to become organized with a tighter coupling between the legs as human infants become older. These studies imply that a wide range of fractionations and exaggerations in early motor performance later give way to more coordinated patterns.

Analogous studies on animals (e.g., rat locomotion; Gruner and Altman, 1980; Gruner *et al.*, 1980) have shown that tracing movement dimensions both separately and in combination provides useful insights. Golani *et al.* (1981) have demonstrated an ontogenetic sequence from lateral to longitudinal to vertical aspects of locomotory movements for rats. Repeated movements along a single dimension prior to coupling into a coherent multidimensional whole were also observed. Cephalocaudal progressions are also common (see, e.g., Ferron and Lefebvre, 1983 on grooming in squirrels and Richmond and Sachs, 1980 on grooming in rats).

3. FORM, ORIENTATION, AND FUNCTION. One aspect of movement dimensionality that has received considerable attention is the separation of form versus function and/or orientation. Many actions can be recognized and classified prior to the time that they serve any obvious function. Barraud (1961) and Nice (1943), for example, observed preening movements in passerine birds prior to the time that the feathers had developed. It is as if these early preening movements "anticipate" subsequent function (cf. Anokhin, 1964; Carmichael, 1970). Analogous observations exist for a number of species.

Eibl-Eibesfeldt (1961) has reviewed his own studies on motor development in squirrels *(Sciurus vulgaris)*, hamsters *(Cricetus cricetus)*, and desert mice *(Meriones persicus)* that suggest the form of many species-characteristic movement patterns develops prior to, and is less dependent upon experience than, the functional orientation of these movement patterns. For example, in the squirrels he studied (Eibl-Eibesfeldt, 1951) gnawing and splitting movements used in opening nuts were at first poorly oriented and gained in precision with practice. Kear (1962) has found, similarly, that the basic movements used in dehusking seeds by finches develop before their effective utilization. Practice, as usually defined, plays a relatively minor role in the initial production of these motor patterns but is much more important in their effective use. To cite a related example, Meaney and Stewart (1981) report that in social development of rats, fighting and mounting patterns are often well coordinated from the perspective of the moving animal prior to the time that orientation with respect ot the partner is perfected. They conclude that the "behavior appears to become more appropriate to the salient stimuli in the environment and especially to stimuli from other animals" (p. 44). Many other ethological examples in a variety of species indicate that proper orientation often occurs after the form of the behavior is well developed, although the full spectrum of interplay among these dimensions deserves further careful study (Fentress, 1984).

Different stimuli may be involved in the elicitation of movement than are involved in its guidance (see, e.g., Hinde, 1970; Wolgin *et al.*, 1980). A common natural history example is that auditory alarms may trigger flight that is directed to the nearest visible cover. The relative importance of various stimuli used in orienting movements may also change during development. The sensory basis for home orientation in kittens and huddling in rats, for example, is initially thermotactile, but by the end of the first weeks it is primarily olfactory (Rosenblatt, 1980).

That proper orientation of motor patterns develops after practice does not necessarily imply that learning is primarily responsible. Szechtman and Hall (1980) have found orientational changes in oral behavior elicited by tail pinching of preweanling rats. At 5 days elicited licks were not directed to any object; by 10 days licks were directed at nonfood objects; by 15 days licks were directed at either food or nonfood objects; by days 20–30 licks were directed primarily at food. Concurrently with the increased orientational specificity of tail pinch elicited licks, the required stimulation threshold decreased, giving further indication of maturational changes.

Cruze's (1935) study of the effects of learning and maturation on the development of pecking accuracy in chicks nicely illustrates how these two ontogenetic

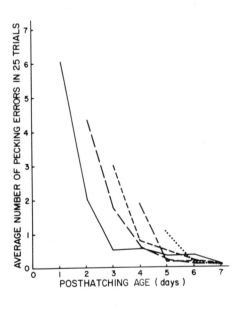

Fig. 2. Data from Cruze (1935) illustrating the interplay between maturation and learning in the development of pecking accuracy in chicks. Graphs depict the average number of pecking errors of five groups of chicks kept in the dark for 1–5 days, until the first test day. (Adapted from Cruze, 1935.)

factors can interact. In Cruze's experiments, chicks were hatched and kept in the dark with no opportunity to peck at food. At 24-hr intervals, groups of chicks were taken out of the dark for the first time and tested for 25 pecks at food daily from then on. The number of errors chicks made on their first exposure to the test situation was inversely related to the amount of time they had been in the dark and therefore their age (Figure 2). In Cruze's words, "maturation, in the absence of practice, very effectively reduces the number of missing reactions to the point where they are almost eliminated" (p. 392). However, when practice was allowed, errors decreased in frequency and were eliminated much more quickly. Furthermore, accuracy improved slightly over the first 25 pecks in groups allowed to peck on the first day. The differential contributions of reinforcers during the establishment of behavioral operants have been reviewed by Hogan and Roper (1978).

B. Spontaneous and Elicited Movements

In this section we move further beyond descriptions of motor patterns in development to potential principles of control. Our concern is with the balance between spontaneous and elicited actions. Three useful introductory perspectives are that (1) tendencies toward spontaneity and responsiveness to eliciting stimuli often coexist, (2) different aspects of the same movement can differ in their relative dependence upon central generation versus elicitation, and (3) within any given dimension abstracted for analysis the balance between spontaneity and elicitation can change with time as well as with a variety of contextual factors.

1. Spontaneity. Experimental approaches to the concept of spontaneity typically involve the removal of inputs that normally impinge upon a defined behavioral or neuronal system. If patterns of activity persist it is assumed that intrinsic circuitry is sufficient for the production of these patterns (i.e., extrinsic information sources are not necessary). In practice, most researchers are willing to allow a *permissive* role for extrinsic variables, such as sufficient muscular strength or body support to permit the expression of intrinsic potentialities. Early human locomotor patterns, for example, are studied in this way (Forssberg and Wallberg, 1981; Lagerspetz et al., 1971; Thelen, 1983; Zelazo, 1983).

Spontaneity can often be traced in a developmental context (see, e.g., reviews in Bekoff, 1981; Hamburger, 1977; Oppenheim, 1982; Provine, 1983; Szekely, 1976). As summarized by Hamburger (1977), this has led many workers to the "conception of the individual as, primarily a system of action" (as opposed to "reaction"). Historically, an emphasis upon spontaneity has not always predominated. For example, in his studies of fetal movement patterns in man, Windle (1940) accepted the reflex as the basic unit of behavior and conducted his experiments accordingly. This was in contrast to the conclusions of Preyer (1855). Based upon analyses of motor patterns in chick embryos, Preyer stressed the "impulsive" nature of early movements. He pointed out that reflexes often could only be elicited after spontaneous movement was observed. Coghill's (1929) many careful observations on behavioral development in the salamander *Ambystoma* led to similar conclusions.

a. Early Evidence (Invertebrates). Prosser's (1933) study of behavioral and neuromuscular development in embryonic earthworms *(Eisenia foetida)* is an elegant early attempt to come to grips with problems of motor spontaneity. Following the general strategy of Coghill (1929), Prosser sought to correlate changes in early motor expression with changes in underlying anatomy. He traced processes of differentiation from anterior to posterior, and laterally, in both behavioral and anatomical terms. Prosser noted that motor integration in embryos of *Eisenia* are present from the beginning of behavioral development, with segmental independence appearing later. Most important, Prosser found that these early movement patterns occurred *spontaneously* in the sense that no special experimental manipulations were necessary to elicit them. It was only later that the first responses to tactile stimulation could be elicited.

Wilson (1961) was the first investigator to demonstrate conclusively that motor patterns in invertebrates (in this case locusts) could to a large extent be accounted for by endogenous central pattern oscillators. Wilson, however, did not examine the development of these motor patterns. Bentley and Hoy (1970) found that flight and stridulation motor patterns are not spontaneously active in larval crickets in the strict sense of the term. These motor patterns could, however, be elicited by tonic stimulation (e.g., windstreams and flight), and/or removal of tonic inhibitory circuits (e.g., mushroom bodies and stridulation). As anticipated by Wilson (1961), the detailed patterning of these movements reflects mechanisms that are to a large extent endogenously organized (cf. Ikeda, 1976; Kristan, 1980; Provine, 1983; Wyman, 1976, for useful discussions).

b. Recent Developmental Emphases. Elsner (1981) has recently reviewed a number of studies of motor development in insects. He stresses that electrophysiological techniques can be used to demonstrate motor circuitry that otherwise would not be expressed in overt motor patterns. Elsner is careful to make the important distinction between exogenous influences in the *elicitation* of behavior from possible exogenous *modifications* of the developmental process. He refers to studies by Weber (1972, 1974), who demonstrated that experience can influence the precision of larval flight motor patterns in crickets (as measured through cycle variability). This effect is very temporary (on the order of 10 min). While recognizing that sensory influences may normally contribute both to the functional integration of motor patterns and to that of their developmental precursors, Elsner concludes that many motor patterns in insects are fundamentally autonomous, even in their early stages of development (see also Elsner and Hirth, 1978; Halfmann and Elsner, 1978; and Lindberg and Elsner, 1977).

Studies by Provine and his colleagues (see, e.g., Provine, 1976, 1983) on the cockroach *(Periplaneta americana)* have clarified problems of spontaneity by severing interganglionic connectives and placing individual ganglia into culture. Electrophysiological analyses indicate that the resulting chain cultures show "spontaneous single unit discharges and multiunit bursts" (Provine, 1976; p. 218).

c. "Impulsivity" in Vertebrates. As noted previously, Preyer (1885) argued nearly a century ago that "impulsive" behavior is critical to an understanding of motor patterns in vertebrate development. He based his conclusions upon a series of elegant studies of chick motor patterns in which motor responses elicited occurred systematically by external stimuli at a later developmental phase than did the first "spontaneous" actions. Hamburger *et al.* (1966) have confirmed many of Preyer's earlier insights. These later investigators, also working with chicks, removed the entire dorsal half of the lumbosacral spinal cord at 2–2.5 days of age (which removes all sensory feedback mediated via the dorsal root ganglia) and also performed a transection of the thoracic cord (eliminating descending influences from the brain). Through careful behavioral observations they concluded that "spontaneous" movements continued to develop normally until 15 days of incubation, after which time degeneration of the lateral motor columns appeared to contribute to a deterioration in motor performance. As Bekoff (1981) has summarized, this study "produced unequivocal proof that normal, spontaneous embryonic motility in the chick was centrally generated and, furthermore, that it could develop in the absence of both sensory and descending input" (p. 138) (cf. Bekoff *et al.*, 1975; Corner, 1978; Hollyday, 1980; Landmesser, 1980; Sedlacek, 1978).

There have been numerous other studies on vertebrate species that make a similar point. Szekely (1976) reviews experiments on amphibian species indicating that histological differentiation of the spinal cord commonly occurs at an early embryonic age in spite of severe sensory restrictions. Many of these studies (and studies with chick embryos; see reviews in Beckoff, 1981; Hamburger, 1977; and Oppenheim, 1981) involve cord transplants and lead to the conclusion that within the spinal cord there are central pattern generators with relatively localized and resistant properties. In this sense workers such as Szekely speak of the underlying circuits as being "predetermined."

d. Retrograde Development. In mammals, motor circuits also tend to develop in a "retrograde" fashion. Output pathways are established prior to the time that functioning sources of activation of these pathways become operative. As illustration, kitten forelimb mechanoreceptors develop an effective influence upon cells within the motor cortex at 60 days postnatally while outputs from the motor cortex can be detected by 45 days (Bruce and Tatton, 1980). Before 40 days, movement patterns can be observed, but these are often inefficient and variable. Circuitry within the cortex, cerebellum, and basal ganglia is still immature at this time. Bruce and Tatton conclude that functional outputs emerge prior to activation by afferents.

The retrogarde progression in motor system development is of special interest in that it places limits upon the role of experience in the initial formation of output circuitry. Retrograde development also relates to the common occurrence of spontaneous activity prior to the time that reliable patterns of evoked behavior can be observed (see, e.g., reviews in Bekoff, 1981; and Oppenheim, Chapter 1, this volume). Furthermore, a similar retrograde progression is found at many levels of the neuroaxis. In Saito's (1979) *in vitro* study of reflex development in the rat fetus, evidence for transynaptic evoked discharges in motoneurons of the 14.5-day isolated spinal cord was found, whereas the first reflex discharge via dorsal roots did not appear until prenatal day 15.5.

e. Examples of Human "Spontaneity." Wolff (1966) has reviewed evidence that mouthing and other movements in postnatal human infants are common when the infants are drowsy. Particular movements have different periodicities, may (within limits) substitute for one another, and lose their apparent spontaneity of expression when alternative actions are elicited by strong stimuli. Thelen and her co-workers (see, e.g., Thelen, 1981, 1983; Thelen and Fisher, 1983) have demonstrated spontaneity of early kicking movements in human infants, together with evidence that these kicking movements can be affected by the infants' states of "arousal" (cf. Fentress, 1976, for a critical evaluation of this concept). Forssberg and Wallberg (1981) argue that in the earliest forms of infant locomotion, spinal pattern generators can act "almost spontaneously" (cf. Grillner, 1981). Each of the preceding authors also points out that during later development, higher CNS circuitry may reduce movement spontaneity. The motor patterns become more precisely attached to complex eliciting events (cf. Prechtl, 1981). These conclusions are compatible with the proposition that during development higher CNS functions take on progressively more important roles in movement control (see, e.g., reviews in Doty, 1976; Gallistel, 1980; Teitelbaum, 1982).

f. Synthesis and Remaining Issues. In sum, these data support the contention of may ethologists that early motor patterns show a high degree of spontaneity (in the sense that they do not depend upon afferent and/or descending pathway activity; see e.g., Hinde, 1982; Hoyle, 1984; Lorenz, 1981). While it is important to qualify the limits of presumed spontaneity, there can be no doubt that early motor rhythms reflect fundamental endogenous tendencies.

Many early motor patterns appear in the absence of causal factors that are of paramount importance at later ages. An example is Kruijt's (1964) ontogenetic study of social behavior in Burmese red junglefowl *(Gallus gallus).* Socially isolated young junglefowl frequently "attack" their own tails. This illustrates problems associated with concepts of spontaneity. The early attack movements are not associated with social interactions and are in that sense spontaneous (cf. Baerends–von Roon and Baerends, 1979; Leyhausen, 1963; Tooker and Miller, 1980). However, they are associated with proximal moving stimuli provided by the animal's own tail. Motor systems must be studied in their expressive contexts, and these contexts can change with age (cf. Fentress, 1984).

There are numerous cases in which the early expressions of motor patterns "surprise" us in that they are not associated with stimuli that play an important role in later life and that we might assume to be necessary. For example, in a recent study employing central nervous system stimulation in young rats, Moran *et al.* (1983; see Chapter 3, this volume) found that motor patterns could be elicited in the absence of sensory cues that later control them (cf. Hall and Williams, 1983; see also preceding examples).

Such evidence does not imply that changes in early sensory events are necessarily without effect. Galef and Henderson (1972), for example, have demonstrated that experiences associated with suckling can have a marked effect upon later food preferences. (For recent reviews also see Galef, 1981, and Hall and Williams, 1983.) During the course of early motor patterns such as suckling, the organism has the opportunity to experience a variety of thermal, tactile, olfactory, and other cues that might contribute to later adaptive functions, including social behavior (Alberts, 1981; Rosenblatt, 1976).

2. Elicited Movements

a. Reflex Actions. Reflexive responses and their development are in many respects more easily studied than are movements that occur "spontaneously" outside of the experimenter's control, and such studies have a long history (see, e.g., Windle, 1944; Hamburger, 1977; Burke, 1981). While the elicitation of isolated reflexes is not our primary concern, reflex elicitation provides a valuable technique for the clarification of a number of issues and principles. Reflexes can be elicited prenatally in a number of animal species. For this to be possible, muscles, motor neurons, and their afferents must be functional. This does not imply that initial reflex patterns are identical to those of later life. Further, as emphasized by Carmichael (1970) and others, many early forms of responsiveness are myogenic in origin; that is, they appear prior to the establishment of sensorimotor connections (cf. Bekoff, 1981; Saito, 1979; and Section II. B.1.d).

Reflexes tend to develop in cephalocaudal and proximodistal directions. One of the earliest studies to demonstrate this in mammals was Angulo Y González's (1932) insightful, albeit qualitative, set of observations on prenatal behavior in rats. These gradients of reflex development were confirmed in a more detailed investigation by Narayanan *et al.* (1971) and reflect the order of ontogeny of spinal motoneurons (Hollyday, 1980; Landmesser, 1980). Similar developmental gradients are found in many areas of the central nervous system (Cowan, 1978; Jacobson, 1978).

Many early elicited reflexes involve widespread movements that in later development become more localized. For example, Bergstrom *et al.* (1962) found that stimulation of the trigeminal nerve in fetal guinea pigs initially produced widespread movements of the neck, limbs, and trunk. Later in fetal development this generalized response was replaced by restricted movements of the head. Gatev (1972) has shown, similarly, that both early monosynaptic and polysynaptic reflexes are widespread in children, with subsequent restriction. As emphasized by Hines (1942) in her classic study of motor development in rhesus monkeys, following the period during which elicited motor patterns become restricted there can be a variety of subsequent re-elaborations that may transcend initial reflex dependencies. (See also reviews by Bekoff, 1981; Hamburger, 1977; Oppenheim, 1981; Provine, 1983.)

The degree of "irradiation" for reflexes seen at any given age can also depend in part upon the strength of stimulation (Corner, 1978; Sherrington, 1906). An elegant study of early irradiation and gradual restriction of reflexes is Saito's (1979) *in vitro* investigation of reflex development in fetal rats. Saito demonstrated that both the development of inhibitory circuits and neuronal cell death sculpt the initially broad reflex excitation. These two processes—(1) inhibitory circuits becoming functional after excitatory circuits and (2) neuronal death—appear frequently in studies of nervous system development (see, e.g., Burke, 1981; Cowan, 1978; Lewis, 1981; Purves and Lichtman, 1980; Wolff, 1981) and undoubtedly contribute to ontogenetic "parcellation" in a number of instances (cf. Ebbesson, 1980). While inhibitory circuits tend to become fully operational after the onset of activity in excitatory circuits, increasingly sophisticated techniques have indicated an earlier onset of inhibitory contributions to patterned movement in various vertebrate species than has often been suspected from earlier studies (Oppenheim and Reitzel, 1975). A recent illustration of developmental segregation is Brown's and Booth's (1983) study on the distribution of motor axons entering the gluteus muscle from segmental roots L4 and L5 in rats. In this study the overlap in distribution at birth was greatly reduced by postnatal day 11 through a process of synapse elimination.

b. Effector Properties. Changes in effectors, as well as in neural circuits, contribute importantly to reflex development. In addition to early polyneuronal and polysynaptic innervation of muscles in development, muscle properties themselves undergo marked changes (e.g., differentiation into fast and slow muscle groups; Lewis, 1981; Navarrete and Vrbová, 1983). Fascinating lines of current research show the intimacy of nerve and muscle properties in a number of species. Two especially interesting invertebrate examples are the demonstrations (1) by Lang *et al.* (1978) that American lobsters housed in smooth-bottomed tanks (without gravel to manipulate) during certain larval stages will grow two cutter claws with predominately fast muscle rather than a single cutter and crusher claw (with slow muscle) and (2) by Mellon and Stevens (1979) that cutting the nerve to the snapper claw in hatchling alpheid shrimp will lead the opposite pincher claw to be transformed into a snapper. Thus changes in peripheral activity can have marked consequences on muscle differentiation on either the same or opposite side of the body. In chicks it is now clearly established that motoneuron cell death can be modulated by muscle activity. By immobilizing chick embryos, Pittman and Oppenheim (1979) reduced the normal decline of motoneuron numbers in both the brachial and lumbar motor columns.

There is evidence in mammals that motoneuron activity can control gene expression in muscle fibers, even in adults. Metafora *et al.* (1980) found marked changes in mRNA sequences of 8-day denervated rat gastrocneumius muscles, and Streter *et al.* (1975) found that cross-innervated fast and slow muscles in rabbits developed changed twitch characteristics as well as protein composition. That skeletal properties not only change in development but can be dependent on muscle activity is also well known (see, e.g., Drachman and Coulombre, 1962—joint development in chicks; Walker and Quarles, 1976—palate development in mice).

c. Control Networks. Changes in reflexes and related processes during development clearly can involve a number of interconnected properties, many of which are just beginning to surface. Because of the richness of potential developmental connections in even relatively simple "elicited movements," elucidation of these processes will have to be sought at a variety of levels, and with an appropriate appreciation for the dynamic maturational processes that accompany them (cf. Thelen, 1983, and Zelazo, 1983, for recent related discussions with reference to human locomotion).

Certainly, routes of reflex *control* can also change importantly during development, as seen in the transition between vestibular and visual dominance in humans for early righting reflexes (Kuypers, 1981) and in changes in response to nociceptive stimuli in rats (Stelzner, 1971). It is also well established that the expression of many early reflexes can be suppressed through the subsequent development of higher CNS regions (see, e.g., DeGroat *et al.*, 1978—reflexive micturition and defecation in kittens; for additional reviews see Bower, 1979; Doty, 1976; Hofer, 1981; Peiper, 1973).

d. Complexities in Elicited Movement. Certainly not all forms of early elicited movement qualify as reflexes in the classical sense of the term (Sherrington, 1906). Because of complex patterns of reflex interaction and descending influences, the precise boundary line between reflex and nonreflex in the intact animal is frequently difficult to determine (Berridge and Fentress, in press). There are, for example, a number of documented cases of sensorimotor responses in prenatal or larval animals that begin with simple movements only to be followed by complex sequences involving central pattern generators. To cite a recent example, Soffe *et al.* (1983) have demonstrated that in newt embryos *(Triturus vulgaris)* a tactile stimulus applied to the head produces turning of the head away from the source of the stimulus, followed by a complex pattern of swimming movements. That there are centrally patterned substrates for these swimming movements can be demonstrated by a variety of means (e.g., through recordings of motoneuron discharge in paralyzed animals). In very young animals head turning and subsequent swimming responses are often disassociated.

Measures of prenatal responses to environmental stimuli can often be used to demonstrate functional sensorimotor pathways that might not otherwise be anticipated. For example, Armitage *et al.* (1980) have shown that in sheep, motor activity can be elicited prenatally by auditory stimuli. Motor activity also occurs in third-trimester human fetuses in response to acoustic events (DeCasper and Fifer, 1980), and as early as postnatal day 1, human infants are said to synchronize movements in rhythm with adult speech (Condon and Sander, 1974). That these early responses may, under certain circumstances, affect later behavior is suggested in a study by Vince (1979). She exposed prenatal guinea pigs to mildly aversive acoustic stimuli and found, using measurements of heart rate change, a subsequent reduction in responsiveness to the stimuli. While these changes are almost certainly not focussed upon motor pathways per se, Vince's studies establish the need to evaluate measures of motor performance in the intact organism as representing products of complex control networks and previous experience.

Elicited postnatal motor patterns in development often display remarkable degrees of sophistication. Perhaps most dramatic among these are early patterns of imitative movements in human infants. Meltzoff and Moore (1977) have reported that as early as 2 weeks of age infants can articulate recognizable facsimiles of several adult facial expressions. Here we have an instance not only of a considerable degree of control by the infants over their own facial musculature, but also an appropriate responsiveness to stimuli of clear social import.

One interesting laboratory approach to motor and behavioral development in animals is to stimulate central nervous system pathways directly. Gorska and Czar-

JOHN C. FENTRESS
AND PETER J.
McLEOD

kowska (1973) applied this methodology in an ontogenetic evaluation of motor cortex function in dogs from birth to 12 weeks of age. Their data include observations of comparatively poorly organized somatotopic representation of body parts, incomplete dominance of contralateral movements, limited repertoire, plus short duration and variable movements with a high stimulation threshold in the youngest (birth to 4 weeks) animals. By 12 weeks adult response patterns were found, but still with elevated thresholds compared to adults.

In a recent study, Moran *et al.* (1983) provided electrical stimulation to the median forebrain bundle at the lateral hypothalamus in 3-, 6-, 10-, and 15-day-old rats. Pups 10 days and younger became behaviorally activated, as expressed through a variety of motor paterns, including mouthing, licking, pawing, gaping, probing, stretching, and even lordosis. As the animals grew older, these motor patterns showed signs of improved organization and often became increasingly dependent upon the presence of appropriate goal objects. By varying stimulus frequency, differential changes in various forms of motor expression could also be obtained. For example, stretch and lordosis responses "appear to represent 'end behaviors,' occurring at the height of behavioral activation" (p. 13). The importance of central stimulus parameters in eliciting different forms of behavior has also been demonstrated by Beagley (1976) in adult rats.

As animals grow older there is commonly a progressive restriction in the movements elicited by a given stimulus, along with an improvement in their organization. The relatively nonspecific nature of early elicited motor patterns deserves further detailed study. It is likely that developmental restrictions in movement often reflect increased specificity of motivational rather than motor pathways. Further, variation of early elicited motor details may in part depend upon the young animal's current posture (Fentress, 1981a,b, 1983b,d). Mature animals with high-level CNS damage are similar to young animals in that they often become "entrapped" by particular postural and exteroceptive sensory events (see, e.g., Deliagina *et al.*, 1975; Teitelbaum, 1982). Conversely, in the study by Moran *et al.* (1983), responses of young rats to central nervous system stimulation were *less* dependent upon sensory events in the environment than were responses of older animals. Here motivational control pathways were activated directly in the young rats, which suggests that these pathways may have a high degree of intrinsic organization, and well-established descending connections to motor pathways, prior to the time that they are addressed by sensory events that later play a critical role in their activation.

There are clearly a wide variety of central as well as peripheral sources of behavioral activation. As stressed by Wolgin (1982), both young animals and human infants may fail to express fully their motor capabilities unless they are appropriately "aroused." Once the appropriate circuits are activated, previously hidden competancies in motor performance are often revealed (see also, Prechtl, 1981). A recent study by Smiley and Wilbanks (1982) on the effects of noise on the precocial elicitation of play and locomotion in young rats indicates the power of such manipulations within a developmental context. Similarly, experiments by Szechtman and Hall (1980) on the ontogeny of oral behavior in rats indicate that various sources of behavioral activation may be used to clarify underlying devel-

opmental (e.g., dopaminergic) circuit properties. Certain of these neurochemically
mediated substrates not only alter responses to specified sensory events but also
activate motor patterns in development more directly (cf. Tamasy *et al.*, 1981, on
dopaminergic and serotonergic pathways in the development of swimming abilities
in rats). Thelen (1981) has argued that activation in human infants may trigger
previously formed motor stereotypes that are relatively independent of sensory factors in their expression. Movement details, in this case, may be relatively "hard
wired" (Thelen and Fisher, 1983) and respond in a characteristic way over a number of activation sources (cf. Turkewitz *et al.*, 1983).

e. Summary. Elicited movements have provided valuable information on motor
development. These movements, examined in a variety of invertebrate and vertebrate species, range from relatively simple reflexes to much more complex patterns
in which environmental activation and central programming are intimately connected. It is often difficult to localize the critical foci of developmental change in
elicited actions due to interactions among sensory, motor, and higher CNS events.
However, careful experiments on animals and detailed observations of early
human movement patterns have begun to clarify some of the issues.

It is important to emphasize that developmental changes in elicited movements can be examined somewhat separately in terms of movement form, causal
antecedents, and functional consequences. The data reviewed in this section also
suggest that while some elicited motor patterns in development exhibit progressive
improvement, other motor patterns may be perfected early in ontogeny, to be
disassembled and perhaps reassembled into novel patterns as development proceeds. These new motor patterns may also come under different systems of control
as the functional needs of the organism change. Other early motor patterns may
be relatively nonfunctional, yet point unambiguously toward future capabilities.
Circuits underlying many motor patterns in development can only be revealed
through special experimental techniques.

3. CONCLUDING COMMENTS ON THE DISTINCTION BETWEEN SPONTANEOUS AND
ELICITED BEHAVIOR. While we have found it convenient in this section of our
review to emphasize the polarities of spontaneous *"versus"* elicited motor patterns
in development, we emphasize that the distinction is not absolute. In "elicited"
motor patterns, fluctuations of internal state may change the effectiveness of any
given stimulus, even in simple reflexes. Concepts of "spontaneity" are necessarily
relative in the dual sense that one must always ask, "Spontaneous with respect to
what?" and "Spontaneous to what degree?" Central and peripheral events necessarily work in concert. The issue is one of relative (rather than absolute)
parcellation.

There is a close conceptual parallel between any forced dichotomy of evoked
and spontaneous behavior and inadequacies of the nature versus nurture perspectives on developmental processes. Emphases upon evoked behavior serve merely
to stress the importance of *measured* system inputs, while emphases upon spontaneous behavior serve merely to stress the limits of specified system inputs. Demonstrating the importance of one source of control does not negate the importance
of another (cf. Hebb, 1955).

John C. Fentress
and Peter J.
McLeod

Basic behavioral and biological functions are often achieved through the simultaneous and sequential articulation of physically distinctive movements. This has led workers in ethology and related disciplines to seek rules by which abstracted actions are joined together. It is "how response units are related" (Hinde and Stevenson, 1969, p. 293) that poses many of the most important problems for developmental analyses.

In this section of our review we examine multiple movements from a perspective that complements the problems of dimensionality in motor behavior already discussed. "Because most natural motor behaviors require complex interactions among multiple movements, conclusions as to the potential contributions of sensorimotor mechanisms based upon unidimensional movement criteria . . . may be premature" (Abbs *et al.*, 1984, p. 196). The importance of examining integrated motor patterns has received growing notice in recent years. Multiple movements may be embedded within a number of higher-order sequences. As summarized by Menn (1981) in the context of human speech, "some motor sequences evidently are easier to coordinate than others, for reasons that we should be starting to study" (p. 132).

1. Integrated Sequences and Ontogeny. Many motor sequences in young animals are incomplete and appear to be of limited functional utility. As animals get older, new dimensions may be added to the sequence and existing processes may be combined in novel ways. In their study of kitten development, Baerends–van Roon and Baerends (1979) subdivided motor development into three phases characterized by (1) exclusive domination of body musculature, as seen in swaying; (2) legs used, but in a manner secondary to the body; and (3) legs used in a manner relatively independently of gross body movements. "During the first four weeks the behaviour patterns distinguished develop progressively, essentially by the addition of new elements. After this period changing continues, but in our opinion these changes have to be considered as adaptive modifications and combinations of the existing patterns; no essentially new patterns appear" (p. 23). Interruptions within functional motor sequences were also observed frequently in young animals.

A recent study by Etienne *et al.* (1982) on the ontogeny of hoarding in golden hamsters also illustrates ethological approaches to the development of sequential motor patterns. These authors conducted observations of four litters, beginning on postnatal day 13. Immature forms of pouch filling and emptying were seen first, but actions were not integrated into coherent behavioral sequences. The young animals frequently exhibited evidence of goal-directed actions even though these actions were in themselves imperfectly organized.

Coordinated hoarding trips in the Etienne *et al.* study were observed from day 21, once motor patterns of inserting and extracting food had gained definitive form. The investigators isolated three ways of inserting food into pouches (one is mature) and five ways of extracting food from the pouches (one is mature). Well-integrated hoarding has four distinct phases that follow each other in a fixed order. The order in which young pups perform constituent acts of hoarding was unpredictable. Early hoarding patterns were uncoordinated in space and time. For example, inserting and extracting were "randomly" associated. Constituent parts of hoarding were intermingled with other classes of action in the young animals, without the adult temporal groupings of functionally related actions. Increases in the frequency of mature motor patterns accompanied decreases in immature motor patterns.

The decreasing performance of immature actions occurred in a regular sequence, with those motor patterns that were most different from mature movements disappearing first. As the motor patterns became more clearly differentiated, so too did the sequential phases within which these motor patterns were grouped. "The fact that the decrease of immature hoarding components coincides with the reduction of less specific, 'clumsy' activities (such as taking and dropping food), shows that the differentiation of particular motor patterns is accompanied by a general progress in the coordination of movements" (p. 42).

A rich nexus of changes can be traced in the ontogeny of such species-characteristic motor sequences. A possibility that has interested many workers is that the sequential order of movement patterns found in adult animals is paralleled by the ontogenic order in which the movements emerge. Whether or not "integration recapitulates ontogeny" is a general rule is questionable at present (although within limits there often appears to be a formal association in motor sequencing over these two very different time frames). While in most of the hamsters studied by Etienne *et al.* movements of food insertion developed before those of food extraction, in socially isolated animals mature forms of insertion and extraction appeared more or less simultaneously. The transition to mature hoarding occurred suddenly in these isolated animals.

2. SEQUENCE "PIECES" AND "RELATIONS." For action properties to be connected they cannot be completely separate. Precisely where the interconnections among otherwise separate motor properties exist and the consequences of these interconnections upon properties that might otherwise be viewed as independent events represent difficult challenges to present research (Fentress, 1984). It is important to determine processes that may transcend abstracted "components" of action (e.g., whether these "components" exhibit regular forms of contextual modification). A number of movement properties, such as the precision of timing, can change during ontogeny and be shared among *sets* of actions. This indicates that the individual actions are either directly influencing one another or share common control processes. Excellent examples are provided by recent studies on speech development of human infants (Cooper, 1977; Kent, 1981). A basic point of these studies is that otherwise individually defined articulatory components change their timing and related properties all together and in this way maintain a relative invariance of overall sequential structure (rather like changing the tempo of an entire melody line in music; Fentress, 1978, 1981b). Kent (1981) compares the articulatory invariances among phonetic attributes in human speech to schema theory in

motor learning, where the individual must develop regularities among performance attributes to achieve functionally adequate sequences.

In a wide variety of human motor skills such as typing (Terzuolo and Viviani, 1980), handwriting (Viviani and Terzuolo, 1980), speech (Tuller *et al.*, 1982; Weismer and Ingrisano, 1979), cycling and piano playing (Whiting, 1980), the *relative timing* of component activities remains invariant even when the overall speed of the sequence changes. This indicates that skilled motor performance is not a simple assemblage of components, but that there are cohesive properties that affect the integrity of the sequence as a whole.

For rapidly performed sequences of movement, anticipatory adjustments can increase the cohesion of elements within the sequence (Bizzi, 1981; Glencross, 1980; Kelso, 1981). Action components are in this sense "coarticulated"; merging together into higher blockings of expression (Liberman and Studdert-Kennedy, 1978). The fragmentation of movement often observed in young organisms reflects an imperfect synthesis of these action components. As Whiting (1980) has summarized, "motor skills . . . do not grow by a simple addition of elements, but by a structural rearrangement—by analysis and synthesis" (p. 543). Even relatively simple motor patterns, such as rodent grooming, are similar in their progressively "coarticulated" expression with age (Fentress, 1983d).

Interesting limitations in sequential motor performance are found in many young mammals, as well as in mammals with various forms of central nervous system disruption. Employing central stimulation in rats, Moran *et al.* (1983) found that infant animals (1) frequently performed motor sequences in the absence of sensory cues necessary for adults and (2) showed sequences of movement that cut across functional categories (e.g., lick, probe, gape, lordosis). Vanderwolf (1983) found that decorticate rats would normally show grooming elements but not full sequences as "inappropriate" acts often intervened (such as walking away). "The behavioral deficit in decorticate or decerebrate rats appears to be largely a matter of failing to perform behaviors in the right place at the right time" (p. 85; Vanderwolf, 1983). Similar phenomena are found in many young animals (Fentress, 1984; Teitelbaum, 1982).

In a recent study, Moss and Hogg (1983) rewarded infants 12–18 months of age for successfully manipulating metal rods in sequences involving the lifting of two rods, connecting them together, placing the connected rods in a hole, and moving the rods around a track. Their study was designed explicitly to test the idea that modules of action are sequenced together in invariant form when higher-level skills are perfected. Contrary to the strict modularity view, variation within modules increased with mastery of the sequential task.

It is clear that any full analysis of motor patterns in development must take into account not only the form of the behavior expressed but also the various "strategies" employed with respect to higher-order goals. In many cases it is useful to make a distinction between *how* an organism goes about doing something and its action details defined in isolation. Related to this problem is the logical distinction between capabilities and strategies of action, to which we shall return. Differential roles of experience in the descriptively separable layers of motor performance constitute a related issue.

3. ROLES OF EXPERIENCE. A common perspective in ethology and related disciplines is that experience during development plays an increasingly important role at higher levels of motor organization (e.g., Eibl-Eibesfeldt, 1956; Lorenz, 1981). Basic motor "components" are viewed as having a strong genetic base, since they take on an invariant form even under diverse developmental conditions (within limits). The *rules of connection* among these "components," as well as the *orientation* of the movement sequences they comprise, are viewed to be more dependent upon particular developmental histories (for an exception see Section II.D.1.b).

An excellent example of this perspective can be seen in the work by Kruijt (1964) on Burmese red junglefowl noted previously. By rearing animals with different degrees of social experience and by documenting the detailed patterns of their motor behavior, Kruijt was able to conclude that a "lack of social experience has little influence on the motor side of behaviour, but much more so on the way in which the motor patterns are released, orientated and integrated with each other" (p. 170). For example, fighting and mating movements in birds reared in social isolation may appear to be well coordinated in terms of the animal expressing them yet fail to be appropriately coordinated with respect to the birds' social partners. Thus, as we have seen previously in the discussion of eliciting factors in motor development, total motor performance must be judged along a number of dimensions and in explicit association with the contexts that may trigger, guide, and modulate the sequences in question.

Prechtl (1983) reviews evidence suggesting that orientational parameters of movement in mammals, particularly those that involve the coordination of visual information with directed expression, are more sensitive to experiential distortions than are the individual "units" of movement per se. A number of disruptions of normal central nervous system processes also have their most obvious effects upon "higher-order" properties of integrated movement sequences (see, e.g., reviews in Teitelbaum, 1977; Vanderwolf, 1983). Not surprisingly, the brain regions that are viewed to have particularly important effects on "higher-order" aspects of motor performance are those regions that are also considered to be "higher-level" (and later developing).

D. SOCIALLY INTEGRATED MOTOR PATTERNS

Interactive motor patterns in social behavior must be properly oriented with respect to specific social and environmental situations and must occur in the proper context to be functional. For the communicative actions of birds and mammals, contextual variations can alter the interpreted "meaning" of the signal by the recipient (Green and Marler, 1979; Goosen and Kortmulder, 1979; Gould, 1983; Marler, 1967; Maurus and Pruscha, 1972; Petersen, 1981; Shalter *et al.*, 1977; Smith, 1965). The social contexts in which signals are given have been shown to affect the probabilities of subsequent behavior by the sender (macaques: Goosen and Kortmulder, 1979) and the elicited response in the recipient (mice: Butler, 1980; macaques: Lillehir and Snowdon, 1978; wolf pups: Shalter *et al.*, 1977). Displays by young animals are often less accurate predictors of their own subsequent behavior than are adult displays (Simpson, 1973).

The environment in which social behavior patterns normally develop includes conspecifics. These social agents differ from other components of the physical environment in their greater mobility, decreased predictability, and ability to act upon their surroundings. More importantly, the information processing abilities of these agents make social feedback and mutual influence possible, providing the infant with a wealth of potential stimulation and information (Mason, 1979). Watson (1981) has argued that behavioral development can be influenced directly by such behavior–stimulus contingencies. The ability of human infants to recognize when facial movements are correlated with speech (Kuhl and Meltzoff, 1982) and to imitate both facial and manual gestures (Meltzoff and Moore, 1977) shows that infants are capable of both perceiving and performing contingent behavior.

Although in some sense all social behavior involves communication, for the purposes of the present discussion we find it helpful to treat communication by relatively discrete signals (and human speech as a somewhat special case) separately from behavioral relations among individuals.

1. COMMUNICATION. The importance of effective communication for survival in a social environment has been recognized since Darwin's (1872) classic work, and many ethological studies have focused on the communication systems used by social animals. In general, the forms of the motor patterns used in social displays of nonhumans have been found to develop with a less appreciable role being played by experiential factors whereas contextual variations in the use of these signals, and the ability to respond properly to the signals of others, is relatively more dependent on experience (Burghardt, 1977; Fox and Cohen, 1977; Green and Marler, 1979). [Some exaggeration of facial expression has been reported in hand-reared monkeys (Bolwig, 1964), and frowning, a display frequently exhibited by rhesus mothers, occurs in mother-raised infants but is not exhibited by terrycloth-surrogate-raised infants (Hansen, 1966), suggesting that the form and expression of these monkey displays are somewhat dependent on environmental conditions.]

Shortly after birth most mammals vocalize. These sounds generally evoke approach or contact by the mother. "Some calls that originate in infancy and appear in the mother–infant context reappear later in life in similar contexts but serve different functions" (Gould, 1983, p. 265). For example, adult female Japanese monkeys use a similar call when sexually soliciting a partner as infants do when soliciting contact from their mothers (Green, 1981). Ontogenetic changes in vocalization and the contexts in which they occur can also differ between the sexes while still sharing some unspecified underlying causal mechanism (Green, 1981).

In some species (e.g., stumptail macaques, Chevalier-Skolnikoff, 1974) the number of different vocal motor patterns increases from the infant to adult with a decrease in vocal activity toward the end of the juvenile stage (Gould, 1983). Specific calls may also undergo some changes in acoustic properties during ontogeny (e.g., Chevalier-Skolnikoff, 1974; Field, 1978, 1979; Lieblich *et al.*, 1980; Newman and Symmes, 1982), generally exhibiting less variability in physical structure (e.g., Field, 1978, 1979; Newman and Symmes, 1982).

Observations that vocalizations usually occur in concomitance with gross head and body movements in neonate kittens (Levine, *et al.*, 1980) and wolves (McLeod, personal observation) indicate that these early vocalizations are controlled less independently of general activity than in older animals. Similarly, early facial expressions in human infants involve flashing "fragments of smiles, grimaces, and frowns and these bouts of facial activity tend to occur when the infant's head turns" (Hofer, 1981, p. 105). At the "local" level, these movements appear as fragmented components of adult expression while exhibiting connectivity with general patterns of movement.

a. Monkey Alarm Calls. In a study of Seyfarth and Cheney (Seyfarth and Cheney, 1980; Seyfarth *et al.*, 1980a,b; reviewed in Seyfarth and Cheney, 1982), adult vervet monkeys were found to give acoustically distinct alarm calls to leopards, martial eagles, pythons, and baboons. Each type of call elicited different, appropriate responses from conspecifics (e.g., look up and/or run into dense brush in response to eagle alarms). Infants and juveniles, however, gave alarms to a much wider variety of species, though the type of alarm an infant gave in response to a stimulus was not arbitrary (e.g., eagle alarms were given by infants to a variety of birds but not to terrestrial animals).

Although there was no significant difference between age classes in the probability of showing a "wrong" response to broadcast alarm calls, there was a trend toward decreasing errors with age. Infants were more likely to make wrong responses if they were more than 2 m from their mothers and generally responded to all alarm calls by looking at their mothers (Seyfarth and Cheney, 1980). Infants also sometimes emitted different calls when they joined in alarm call "bouts" initiated by other group members, whereas adults and juveniles always called with the same type of alarm (Seyfarth *et al.*, 1980b).

These findings suggest that the association between predatory species and alarm call type sharpens as infants grow older. The stimuli that elicit each of the different signals become more specific as do the developing individuals' responses to the alarms, possibly because of "subtle reinforcement" (Seyfarth and Cheney, 1982). This is consistent with the statement by Newman and Symmes (1982) that the "primary role of learning in the acoustic behavior of primates, including humans, may be gradually to restrict the variability of vocal utterances and behavioral responsiveness with age. By this process, the individual would acquire, at maturity, a considerable degree of communicative predictability" (p. 272).

A similar process of reinforcement by the human infant's social environment has been proposed to account for the emergence of social smiling from the nonspecific smiling that occurs in response to a wide variety of stimuli prior to 3 months of age (Emde *et al.*, 1976).

b. Bird Song. Studies on the ontogeny of bird song have been most productive in determining the combined contributions of genetic and epigenetic factors in the development of complex communicative motor patterns in a variety of species. The neuroanatomical regions involved in vocal learning, hormonal effects on these areas, and the roles of both auditory and proprioceptive feedback have also been studied in birds (see, e.g., Nottebohm, 1980). As this literature is covered in Chapter 4 of this volume, we will mention only a few studies that illustrate issues relevant to the present discussion.

Song development in several bird species starts as an "amorphous, highly variable and unstructured subsong" (Marler and Peters, 1982a, p. 445). As described in detail for swamp sparrows (Marler and Peters, 1982a) syllables then appear in a very rudimentary form of unidentifiable sequences. Note sequences, though variable, can then be identified and syllables can start to be repeated. The syllables then become less variable, are produced as trills, and occur in a progressively more stable order. There is a trend that the "acquisition of the smaller acoustic unit [is] accomplished before a complete commitment is made to higher levels of temporal organization" (Marler and Peters, 1981, p. 86). At the same time the variability in song duration is greatly reduced and the number of phrases per song decreases (as does song duration). Subsong also exhibits a greater frequency range than crystallized song and tends to be quieter (Table 1).

During plastic song, swamp sparrows produce four to five times as much song material as is needed for normal adult song (Marler and Peters, 1982a,b). The excess is then subjected to selective attrition (Marler and Peters, 1982b). When young swamp sparrows are exposed to tape recordings of both swamp sparrow syllables and syllables of the closely related song sparrow, they later produce many more conspecific syllables than song sparrow syllables (Marler and Peters, 1982b) in part (at least) because of early perceptual selectivity (Dooling and Searcy, 1980). During crystallization of song production there is also a trend for any song sparrow syllables that might be present to be selectively rejected. Marler and Peters (1982b) suggest that "species-specific motor constraints [possibly] contribute to this 2nd selective phase" (p. 375) as components that fit more comfortably with the "motor predispositions" of the species are selectively retained. Both perceptual and motor

TABLE 1. STAGES IN THE SONG DEVELOPMENT OF SWAMP SPARROWS[a]

	Syllables				Songs	
	Stage[b]	Number[c]	Form	Repetitions	Form	Duration
Crystallized song	I	2.9	Stereotyped	Clear trills	Stable order	Short
	II		Stereotyped	Clear trills	Variable order	Short
Plastic song	III	9.5	Minor variations	Clear trills	Stable order	Longer
	IV		Variable	Clear trills	Variable order	Long and
	V	12–13	Rudiments	Some	Variable order	variable
Subplastic song	VI		Rudiments	None	Variable order	Variable
Subsong	VII		None	None	None	Variable

[a]Adapted from Marler and Peters, 1982a.
[b]Stages II and III may represent alternate rather than sequential stages.
[c]Average number of syllables per individual per session.

constraints may therefore influence the development of motor patterns used in song production. Baptista and Petrinovich (1984), however, have recently argued for the importance of social factors in the development of avian song. They review studies showing that several species will copy whole songs from a live tutor but only limited copying occurs when exposed to tape recorded songs. Baptista's and Petrinovich's data demonstrate that white-crowned sparrows could learn the alien song of a strawberry finch social tutor even when conspecifics could also be heard but not seen. These authors suggest that social factors can override predispositions to learn species-specific songs.

Other issues in motor pattern development are illustrated by Guttinger's (1981) study of song development in deaf canaries. Canary songs exhibit three main levels of temporal organization but a highly plastic choice in the combination of notes. Briefly, these levels of organization are as follows: (1) Single notes or syllables are the shortest components with individual repertoires of 30–60. (2) "Tours" or trills exhibit a correlation between note duration and repetitions and systematic changes in repetition rate of the tour duration. (3) Phrases or songs exhibit characteristic intervals between syllables, between tours, and between phrases; a restricted number of vocal patterns that can occur after pauses longer than 1.2 sec; and a relation between tour and succeeding interval durations (short tours being followed by longer pauses). Following deafening of juveniles and adults these hierarchical levels of organization were differentially affected. In contrast to the view that sequential organization is in general more susceptible to experiential modifications than the form of motor components (see Section II.C.3), in this case: "The higher orders, the syntactic rules, develop quite normally whereas at the lowest level there is a dramatic reduction not only in repertoire size but also in the diversity of tonal quality" (p. 338). Although auditory monitoring and learning both play important roles in the differentiation of individual notes, Guttinger suggests that the two motor patterns involved (determining the pattern of respiratory movements and the impedance of the syrinx) might be either centrally programmed or monitored by a "proprioceptive template."

Another species that exhibits a complex and variable song structure (as opposed to the relatively simple, stereotyped songs of sparrows), the parasitic widow bird *(Tetraeunura fisheri)* sings both species-specific phrases and vocalizations learned from the host species. In this species, both the species-specific phrases and the syntactic rules controlling switches to learned contraspecific vocalizations are under clear genetic influence and are to a large extent model independent (Nicolai, 1973; cited in Guttinger, 1981). These parallel findings have suggested that in species that exhibit a variable song, in terms of the component syllables, the syntactic rules of connection between syllables may encode species identification (Guttinger, 1981).

c. Human Speech. The motor system used in speech is a complex one involving approximately 100 muscles (versus 7 in song birds). Parallels between the ontogenetic development of song in birds and speech in humans, however, have attracted much interest (see, e.g., Marler and Peters, 1981; Studdert-Kennedy, 1981). The issues arising require consideration in any study of motor development.

The first sounds produced by human neonates have been described as discomfort sounds or crying (Cooper, 1977; Emde *et al.*, 1976). Stark (1980) reports that vocalizations recorded in the first 15 weeks of life can be classified as cry, discomfort, vegetative, and comfort sounds. Prior to 15 weeks these vocal categories were highly correlated with nonvocal behavior. After this time the relations to nonvocal behavior decreases as the infant gains control over "reflexive" sound production. As with other developing motor systems, it is important to note that many anatomical and neuromuscular changes in the speech apparatus are occurring during the ontogeny of speech and that these changes may alter the physical constraints acting upon the system (Bosma, 1975; Kent, 1976, 1981).

Between the ages of 16 and 30 weeks, which Stark labels the "vocal play" stage,

> The infant takes apart the elements, of which his utterances are formed, selects simple elements, produces them over and over, prolongs them, elaborates them, and divides the elaborated versions into new segment types by means of glottal stops and silent intervals. The infant then shortens them again and puts the new segment type or new element back in series with others that are of earlier origin or are more highly practiced. . . . This disassembly of parts and their reassembly in new series is an essential characteristic of all subsequent development in speech production [Stark, 1980, p. 86].

By 10 months of age, consonant–vowel transitions are handled efficiently in sequences (Stark, 1980) and the relatively rigid timing characteristics of syllabification that conform to natural language restrictions are apparent. Reduplicated babbling (e.g., "bababa") is common "without communicative intent" (Stark, 1980) or any evidence of a denotative meaning associated with the babble (Oller, 1980). The infant frequently focuses on particular syllables at this stage to the exclusion of others (Oller, 1980). The similarities between babbling and rhythmical stereotypies (cf. Thelen, 1981) in their unidirectionality and functional isolation have been pointed out by Kent and Murray (1982), who also note that both reduplicated babbling and nonspeech stereotyped rhythms peak in frequency at roughly the same age.

Over the longer period during which the child develops his referential speaking abilities, speech is marked by severe restrictions in the range of sounds produced (Cooper, 1977; Locke, 1983). Even after the child's phonetic repertoire is complete, the motor skills of speech appear to continue being perfected (Kent, 1976). These changes in speech production are characterized by diminished variability of segment durations and spectra, a general reduction in segment durations, and increased anticipatory articulation (Kent, 1981). As we have seen in previous sections, a similar separation of articulatory profiles into attributes of relative restrictions, variability, duration, and anticipation can be applied to many forms of animal and human motor performance during development. Potential parallels among these diverse examples during development can lead to the generation of formal rules and hypotheses of control that transcend particular cases (cf. Fentress, 1983c, 1984).

In this spirit, Kent (1981) proposes a motor schema model for speech production involving comparisons between the expected or desired outcomes (phonetic targets) of the speech "motor program" and proprioceptive and auditory feedback.

> Depending on the stage of acquisition and the articulatory requirements of a particular phonetic target, the emerging motor schema must deal with at least three major types of feedback . . . tactile, kinesthetic and auditory. In the early stages of speech acqusition, probably all three types of feedback are closely monitored. But as skill develops (i.e., the motor schemata are well established), the child may lessen dependence on the slow-loop auditory channel and rely to a greater degree on tactile and kinesthetic feedback [pp. 176–177].

Relatively few data are available on the sounds produced by children during the transition from babbling to word acquisition. Although studies such as Stark's (1980) and Oller's (1980) document a developmental progression in sound production from "reflexive" to "nonreduplicated babbling" and "expressive jargon," studies concentrating on speech produced by children as their lexicon develops (usually with subjects of 10 months of age and older) often conclude that words, and not phonetic elements, are the "units" being manipulated (see, e.g., Labov and Labov, 1978; Leonard *et al.,* 1980; Macken and Ferguson, 1981; Menn, 1981; Menyuk and Menn, 1979). One of the most important developments that occurs during this period is that the child learns that phonological sequences have meanings apart from the contextual situation in which they are usually perceived.

As words are broken into phonetic components, the child discovers or invents rational rules between sounds in different contexts (Macken and Ferguson, 1981). As new words are learned they are often modified so that they will comply with the sound-patterning rules the child has for the existing words in his or her lexicon (Menyuk and Menn, 1979). This overgeneralization of rules may even result in the mispronunciation of words that had previously been articulated properly (cf. Macken and Ferguson, 1981; Menn, 1981; Menyuk and Menn, 1979).

Peters (1977) points out that although it is reasonable to assume that a child's speech is initially simple and gradually becomes "more complex," this progression can occur in either of two ways. The developmental strategy emphasized in most studies involves forming short, clear utterances that are later integrated into long, clear ones. In contrast to this, Peters' (1977) study of the vocalization of one subject indicated that the number of syllables and their intonation (including pitch and amplitude contours) was often more consistent than his articulation of individual segments. This illustrates a second, "gestalt" strategy of speech development whereby longer utterances are first sketchily approximated and later the details are filled in. Individual children vary in the relative degree to which these strategies are employed (Menn, 1981). These strategies therefore should be considered as extremes of a continuum (Peters, 1977).

2. INTERINDIVIDUAL BEHAVIORAL RELATIONS. Many ethological studies of development have been directed at accounting for individual variability in the expression of adult behavior patterns. Even minor differences in experience, such as the position of a preferred nipple (Pfeifer, 1980) or slight temporal shifts in physiological development (Cairns, 1976) have been found to correlate with a number of

social behavioral differences. (Reite and Short, 1980, however, have tried unsuccessfully to account for individual behavioral differences in pigtailed monkeys by measures of 23 physiological variables.)

After 180 days, "virtually the same experiences have apparently opposite effects on the behavior of cats depending on their early defensive biases" (Adamec *et al.*, 1980). In cichlid fish, social development in one individual may influence the rate of development of the same behavior patterns in other same-aged members of the social group (Fraley and Fernald, 1982). Socially isolated chimps raised in impoverished (less variable) environments developed more highly variable idiosyncratic behaviors than chimps raised in a normal social environment (Rogers, 1973). Social factors can therefore both facilitate and reduce individualization. Social factors have also been found to suppress the expression of adultlike hoarding behavior in hamsters (Etienne *et al.*, 1982) and to decrease the effectiveness of songs developed by male cowbirds in eliciting female receptive posturing (West *et al.*, 1981).

Descriptively, rules of relations between two animals' behavior may be simpler than the component movements of either individual (cf. Fentress, 1981b; Golani, 1976, 1981; Moran and Fentress, 1979; Moran *et al.*, 1981; Simpson, 1973). Most analyses of movement relations in social behavior have, however, focused on interactions between adult animals—for example, "supplanting" (Moran, 1978; Moran *et al.*, 1981) and courtship movements in canids (Golani, 1981; Moran and Fentress, 1979; Golani, 1981). Relevant developmental data from our lab are discussed in the following section.

III. Some Personal Explorations

Our work has concentrated largely upon the detailed patterning of movement in rodents and canids. As a supplement to these focal studies we have also explored "higher-order" phenomena such as the development of skilled performance in humans (Fentress, 1983d) and "lower-order" problems of CNS structure (Fentress *et al.*, 1981). These latter observations are touched on only briefly here.

The primary goal in our work has been to determine *rules of order* in the sequential and hierarchical organization of mammalian behavior (Fentress, 1981b). In particular, we have been concerned with "boundary conditions" (Fentress, 1976) that delimit adaptive performance. Analyses of this problem involve the search for *both* separations and cohesions in behavioral expressions, including the developmental substrates of these expressions. The dynamic, relational, and multileveled properties of behavior have been a major concern (Fentress, 1983a, 1984).

A. Rodent Motor Patterns

Rodent motor patterns provide a convenient assay of many of the issues discussed in earlier sections of this paper. Grooming, for example, is a motor pattern that occurs in predictable contexts, has recognizable actions within a defined hierarchical structure, permits evaluations of both central and peripheral determinants, and reflects both genetic substrates and developmental transitions in expression and control (Fentress, 1972, 1978, 1981a,b, 1983b). As the details of our studies of rodent grooming and associated motor patterns have been reviewed previously, we shall restrict ourselves here to representative highlights.

1. CONTEXTUAL AND DIMENSIONAL ANALYSES. In adult rodents, grooming occurs reliably under specified contexts such as during transitions between locomotion and resting states. Once the animals are in these transitional states, moderate stressors in the appropriate context can increase the probability of grooming behavior and later the distribution of movements employed. Strong stressors block grooming. These data indicate subtle interplay among underlying motivational systems. Control processes are more usefully conceptualized as being dynamically organized and partially overlapping, rather than being either totally specific or nonspecific (Fentress, 1973a). Similar conceptualizations are now being applied to a number of higher CNS functions (see, e.g., Mountcastle, 1979).

In young rodents, transitions into grooming can often be predicted in terms of lower-level sensory and motor processes. For example, if a mouse approximately 1 week of age moves its paw in proximity to its face during locomotion (or even swimming!), it is likely to go into a bout of grooming behavior. Similar transitions can be seen in adult animals with disrupted higher CNS functions (such as following KC1 applications to the cortex; Fentress, 1977).

When mice even a few hours old are supported in an upright sitting posture, they frequently exhibit grooming-like movements. Further, once in the appropriate posture, light pinches of the tail can elicit grooming. Our data indicate that the motor apparatus relevant to grooming can function (albeit imperfectly) at birth. When the neonatal mouse is placed in an appropriate sitting posture, various extraneous sources of activation can generate grooming-like movements. (For a recent review of activation in behavioral development see Wolgin, 1982. A complementary model, supported by a number of ethological and related studies, is that activation is often indirect in that it operates by releasing high-probability motor patterns from inhibition by normally competing activities. For reviews of the "disinhibition" model in the context of "activation" see Fentress, 1976, 1983a).

Adult grooming sequences include a number of recognizable stroke types, based upon contact pathways between the forelimbs and the face. These strokes can be classified into several higher-order groupings that are temporally organized by definable rules (Fentress and Stilwell, 1973; Figure 3). While recognizable by the trained observer, these stroke types and their higher-order groupings are not invariant in form. When individual dimensions of grooming movements such as velocity and amplitude are examined for a number of grooming bouts, they form a continuous distribution (Woolridge, 1975). Within a bout, repeated movements (e.g., "overhands") tend to increase in their amplitude and duration. Further, the form of a defined movement may vary in detail as a function of other movements that either precede or follow (as in the phenomenon of coarticulation among successive phonemes in human speech). This incomplete separation of movement profiles over time contributes to the smooth appearance of adult grooming behavior.

JOHN C. FENTRESS
AND PETER J.
McLEOD

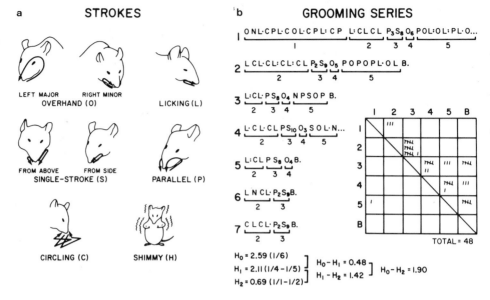

Fig. 3. (a) Cartoon representation of some major stroke "types" in mouse facial grooming. (b) Depiction of seven individual grooming series. The letters refer to stroke types while the numbers beneath the brackets refer to "higher-order" groupings of these strokes. The table to the right represents sequential transitions among the groupings, with information statistical evaluation of groups given by Ho (number of items), H1 (different probabilities of items), and H2 (sequential dependencies among item dyads). The progressive lowering of values from Ho through H2 indicates that grooming units have differential probabilities and are also linked by stochastic rules. (Modified from Fentress and Stilwell, 1973.)

It is not until mice are approximately 10 days of age that individual actions become readily definable by criteria derived from adults. As the stroke types take on a characteristic adult form they also become arranged into higher-order groupings. Both in adult grooming sequences and in the development of grooming motor patterns one can also see an overall cephalocaudal progression (see also Richmond and Sachs, 1980). This is true only in a statistical sense, however. When more detailed analyses are employed, a number of individual exceptions can be found (Fentress, 1984).

An example of individual stroke development and the arrangement of these strokes into higher-order groupings is shown in Figure 4. During the central phase of facial grooming sequences adult mice perform an uninterrupted series of "single strokes," followed by a series of "overhands." During the single-stroke series the forelimbs move simultaneously but alternate between larger and smaller movements along the sides of the face. Each movement is approximately 100 msec in duration. The overhand strokes are larger in amplitude and cross over the midline of the head.

High-speed film records show that 10-day-old mice articulate single strokes with adult duration, but these strokes are separated from one another by momentary pauses. The forelimbs in young mice also frequently fail to alternate reliably between larger- and smaller-amplitude strokes. The single strokes as a group tend to precede the overhand series, as in adult higher-order groupings, but other strokes intervene between these groupings. During development the invariant adult form of the behavior appears rather suddenly between postnatal days 16 and 19 (Figure 4). This suggests that during development capacities for higher-order sequential parameters of movement may crystallize in a relatively discontinuous manner, a position that is compatible with a number of other data in both the neurobiological and behavioral literature (Gierer, 1981; Immelmann *et al.*, 1981).

There are thus a variety of developmental changes in grooming that occur more or less together. Our data also indicate limitations to the use of simple "unit" concepts. For example, "stroke types" change their form as well as their rules of combination as the animals grow older. In certain cases it can be rather arbitrary to designate a movement either as an immature version of a stroke defined for adults or as a "different" stroke. For very young animals even the designation of grooming versus nongrooming can be troublesome. Infant mice often go through forelimb movements that are similar to those found during adult grooming yet fail to make contact with the face. In other (more rare) instances, facial contacts can be made but via strokes that resemble locomotor patterns of alternating limb movements rather than the simultaneous movements of the forelimbs found in adult grooming. Immature grooming movements are also characterized by jerky limb trajectories, imperfect coordination between rotational and planar movements, and so on.

From such considerations Golani and Fentress have recently completed a detailed analysis of the structure of grooming behavior in mice from birth through postnatal day 14. Grooming was examined from the three complementary perspectives of forepaw to face contacts, forelimb trajectories, and limb segment kinematics. Rather than dividing the grooming behavior into stroke types as in our previous work, the movements were described in a continuous fashion from each

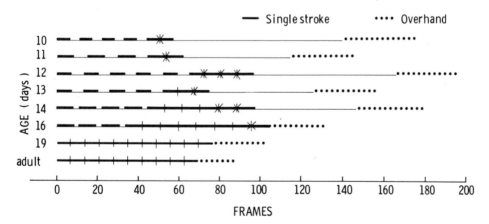

Fig. 4. Summary representation of single-stroke (solid bars) and overhand (dotted bars) grooming clusters in 10-day through 19-day and adult DBA/2J mice. Film frames @ 64 fps are given on the abscissa. Thin lines during single-stroke series represent pauses in movement. Thin lines between single strokes and overhands represent pauses and miscellaneous stroke types. X's in single-stroke series indicate failure to alternate between contralateral major and minor stroke trajectories. (From Fentress, 1981a.)

of these perspectives. This was done by applying a form of dance notation, the Eshkol-Wachmann choreography system (Golani, 1976). This system permits one to perform a fine-grained dissection of both simultaneous and sequential movement dimensions, and also to determine the rules by which these movement dimensions are combined.

From a few hours following birth, individually marked mice were placed within a mirrored filming chamber that allowed orthogonal viewing perspectives of their movements. Films were taken at 100 fps via a combined camera and strobe system that provided fine resolution of the movements. The mirrored chamber also served to support the young mice in an upright sitting posture, thereby facilitating the expression of grooming-like movements. These movements were documented for individual limb segments both in terms of their relations to the camera and in terms of their relations to one another. Illustrations of some of the early movements from the perspective of forepaw trajectories and contacts with the face are shown in Figure 5.

Early grooming development can be divided into three descriptive phases.

1. During the first 4–5 postnatal (0–100 hr) days contact paths along the face are variable, ranging from momentary point locations on different parts of the face (Figure 5A–E) to long traces that extend above the eyes (Figure 5F). Contact may be missed altogether (Figure 5B and D, left paw), and when contact does occur a forelimb may remain as if "stuck" on one part of the face for a considerable time (see, e.g., Figure 5A, right paw) or "bounce" from the face in the horizontal plane (see, e.g., Figure 5D, right paw). When contact pathways do occur they are seen during the downward phase of forelimb movements, as in adult grooming. During early grooming the head is held in a relatively fixed position, whereas in adults head and arm movements work together in producing the contact pathways. The forelimb trajectories during these first few days of life are often highly irregular, with frequent excessive displacements both vertically and horizontally (see, e.g., Figure 5B). Individual forelimb segments are also often poorly coordinated with one another, and within a limb segment rotational and planar aspects of the movement may be temporally disconnected. This is in contrast to adult movements where a limb is rotated as it is lifted and lowered, thereby bringing the thumb region of the paw into smooth contact with the face.

2. Between postnatal days 4 and 7 (100–200 hr) the contact pathways between the forepaws and face become consistent but are restricted to the rostral aspect of the snout. Also at this time, the limbs follow a restricted trajectory and the components of movement become more tightly coordinated with one another. This second developmental phase can be summarized in terms of improved precision of movement with a concomitant loss in the variety of movements seen earlier.

3. From approximately postnatal day 8 (200 hr +) the movements become re-elaborated. By postnatal day 10 or so stroke profiles can be classified in terms of adult criteria. Occasional poorly coordinated movements can still be observed through the first 2.5 weeks, over which time the sequential structure of grooming strokes is also becoming elaborated. The movements soon take on a rich and smoothly flowing appearance, with the head, body, and forelimbs contributing to

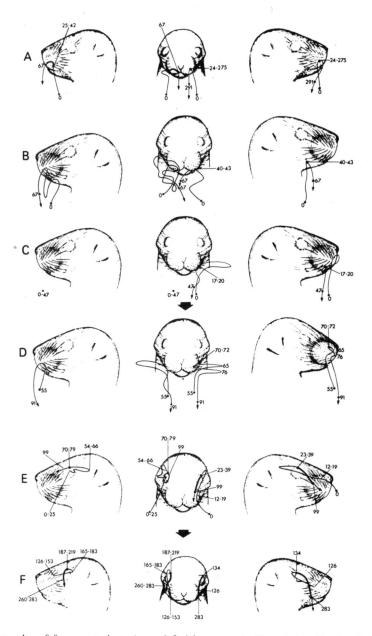

Fig. 5. Examples of forepaw trajectories and facial contacts in 1- to 4-day-old mice during "facial grooming." Numbers indicate film frames at 100 fps. Note irregularity of trajectories and contacts. Forelimb trajectories without facial contact are indicated by the thin lines; contact pathways are indicated by the heavy lines; static forepaw to face contacts are indicated by the small filled circles on the mouse's face. Successive strokes in a grooming sequence are shown by the heavy connecting arrows between rows C + D and E + F. (From Golani and Fentress, in press.) These drawings do not illustrate articulation of individual limb segments, which are also quite irregular during this phase. Note that mirror images are shown for frontal views, thus preserving normal orientation for side views.

the production of contact pathways. Consistent contact pathways can now be accomplished by a wide variety of individual movement combinations.

In summary, these observations emphasize the rich multidimensional nature of motor development and demonstrate the value of examining the same movement from several complementary perspectives. They also illustrate that movement may show a number of regressions as well as progressions along various dimensions (e.g., paw-to-face contact pathways). Further, they suggest that as the animal matures it gains the ability to accomplish the same "goal" in movement (here contact pathways between the forepaws and face) through an increasing variety of movements. This can be accomplished only when there are mutually compensatory relations among the individual aspects of movement (cf. Lashley, 1951). Compensatory relations imply the operation of complementary layers of motor control.

There are a number of parallels between these data on rodent grooming and other areas of motor development. That the multidimensionality of movement necessitates a variety of analytical perspectives, for example, has been emphasized by Golani (1981) in terms of both bodywise and external attractors. Bower (1979) has reviewed considerable evidence on human motor development that indicates that regressions as well as progressions along a number of expressive dimensions may be the rule rather than the exception. Kelso (1981) has argued that the mature expression of most forms of human motor behavior are characterized by invariants among individual movement properties, and Hofer (1981) has provided an excellent review concerning the balanced developmental tendencies among differentiation and integration that contribute to these adaptive characteristics of motor behavior in the mature individual (cf. Fentress, 1984).

2. COMPENSATORY RESPONSES TO PERTURBATIONS. An important implication of the preceding considerations is that individual aspects of an integrated motor sequence often exhibit mutually compensatory relations. Thus if one reaches toward an object it is possible to adjust the relative tensions of muscle groups in response to unexpected perturbations (e.g., phasic loading of a limb). In this way the limb can maintain its overall ("goal-directed") trajectory. A question of considerable interest is whether immature organisms are limited in this adaptive capacity. Even in adult organisms the degree of adaptive response to experimentally applied perturbations may vary with the broader contexts within which motor performance is evaluated (Berridge and Fentress, in press; Fentress, 1976, 1983a).

By applying both tonic and phasic disturbances to an animal, properties of motor organization, such as central patterning mechanisms, can be clarified. It is clear that adult rodent grooming is to a large extent centrally organized. This can be shown by removing tactile input from the face (via section of the sensory branches of the trigeminal nerve) and by eliminating normal routes of proprioceptive input (via sections of the appropriate dorsal roots). Following each of these manipulations (see, e.g., Fentress, 1972) grooming patterns can still be observed. Also, when mild peripheral irritants are applied to the backs of mice, they still initiate grooming sequences with their faces.

However, simple dichotomies between central and peripheral control can rarely, if ever, be justified. For example, following dorsal root lesions both the speed and precision of (especially large) grooming movements are often reduced.

The strength of peripheral disruptors is also an important consideration (Fentress, 1972, 1981a). Most interestingly, the consequences of sensory manipulations can vary as a function of the context, phase, and/or details of motor performance. After trigeminal nerve sections the rapid grooming movements that occur in strange environments are minimally affected, whereas more slowly performed movements in an animal's home cage can be quite markedly altered. Berridge and Fentress (1984) have recently completed an assessment of facial grooming in rats with sensory trigeminal lesions. These data emphasize the importance of evaluating both contexts of expression and the various dimensions over which grooming expression occurs.

By applying phasic loads to the forelimbs during various phases of grooming, Michael Woolridge (unpublished) has demonstrated the *dynamic balance* between central and peripheral contributions (Figure 6). When mice are engaged in large-amplitude and relatively slow grooming movements (e.g., "overhands"), pulling the limbs out from the face terminates the grooming sequence. However, when engaged in rapid and smaller grooming strokes (e.g., "single strokes"), the same animals continue to perform that phase of the grooming cycle even though the paws are no longer able to contact the face. Removal of the vibrissae increases the probability that rapid grooming phases will be interrupted, and even when the cycle is not terminated the animals employ stronger inward movements of the paws toward the face. Thus, though rapid grooming has a strong central component, peripheral events may still exert some influence.

The normal course of grooming development can be relatively unaffected even after dramatic changes in an animal's sensory experience. Even when the lower forelimbs are surgically removed at birth, one can trace a developmental progression in upper arm rotations that are very similar to those found in normal ontogeny (Fentress, 1973b). Even details such as rhythmic eye closure and tongue protrusions occur at the appropriate age and grooming phase in animals who have not had the opportunity to contact their faces with their forepaws since birth. Animals amputated at birth do show some postural adjustments as well as slight modifications in the temporal structure of upper arm movements, however. In rats, grooming-like movements have been observed prior to birth (Angulo y González, 1932; Narayanan *et al.*, 1971). Much of the grooming motor circuitry is therefore already present *in utero*.

In our laboratory, Ilan Golani has filmed young mice huddled among littermates in their nests. By 10 days of age mice that are unable to move one forelimb in a normal manner because of contacts with littermates will often continue to make symmetrical contact pathways between their forelimbs and face during grooming. This is accomplished by adjusting other body segments, including the head, trunk, and lower legs. In younger animals such adjustments are less apparent.

These data allow one to evaluate relative control priorities in motor performance. By the time mice are 10 days old, *invariance* in grooming function is accompanied by *flexibility* in the motor patterns used to accomplish this function. Our

JOHN C. FENTRESS
AND PETER J.
McLEOD

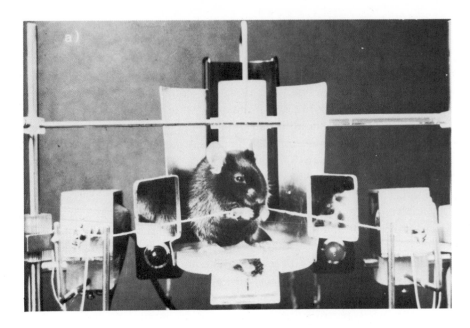

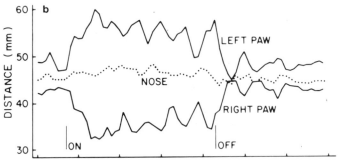

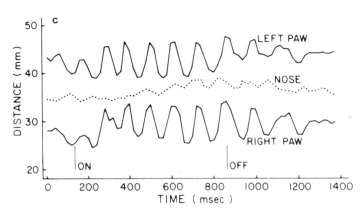

Fig. 6. (a) Mouse in filming apparatus that permits application of phasic loads to the limbs during different phases of grooming behavior. (b and c) Example of grooming strokes during rapid phase before and after pulling of the forelimbs from the face. "Horizontal paw/nose movements" (b) is view of two limbs and nose of mouse from above; "vertical paw/nose movements" (c) is view from the side, with the two limbs separated for visual presentation. (From Woolridge and Fentress, unpublished.)

preliminary data suggest that the invariance in performance shifts during development from individual details in motor performance to *relational properties* among these details. As noted by Golani (1981) a valuable perspective in studies of behavioral development is to assess *relative degrees of invariance* along specified dimensions of performance (including combinatorial dimensions).

A reasonable assumption is that properties of movement that are simplest in their description (i.e., relatively invariant) represent the functional "goals" of the system. Variations represent the strategies by which these goals are accomplished. Both goals and strategies can be scaled in terms of their relative priorities, often with counterintuitive results. For example, in one set of observations Golani and Fentress found that when a single forelimb was pulled outward from the face (via an attached thread) during grooming behavior in an adult mouse, the other limb followed, breaking all contact with the face. The limbs continued to oscillate in unison without contacting the face for several cycles. In this instance 'the priority was maintenance of forelimb coupling over facial contact. Tracing the developmental changes in such priorities is likely to provide a productive line of future research.

3. SUMMARY AND CONCLUSIONS. Rodent facial grooming involves richly orchestrated movement sequences that can be described and analyzed at several levels. In adult animals it is clear that a number of movement dimensions are orchestrated together. During development *both* differentiation of individual motor properties and their improved integration can be observed. We have previously reviewed many analogous cases of motor development, which range from relatively simple elicited actions to the much more complex phenomena of human speech.

Our data indicate that the sequential structure of grooming is to a large extent centrally controlled, but the animals are also able to adjust (within limits) to changes in sensory information. The data also indicate that normal maturational processes can result in grooming movements even after major disruptions of experience, although there can be adjustments in the detailed performance of such movements. *Thus in both the integration and development of grooming behavior there is a dynamic balance between tendencies toward intrinsic organization and interaction with the surround.* This dynamic balance between intrinsic organization and surround interaction is a principle that can be applied to virtually all documented cases of motor, as well as neural, development (Fentress, 1984; Gierer, 1981).

Relative invariances in grooming movements are often most obvious when rules of relation among motor dimensions are examined over several complementary levels of organization. Motor sequences do not necessarily become descriptively more simple at lower levels of organization. Developmental analyses of changes in the foci of relative invariances provide a valuable analytical tool. Through such analyses it is possible to determine changes in control priorities over various aspects of motor expression.

For a behavior as rich in its detailed structure as is grooming, one can anticipate the intimate participation of a number of control (e.g., CNS) mechanisms. The differential contribution of these mechanisms during ontogeny can also be anticipated. Analyses of such mechanisms are very much needed in developing systems, but they will only become fruitful after a sound descriptive base of the various aspects of movement is established.

We have just begun to examine central nervous system mechanisms in grooming development. Potentially fruitful approaches will be to employ neurological mutant animals (e.g., Northup, 1977) in combination with histological evaluations of motor regions (e.g. Fentress *et al.*, 1981) and electrophysiological stimulation and recording techniques (Fentress, 1977; cf. Moran *et al.*, 1983).

B. Conditioned Movement Sequences and Stereotypies

1. CORTICAL CONDITIONING. One relatively unexplored approach to the establishment of motor patterns is to activate neural loci relevant to individual modes of motor expression in a protracted and sequential manner. Fentress (unpublished) provided repeated sequential stimulation of motor cortical loci in squirrel monkeys *(Saimiri sciureus)*. Stimulation of one right motor cortical site, for example, might characteristically lead to a left forelimb movement, while similar electrophysiological stimulation of a left motor cortical site might lead to a specific right arm movement (Figure 7). The question of immediate interest was whether the sequential stimulation of contralateral motor loci could lead to functional connection between them, as measured by the systematic occurrence of movement "B" after it had been repeatedly paired with movement "A." For test purposes, cortical locus A was stimulated alone (with controls for pseudoconditioning, sensitization, etc.).

In summary, there are five basic points from these experiments. (1) After repeated sequential stimulation of motor cortical loci in squirrel monkeys, stimulation of the first locus alone began to elicit both arm movements (Figure 7A). (2) These arm movements not only occurred in the proper overall sequence, but also occurred with the appropriate temporal relations. (3) The production of limb movements was often preceded by characteristic patterns of EMG activity (Figure 7B). (4) There were obvious constraints on movement sequences that could be produced by cortical conditioning. Unnatural movements not only were difficult to establish, but also frequently became transformed into movements that were more characteristic of the species. (5) Once well established through cortical conditioning procedures, either one or both of the motor actions could be elicited by a number of environmental disturbances. For example, a brief exogenous startle stimulus might lead to the performance of either one or both of the movements previously elicited by stimulation of motor cortical loci. This last fact raises questions of the processes that occur when motor patterns are activated in a perseverant manner during development.

2. PERSEVERANT STEREOTYPIES. Severe environmental restriction can lead to the perseverant activation of certain motor patterns at the expense of others (Fentress, 1983a; Fox, 1968). These motor patterns differ for different species, and once well established, they can be activated by a variety of environmental disturbances (Fentress, 1976). They also show a high degree of central programming in the sense that they can persist under changed conditions in the environment to

**STIMULATION: RIGHT CORTEX
RECORD: RIGHT BICEPS**

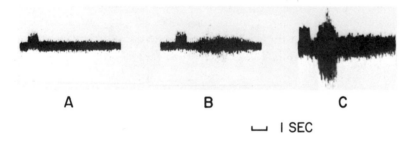

A   B   C

└─┘ I SEC

Fig. 7. (a) Both a relatively normal (left) and abnormal (right) arm movement elicited by electrical stimulation of the motor cortex in a squirrel monkey. (b) EMG responses in unstimulated limb (A) before pairing with contralateral stimulation, (B) during early conditioning, and (C) after full conditioning. The EMG of the previously paired limb grows in amplitude and becomes restricted to the appropriate temporal frame.

which the animals normally adjust. This perseverant rigidity is particularly marked in animals who are strongly activated.

When field voles *(Microtus agrestis)* and woodland voles *(Clethrionomys britannicus)* were housed in closed containers for several months, each developed perseverant movements around a glass water spout. The movements shown by *Microtus* were smooth "weaving" actions around the spout, while those shown by *Clethrion-*

omys involved repeated jumping of the spout. Each of the motor patterns was an exaggeration of actions normally observed in these ecologically distinct species. Each of the perseverant motor patterns could be activated by a number of environmental disturbances, and when the patterns were strongly activated, each persisted even after the water spout was removed.

A Cape hunting dog *(Lycaon pictus)* observed at the London Zoo (Fentress, 1976) developed a highly regular stereotyped pace pattern that involved the tight coordination of a number of exaggerated motor components. The probability of these movements and the tightness of their coupling could be enhanced by various environmental distrubances, such as impending feeding or the approach of schoolchildren. As illustration of the extreme central determination that can operate in perseverantly exaggerated movements, when highly activated this animal continued to jump a chain that had been lowered to the ground for several days.

3. COMMENT. Both cortically conditioned movement sequences and perseverant motor sequences that result from environmental restriction illustrate in simplified form processes that can be assumed to contribute to the ontogeny of normal motor patterns. (1) The repeated activation of central circuits, either directly or through environmental contingencies, can exaggerate predispositions that the animal has at the time. (2) Motor components can be linked in novel ways, but these ways are limited by the animals' inherent predispositions. (3) Once such motor sequences become well established, they may occur at the expense of alternative forms of movement and may be activated by a number of environmental disturbances. (4) They may, under such circumstances, show an extreme rigidity in their form, even to the point where the animals do not adjust to changes in the environmental stimuli toward which the motor sequences are normally oriented. We have seen analogues to each of these phenomena in the literature reviewed previously. We turn now to more subtle aspects of normal motor development that we have observed in captive wolves.

C. CANID MOTOR PATTERNS

We have extended our analyses of motor development to include both individual and socially coordinated movements in wolves *(Canis lupus)* and coyotes *(Canis latrans)*. Through close observation of the development of a number of animals for two decades (see, e.g., Fentress, 1967; Fentress *et al.,* in press) we have sought possible parallels to the principles of motor development that have emerged from our laboratory research.

1. DIMENSIONS OF INDIVIDUAL MOVEMENTS. Wolf pups frequently express precocial motor components that are poorly oriented with respect to the environment. Pelvic thrusting movements, for example, were observed in a hand-reared male at several weeks, and early pouncing movements could be elicited by objects moving through the grass even though the orientation of these pounces was far from perfect (Fentress, 1967). Improved coupling among the four limbs during locomotion is also apparent during development. Havkin (1981) observed that when running

across a raised surface, one leg (usually a hindlimb) would often be "left behind," causing the animal to trip. When films of these early locomotor movements were examined, it was clear that (as in young mice, Section III.A) the individual limbs frequently failed to work as a closely integrated unit. Further, motor components such as raising and rotation of a limb often operated in a fragmented fashion. Other changes in motor coordination appeared to be due to an overall increase in muscular strength and balance.

Relations among movement and sensory factors can also change with age. It is not uncommon to observe one wolf pup chewing upon another and then redirecting its chewing to one of its own forepaws or an environmental object. Such observations contribute to the impression that motor activity in young wolves operates somewhat autonomously from what later becomes a goal direction. At a somewhat broader level of discourse, many early motor patterns appear to exist without later developing separations between the individual organism and its surround (e.g., "finding out about yourself"; Blass, personal communication, 1984).

We have made similar observations in children (as well as rodents) of surprising transitions between different functionally defined classes of behavior that share common movements. It is as if the young organism is "trapped" by the movement, with little regard for the movement's functions (Fentress, 1981b, 1983a,c,d) or for environmental constraints.

Young wolves can often be dramatically "trapped" by unexpected sensory events as well. Havkin (1981) has observed wolf pups engaged in running play suddenly drop to a sleeping mat that is in their path. Sensorimotor events can also be used to manipulate the behavioral states of young animals. Heather Parr (unpublished) has observed that pups who have attempted to flee from a human handler can be lured into play by roughly pulling upon a forelimb (as pups do to one another in play). In these circumstances it is as if the motor actions themselves (plus their sensory consequences) reset the young animals' motivational states.

2. VOCAL MOTOR PATTERNS AND RESPONSES. The vocal motor patterns of adult canids serve important social functions. Through the use of spectrograms and various methods of digital analysis, it is possible to classify these vocalizations into a small number of groupings. In wolves, vocalizations can be classified broadly into howls, squeaks, whines, barks, and growls (Fentress, 1967; Field, 1981). These vocalizations occur in specified contexts and have definable effects upon the behavior of other animals.

Within these vocal categories there is considerable heterogeneity, some of which can be accounted for by the specific context(s) within which the sounds occur. Squeaking sounds directed toward pups or adults, for example, can be distinguished (Fentress *et al.,* 1978). Within a protracted series of squeaks, systematic transitions in sound properties (intensity, duration, frequency modulation) may also occur. In this sense the production of individual squeaks is not an autonomous event; the vocalization details depend in part upon contextual associations. The vocalization patterns of pups, however, also exhibit greater variability (in syllable duration as well as greater frequency fluctuations) within a given context than is found for adults (Field, 1981). Research by Marler and his associates [see, e.g., Marler and Peters, 1982 (birds) and Seyfarth *et al.*, 1980b (primates)] supports the generality of this contention.

3. SOCIALLY COORDINATED ACTIONS. An approach that permits a logical extension to the observations and issues raised earlier is to observe two or more interacting animals. Here the task of the animal is not only to coordinate its own motor patterns from an internal or even stationary environmental reference, but to adjust its movements to those of a moving social partner (Golani, 1981). Moran (1981; see also Moran *et al.*, 1981) has shown that for adult wolves certain properties of "ritualized fighting" can be described best in terms of the dynamic relations among the movements of the individual participants. As in a dance, each animal compensates for movements of its partner. This compensation maintains consistent relations between the animals even though each individual may be performing highly variable motor patterns. The partners become part of a common system of expression. These movement relations can be described continuously and across a number of dimensions. The resulting descriptions reveal a relatively small number of stabilities between the two animals, as well as systematic transitions between these stabilities (a "social interaction space").

Developmentally, Havkin (1977, 1981; Havkin and Fentress, 1985) has looked at the relational aspects of wolf–pup interactions in terms of the degree of symmetry in their reciprocal movements (e.g., simultaneous snout-to-body contacts) and the combative strategies used at different ages. Havkin's (1977) quantitative measurements of contact symmetry between pups 4–6 weeks and 10–12 weeks old indicated that while younger pups made "loosely symmetrical" reciprocal contacts, the older pups' contacts were either highly symmetrical or highly asymmetrical. In adult wolves, many interactions are asymmetrical in that two individuals often perform opposite and compensatory movements *in relation to* their social partner(s). An adult may play the same "role" in these interactions over periods of weeks or even years (cf. Moran, 1978; Moran *et al.*, 1981). These data suggest that as the animals grow older their actions become increasingly constrained by the actions of their social partner(s).

In combative social interactions wolf pups frequently topple one another over (Figure 8). Havkin (1981; Havkin and Fentress, 1985) documented the developmental progression of such losses of balance. Pups 3 weeks of age usually fell sideways, absorbing all the momentum of the fall in a very short period of time. By 40 days, back falls occurred whereby the pups rolled to rest more slowly and in a more controlled manner. A third type of fall, the "forward fall," was first observed at 53 days of age. This fall involved postural changes in a falling animal that could assist in projecting itself over its head and shoulders. Forward falls may be initiated prior to contact with the partner. They result in a rapid change in the falling pup's orientation relative to the standing pup, and release of the standing pup's bite. In addition, the predictability of both the falling pup's final head position (and weaponry) and the fall duration could further benefit the falling pup, suggesting that this strategy may be used in active defense (Havkin, 1981).

These data show that some developmental changes in a young wolf's behavior can be explained most readily in relation to another animal's behavior. During the very earliest contacts between wolf pups there is relatively little reciprocity in their motor behavior. One pup may "attack" its partner and go through a variety of biting and pawing movements while the recipient animal acts as if it is oblivious to the initiator's actions. Later, three ontogenetic phases of social interaction (analogous to those found for interactions between limb segments in the individual animal; see Section III.A.1) may appear. In the first phase, the repertoires of motor patterns are rich, but with only loose coordination among the animals. In the second phase, the animals achieve interactions that are stylized, but impoverished in their form. In the third phase, compensatory relations in response to the animal's partner are orchestrated smoothly. Function gains ascendance over form.

These perspectives on social motor development do not involve concepts of static behavioral "units," as usually defined, and allow one to examine movement from several complementary perspectives. The approach has shown that the dynamics of two animals together may preserve a relatively fixed relation between them (e.g., "T" appositions, where one animal stands perpendicularly to the other and makes contact with its partner approximately midway along the partner's body axis; Havkin, 1981) or lead to systematic transitions in the relations between animals (e.g., "circles," "follows," and "twists and turns" described by Moran, 1981). "Units" become not only relative, but explicitly relational and dynamic in their

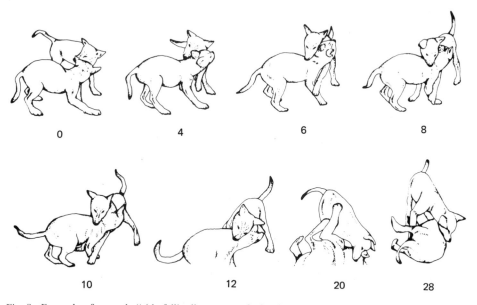

0   4   6   8

10   12   20   28

Fig. 8. Example of an early "side-falling" sequence during interactions between two wolf pups. Numbers indicate film frames at 16 fps. In later development more controlled "backward" and "forward" falls occur. (From Havkin and Fentress, 1985)

organization. Levels of relative stability and change can then be subjected to developmental dissection.

4. SUMMARY AND CONCLUSIONS. The development of canid motor patterns can be studied fruitfully at both the individual and social levels. In each instance differentiation of action and the improved integration of action into functional ensembles is apparent. That many motor *capacities* develop within a changing social context emphasizes the importance of evaluating the *strategies* of action employed by animals at complementary levels of organization.

D. EMERGENT ISSUES

Two basic issues of relevance emerge from both the observational studies on motor development in canids and the more fully controlled (and limited) laboratory studies previously discussed.

1. CENTRAL–PERIPHERAL DYNAMICS. To classify a behavior as social is to imply that the behavior (in our case, motor patterns) of one animal is constrained by the behavior of one or more other animals. The extreme alternative is that individual animals "program" their motor patterns independently of conspecifics. This appears to be the case in very early "social" motor patterns, as it is also the case in individually separable movement dimensions within the isolated organism. More directly, one can ask whether the motor patterns expressed during the development of an organism reflect influences from that organism's environment.

Our observations suggest two facets to this question. The first concerns the dynamics of motor behavior for any given phase of development. The second concerns the possibility of systematic changes during ontogeny.

With respect to the first perspective, in terms of individually coordinated and socially coordinated actions, animals show dynamic relations between tendencies toward interaction with the external world and self-organization (Fentress, 1976). More specifically, when animals are engaged *either* in individual activities *or* in social behavior, an increase in the intensity of expression may reduce the animal's reliance upon, and sensitivity to, otherwise relevant extrinsic cues. We have on several occasions been surprised, for example, at how strongly motivated fighting behavior in wolves can persist in contexts that would interrupt less strongly activated behaviors (Fentress *et al.*, 1978). This provides an analogue to the self-organizing properties of strongly activated motor patterns within the individual in that strongly activated pathways may block potential activity among other pathways, at a number of levels (Fentress, 1976, 1983b, 1984; Mountcastle, 1979). Reduced sensitivity to certain stimuli may, of course, accompany an increased sensitivity to others.

The next question is whether there are developmental transitions in the balance between central and peripheral control. For both individual and social motor patterns we have found that this is the case. Adjustment to, and reliance upon, environmental cues have often been reported to change during development (see, e.g., Moran *et al.*, 1983). These transitions can occur in either of two directions, depending upon the species, motor pattern studied, experimental context(s), and so on. Thus CNS stimulation elicits certain motor patterns in young rats more reliably in the absence of environmental cues that facilitiate elicitation in older rats (Moran *et al.*, 1983), while bird songs and many human motor skills become less dependent on afferent information with practice (Bottjer and Arnold, Chapter 4, this volume; Keele, 1981).

2. HIERARCHICAL COHESIONS. Comments in the previous sections suggest that it is useful to view motor patterns in development as hierarchically ordered in time. Again, there are two perspectives that can be offered; those of motor integration per se and those of development.

The first perspective rests upon the idea that in both social and individual behavior one commonly finds a regression toward more simple and stereotyped forms of motor expression during periods of stress, fatigue, neurological dysfunction, and so on (Fentress, 1983b). The assumption of most reasonable models is that these simplifications in motor output reflect increased focus upon lower levels of organization, those that are normally predominant during early phases of ontogeny (cf. Doty, 1976; Teitelbaum, 1982). The second, developmental perspective has been documented in many places during the present review. Complementary layers of organization can occur throughout ontogeny, and the ascendance of higher functions may produce important modulations of more basic movement properties.

Our observations on canids add important impetus to the argument that higher levels of motor cohesion can exist in the face of lower-order fluctuations. A wolf skilled in fighting can adjust its individual motor actions to compensate for both actual and anticipated motor patterns of an advisary. Younger wolves are rarely able to perform such adjustments skillfully. The important point in these observations is that what one might call "units" of motor organization develop an increased hierarchical dependence. Properties of functional coherence take over the details of individual motor actions. The challenge for future research is to be able to define these various levels of motor organization in detail to allow quantitative evaluations. Here we see that what one might wish to define as a "unit" of motor behavior takes on increasingly dynamic and (most important) relational properties as the organism grows older. Units in this sense become reliable collective configurations that can include not only individual body parts but also other animals.

IV. DEVELOPMENTAL PROCESSES: A MOTOR PERSPECTIVE

A. CAPABILITIES AND STRATEGIES

Throughout this review we have encountered numerous dichotomies that are of both questionable validity and limited heuristic value. It is with some hesitation, therefore, that we now introduce another distinction—between two possible

JOHN C. FENTRESS
AND PETER J.
McLEOD

sources of developmental changes in overt behavior. A fundamental problem in motor pattern development concerns the extent to which age-related changes are attributable to (1) changing *capabilities* of the animal, or "internal" developmental changes, and (2) switching behavioral *strategies* in response to changes in both the animal's needs and its perceived environment. Although these two categories are not mutually exclusive, the distinction is conceptually useful. Fundamental concerns that arise, for example, include the changes in expressive strategies (both ontogenetically and phylogenetically) that might occur against a given substrate of motor capabilities (Blass, personal communication). Stated in a somewhat alternative form, if "capabilities" are already present, in either an ontogenetic or phylogenetic sense, how do we best acccount for strategic changes in the utilization of these capacities?

Changing capabilities of the developing animal can correlate with a number of events, including (1) changing neural organization (e.g., as new pathways are developed or as existing pathways degenerate), (2) physiological development (e.g., as independent ingestion, egestion, and thermoregulation become possible), (3) biomechanical changes (e.g., in the mass/strength ratio of limbs), (4) perceptual development involving changes in capabilities within systems and in the relative salience of difference modalities, and (5) developing "cognitive" and attentional abilities. As the aging individual's capabilities are changing (and these changes may also involve losing abilities), so too are the surroundings with which it must interact. Changes can occur in both the animal's social environment (e.g., as it leaves the nest or den) and nonsocial environment (e.g., seasonal changes in food availability, weather, etc.). The changing requirements of developing animals could also influence interactions within a stable environment. Postweaning mammals, for example, will behave differently to both solid food and their mothers because of different physical and psychological needs (cf. Tanner, 1970). Changes in the salience of various contextual factors are also observable in the "motor trap" phenomenon we have discussed.

The specific context in which an animal is found will determine, to a variable degree, the motor patterns that are observed. This context-specific repertoire will be a subset of the organism's capabilties at that phase of development. With changing contexts, different motor patterns may be used for the same or analogous functions. Such strategic changes might also occur in a static context in that a functionally unsuccessful behavior pattern could be replaced; that is, within the context-specific repertoire the animal may choose among several distinct motor patterns that can serve the same function. Under normal circumstances only a "preferred" behavior might be performed, leading to the false conclusion that the animal is only capable of performing this one behavior pattern. Without experimental manipulation of the functional effectiveness of motor patterns displayed by developing (and adult) individuals, and provision of necessary contextual supports that permit the expression of existing motor circuitry (see, e.g., Bekoff, 1981; Fentress, 1981a,b; Thelen, 1983; Zelazo, 1983), a complete repertoire of capabilities cannot be determined. At present such experimental data are largely lacking (but see Chapter 3). Reed (1982) has argued that the functional circumstances of an action will control the amount of variability observed in action patterns. An individual's true capabilities therefore might be expressed only under demanding task conditions.

The "role switching" observed in many mammalian social orders illustrates this concern (cf. Fentress, 1983b). For example, a change in the "dominance" relationship between wolves involves extensive changes in each individual's repertoire of *expressed* behavior (Fentress *et al.,* in press). Similarly, the performance of caregiving behavior by adult female wolves (to other adult females, juveniles, and pups) depends on that individual's breeding success for that season (Fentress and Ryon, 1983).

This distinction between "capabilities" and "strategies," though seldom explicitly addressed in the literature, is important in behavioral assessments of neuromuscular maturation, the contribution of developmental events to individual variability (both within and between individuals) in adults, and the continuous–discontinuous nature of ontogeny (cf. Thelen, 1983).

The issue of continuity in behavioral development has received much attention (see, e.g., Hall and Bryon, 1980; Immelmann *et al.,* 1981; Kugler *et al.,* 1982). In general, discontinuities are more likely if the environment changes (Sackett *et al.,* 1981). More precisely, discontinuities (in this restricted developmental sense) in overt behavior would be expected when the *perceived* environment changes or when strategic shifts occur.

In cases where the differential utilization of motor capabilities affects subsequent expressive profiles, the *link* between capabilities and strategies could determine later behavior to an unknown extent (see Section IV.A). By separating the issues of capabilities and strategies, rules of interconnection that relate to developmental processes may become apparent.

B. Perspectives on Mechanism

Our review has emphasized phenomenological properties of motor patterns in development. With respect to developmental mechanisms there is surprisingly little known. In terms of lower levels of motor control a growing consensus is that genetic predispositions buffer the organism from modifications by experience (see, e.g., Grillner, 1981). Genes must work in the context of their own environment, however, and this interplay between intrinsic and extrinsic properties of developing systems remains poorly understood. In a recent review Stent (1981) has argued convincingly that simple views on genetic "programming" cannot in themselves provide an adequate account for the dynamics of developmental processes. Our goal here is to highlight some of the problems we see in thinking about developing systems that are *both* self-organizing and interactive with their surround. This complements previous reviews where similar issues have been considered in the context of integrative functions (Fentress, 1976, 1983a).

Many early events in motor development occur with a high degree of specificity in the absence of environmental influences that contribute to later performance (see, e.g., Bruce and Tatton, 1980; Hollyday, 1980; Landmesser, 1980; Chapter 1, this volume). For example, neuromuscular connections form before the motoneurons connect to interneurons, with sensory neurons connecting upon interneurons still later. Such "retrograde" sequences have been observed in the human spinal cord (Okado *et al.*, 1971) and have been subjected to precise *in vitro* analyses in laboratory animals (see, e.g., Saito, 1979).

The absence of sensory influences does not imply that "experience" in a more broadly defined sense is unimportant. Neural influences upon muscle differentiation are a case in point (Brown and Booth, 1983; Metafora *et al.*, 1980). That unusual sensory events can have important ramifications upon subsequent motor performance is revealed dramatically in human breech infants (Prechtl, 1965).

A number of current investigators have attempted to specify different roles and limitations of experience for particular dimensions of behavioral expression (for reviews see Aslin, 1981; Bateson, 1981; Gottlieb, 1981). As sketched in Figure 9 one can, in principle, have either a narrow or wide range of environmental inputs affecting either a narrow or wide range of motor inputs. Further, these inputs may serve roles such as induction, facilitation, or maintenance (Gottlieb, 1981). The degree of influence resulting from any given environmental event or condition can vary, in principle, over a wide range and can involve a number of mechanisms. Developmental connections among motor patterns in ontogeny may be either weak or strong, and may depend either upon successive phases of motor performance or merely upon the "developmental scaffolding" of underlying neural circuitry.

In spite of increasingly sophisticated approaches to behavioral development, a persistent view is that organisms (and their nervous systems) develop with few

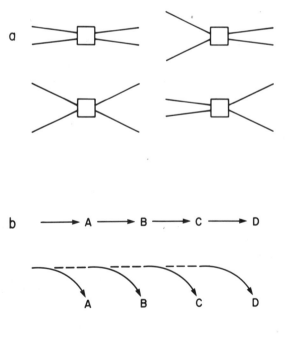

Fig. 9. (a) Schematic representation of the fact that either the antecedents or consequences of experience in development may be relatively specific or nonspecific, thus yielding four logical possibilities. (b) Performance of motor patterns [A, B, C, D] may contribute to the development of subsequent motor patterns more or less directly (above), or there may be a continuity among the precursors of these activities (e.g., neural machinery) even though motor performance itself is of little developmental consequence (below).

intrinsic biases. The presumption follows that the environment provides detailed "instructions" that are accepted more or less at face value. In this sense, the developing organism is viewed as a *mirror* of the experience to which it has been exposed. Numerous limitations to such models can be articulated. Perhaps the most telling of these is that they place the "brain too much at the mercy of the outside world" (Edelman, 1978, p. 54).

In nature, animals have variable and imperfect environments. Chomsky (1980) has written thoughtfully about the "poverty of the stimulus" in human language development. Cromer (1976) has reviewed data on systematic errors in children's speech that reflect their internalization of general linguistic rules that were previously contradicted by numerous exceptions they experienced. Such data indicate that *organisms utilize environmental information within the context of their existing developmental structures.* They do not mimic the details of their environment in a slavish fashion.

Given that animals frequently develop within environmental contexts that provide only partial and imperfect information, it may appear surprising that they do so well. Here we see two avenues of research that have to date been imperfectly exploited. What, for example, happens when an animal is given only *partial information* during its development? Might it, so to speak, be able to "fill in the gaps"? Or suppose an animal is exposed to information that is broadly correct but *distorted in detail?* Might it be able to utilize this environmental information by recoding it in a way that permits adaptive phenotypic adjustments?

There are few data relevant to these questions. Consider first the mismatch between the details of environmental stimulation and the details of subsequent motor performance. As illustration (Figure 10), the characteristic distinctions between song sparrow songs and swamp sparrow songs increase when the birds have normal auditory access to their own immature vocalizations. This indicates that the birds are able to utilize their own immature vocalizations to increase species specificity. If they simply "copied" what they heard, such progressions would not be expected. Gottlieb (1981) has shown that for wood ducks, exposure to their own or sibling alarm distress calls can affect subsequent preferences for species-characteristic maternal calls. As Bateson (1981) has pointed out, this may represent a developmental utilization of sensory events that are relatively nonspecific. Gottlieb (1981) complements this perspective by emphasizing the apparent ability of the young ducklings to abstract out, and then apply, relevant frequency modulated components from these calls.

Interesting similarities can be found in the neurobiological literature. For example, it is well established that individual visual cortical cells in most mammals respond to visual stimuli in specific (e.g., horizontal as opposed to vertical) orientations (Wiesel, 1982). Numerous studies have demonstrated that exposure to *either* horizontal or vertical visual stimuli selectively enhances the responsiveness of cortical cells to these orientations. Of specific interest is the study by Leventhal and Hirsch (1975) on the consequences of early exposure to diagonal lines in otherwise visually deprived kittens. Normally existing biases toward horizontal and vertical sensitivities in cortical neurons was greater than in visually deprived controls. That is, even though the exposure involved diagonal lines to which the animals are normally not especially sensitive, restricted exposure to diagonal lines magnified the selective responsiveness of cortical neurons to visual stimuli in vertical and horizontal orientations. The only way that this can occur is for experiential inputs to magnify preexisting developmental predispositions (cf. Aslin, 1981). Fregnac and Imbert (1978) have shown that there is a significant increase in binocularly driven neurons with age, and hypothesize that the binocular cells are the units that can have their orientational specificity altered by visual experience. An interesting possibility (but to our knowledge unexplored) is that moderate degrees of exposure to distorted stimuli may accentuate the development to preexisting biases, whereas more pronounced exposures would serve to respecify these biases (cf. Fentress, 1976, 1983b).

The term *equifinality* refers to corrections in developing systems, where initial distortions in expression (caused by inadequate environmental input over a given phase of development) are followed by normal expression (Bateson, 1981). In most cases these developmental compensations have not been related precisely to the details of distorted and subsequent experiences. However, they do reflect important intrinsic predispositions. The developmental compensations, which must also be contextually defined, result from still poorly understood genetic and experimential interactions that have occurred during earlier developmental phases.

Within the context of motor patterns in development it is not clear how broadly reaching the consequences of any given set of experiential events should be (Figure 9). Lagerspetz *et al.* (1971) have argued that early locomotory experience in human infants can have widespread consequences, while Loeb (1983) has

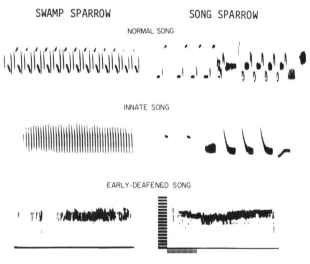

Fig. 10. Consequences for both swamp and song sparrows of hearing their own immature songs, as opposed to being deafened in infancy. The structural improvement in so-called innate over deafened song indicates the ability of the birds to utilize incomplete environmental information. (Subsequent deafening of birds with "innate" and "normal" songs can further clarify the extent to which these distinctions are primarily those of performance or development per se.) (From Marler, 1983.)

argued from a number of theoretical as well as empirical perspectives that the
degree of generalization for different facets of motor performance should be
severely limited.

An especially interesting question, for which there are few data at present,
concerns the *rules* by which the consequences of a given experience generalize during development. When generalized effects have been implicated in motor development research, the generalization has usually been assumed to occur between
movements that are physically similar in *form*. Thus practice in biting during feeding behavior might have an influence on biting used during fighting or mating. An
alternative possibility is that generalization occurs within a functional system, even
across motor patterns that are physically distinct. In this framework, improvement
in biting during aggression might also affect the later production of physically distinct movements such as kicking or scratching (also used in fighting). The level at
which given developmental events generalize may also change during the course of
development. If this is true, then interpretations of function at different developmental periods would be especially important to a full clarification of mechanism.

That an organism's susceptibility to environmental influences can vary during
development has long been appreicated in the ethological, neurobiological, and
psychological literature. These "sensitive periods" (Bateson, 1976) can affect both
the range of stimuli that influence the developing system and the extent of that
influence (cf. Aslin, 1981). Gene regulation through successive phenotypic changes
is a common hypothesis for such developmental profiles (see, e.g., Ede, 1978; Edelman, 1984).

Rather than simply "adding" new information, experience (in the broad sense
of the term) may serve to "select" among existing inherent potentialities, rather
like a key that opens appropriate file drawers at the appropriate times. By this analogy the details of the instructions for subsequent developmental constructions are
contained within the files, rather than in the key itself (Fentress, 1981a, 1983a). It
is for such reasons that "mirror models" can be expected to prove inadequate. This
does not deny a role for specific forms of experience during development, as the
appropriate keys must be employed in an appropriate sequence.

The gap between behavioral and cellular analyses is obviously very large, and
attempts to link these levels of analysis must be made with extreme caution. However, conceptually the fields of motor behavior and cellular biology do appear to
be moving closer together. In each case the importance of interrelations among
networks of events has become increasingly appreciated. Isolated processes or structures, whatever their form, are inadequate for a complete account. Thus Cooke
(1980) has stressed the importance of "dynamic intercellular communication systems" (p. 398) for the construction of nervous system patterns. As stated by Tran
Thanh Van (1980): "One is tempted to reduce the multicellular systems to isolated
cells; unfortunately, however, such isolated cells behave neither as unicellular nor
as multicellular entities" (p. 175).

It is important to bear in mind that the control networks underlying different
aspects of motor behavior may be intimately linked, in both an integrative and
developmental sense. The production of one motor activity may interfere with the

production of another. Turkewitz and Kenny (1982) have suggested that in sensory system development, early stages involve competition within one system and as new pathways become functional the new competition "results in the disruption of a stabilized organization and provides the basis for more advanced organization" (p. 359). They also suggest that excessive input from several modes simultaneously may disrupt systems that are not completely developed.

Goldman (1976) has addressed the same issue in her review of the development of the nervous system. She suggests that: "One function of a stepwise maturational progression may be to regulate the order and impact of internal and external stimuli and experience on the developmental process itself" (p. 70; cf. Cowan, 1978). Of special interest to future studies on motor patterns in development would be the extent to which the selective practice of motor patterns (1) inhibits or retards the development of similar (in form, or function) motor patterns [as is suggested by Labov and Labov (1978) for children "concentrating" on the production of single words and by Bower's (1976) suggestion that mastery of one task can interfere with learning another]; (2) facilitates the development of formally or functionally related patterns that involve diverse motor dimenions (or conversely, inhibits functionally antagonistic motor patterns). It is also conceivable that overpractice at a particular stage in development could have either of two effects on developing motor patterns: (1) accelerate the normal rate of development or (2) maintain the behavior at the practiced stage for a prolonged period of time.

There are a number of ethological observations that show interesting parallels to the hierarchical ordering and dismantling of motor sequences. One of the most interesting lines of evidence comes from the "dismaturation" of reproductive motor patterns at the end of breeding seasons. In a classic study on cormorants, Kortlandt (1955) examined in detail both the maturation and "dismaturation" of nest building movements. It was his conclusion that "ontogenetic development of behaviour as a rule proceeds in ascending stages" (p. 222) and that at the conclusion of a breeding season these "stages" are stepped through in a descending sequence—only to "redevelop" in subsequent years. Illness produced similar reductions in higher-order coordinations in several instances. These observations provide a fascinating parallel to many of the observations reported by Golani *et al.* (1981) on "warm-up" along movement dimensions in both young animals and animals recovering from CNS lesions. In their terms: "The warm-up principle is evident throughout early development. It involves a buildup in movement amplitude along at least two relatively independent dimensions, with partial coupling between them" (p. 7227).

Cortical circuitry is involved in the construction of motor details and in coordinating functionally integrated action. Once the motor capabilities have developed (i.e., in normal adults), cortical damage differentially affects these levels of performance. For example, Vanderwolf (1983) has noted that, "The behavioral deficit in decorticate or decerebrate rats appears to be largely a matter of failing to perform behaviors in the right place at the right time" (p. 83). Here one finds interesting parallels between the effects of CNS insults (plus the recovery from these insults), and normal motor development—a point that has been emphasized especially by Teitelbaum and his colleagues. (For a review see Teitelbaum 1982. See also Prechtl, 1981, for potential limitations in such conceptualizations.)

85

MOTOR PATTERNS IN
DEVELOPMENT

C. Summary Themes and Conclusions

Two complementary themes have emerged in theoretical considerations of motor patterns in development. The first concerns pattern formation. As we have seen, pattern formation implies both differentiation and integration. The problem of coherent behavioral function is to a large extent one of being able to isolate parameters of movement from one another while also being able to join these parameters together into functional combinations.

The second theme stresses the source of controlling events. Here one finds the dual tendencies of self-organization and interaction among systems (which can be defined at a number of complementary levels). It is the difficulty in thinking of systems that are *both* self-organizing and interactive that has led to such sterile dichotomies as nature versus nurture, central versus peripheral and so on. We have attempted to show that such dichotomies can lead to serious conceptual (and thereby, analytical) limitations.

We therefore have stressed the dynamic, relational, and multileveled nature of motor patterns in development. We have also found it useful to supplement the usual employment of "classes" of movement with a focus upon the "dimensionality" (or more properly, "multidimensionality") of movement.

The fact that a number of workers in other fields are devoting their energies to similar problems is encouraging to those of us searching for a closer synthesis among biological and behavioral disciplines. As stated by Ede (1978, p. 142): "How pattern and form are established is the most fundamental problem of developmental biology." It is our view not only that careful studies of motor patterns in development are of interest in their own right, but that they may help us to develop perspectives that will further the link between different levels of behavioral and biological science.

Acknowledgments

We thank Elliott Blass for his invitation to write this review, and for his many valuable comments on earlier drafts. Many colleagues contributed importantly to our literature search, among whom Steven Keele, Ron Oppenheim, Josef Rauschecker, and Janet Werker were especially helpful. Kent Berridge and Heather Parr contributed through a number of discussions. Wanda Danilchuk translated our scribbles, sorted out our note cards, and put the various manuscripts into order. For technical assistance at the final output stage we also thank David Conrad and Gordon Troop. We express with gratitude support from Medical Research Council and Natural Sciences and Engineering Research Council operating grants (J.C.F.), as well as a Natural Sciences and Engineering Research Council postgraduate scholarship (P.J.M.).

REFERENCES

Abbs, J. H., Gracco, V. L., and Cole, K. J. Control of multimovement coordination: Sensorimotor mechanisms in speech motor programming. *Journal of Motor Behavior* 1984, *16*, 195–231.

Adamec, R. E., Stark-Adamec, C., and Livingston, K. E. The development of predatory aggression and defense in the domestic cat *(Felis catus)*. II. Development of aggression and defense in the first 164 days of life. *Behavioral and Neural Biology*, 1980, *30*, 410–434.

Alberts, J. R. Ontogeny of olfaction: Reciprocal roles of sensation and behavior in the development of perception. In R. N. Aslin, J. R. Alberts, and M. R. Peterson (Eds.), *Development of perception: Psychobiological perspectives*. New York: Academic Press, 1981, pp. 322–357.

Altman, J. S. Changes in the flight motor pattern during the development of the Australian plague locust, *Chortoicetes terminifera. Journal of Comparative Physiology*, 1975, *97*, 127–142.

Angulo y González, A. W. The prenatal development of behavior in the albino rat. *The Journal of Comparative Neurology*, 1932, *55*, 395–442.

Anokhin, P. K. Systemogenesis as a general regulator of brain development. *The Developing Brain, Progress in Brain Research*, 1964, *9*, 54–86.

Armitage, S. E., Baldwin, B. A., and Vince, M. A. The fetal sound environment of sheep. *Science*, 1980, *208*, 1173–1174.

Arutyunyun, G. H., Gurfinkel, V. S., and Mirsky, M. L. Investigation of aiming at a target. *Biophysics*, 1981, *13*, 536–538.

Aslin, R. N. Experiential influences and sensitive periods in perceptual development: A unified model. In R. N. Aslin, J. R. Alberts, and M. R. Peterson (Eds.), *Development of perception: Psychobiological perspectives*. Vol. 2. New York: Academic Press, 1981, pp. 45–93.

Baerends-van Roon, J. M., and Baerends, G. P. *The morphogenesis of the behavior of the domestic cat, with a special emphasis on the development of prey-catching*. Amsterdam: North-Holland, 1979.

Baldissera, F., Hultborn, H., and Illert, M. Integration in spinal neuronal systems. In J. M. Brookhard and V. B. Mountcastle (Eds.), *The handbook of physiology, Section I: The nervous system*. Bethesda, Md.: American Psychological Association, 1981.

Baptista, L. F., and Petrinovich, L. Social interaction, sensitive phases and the song template hypothesis in the white-crowned sparrow. *Animal Behaviour*, 1984, *32*, 172–181.

Barlow, G. W. Modal action patterns. In T. A. Sebeck (Ed.), *How animals communicate*. Bloomington, Ind.: Indiana University Press, 1977, pp. 98–134.

Barlow, G. W. Genetics and development of behavior, with special reference to patterned motor output. In K. Immelmann, G. W. Barlow, L. Petrinovich, and M. Main (Eds.), *Behavioral development*. New York: Cambridge University Press, 1981, pp. 191–251.

Barraud, E. M. The development of behavior in some young passerines. *Bird Study*, 1961, *8*, 111–118.

Bateson, P. Ontogeny of behavior. *British Medical Bulletin*, 1981, *37*, 159–164.

Bateson, P. P. G. Specificity and the origins of behavior. *Advances in the Study of Behavior*, 1976, *6*, 1–20.

Beagley, W. K. Grooming in the rat as an aftereffect of lateral hypothalamic stimulation. *Journal of Comparative and Physiological Psychology*, 1976, *90*, 790–798.

Bekoff, A. Ontogeny of leg motor output in the check embryo: A neural analysis. *Brain Research*, 1976, *106*, 271–291.

Bekoff, A. Embryonic development of the neural circuitry underlying motor coordination. In W. M. Cowan (Ed.), *Studies in Developmental Neurobiology*. New York: Oxford University Press, 1981, pp. 134–170.

Bekoff, A. Behavioral embryology of birds and mammals: Neuroembryological studies of the development of motor behavior. In K. Immelmann, G. W. Barlow, L. Petrinovich, and M. Main (Eds.), *Behavioral development*. New York: Cambridge University Press, 1981, pp. 152–163.

Bekoff, A., and Trainer, W. The development of interlimb coordination during swimming in postnatal rats. *Journal of Experimental Biology*, 1979, *83*, 1–11.

Bekoff, A., Stein, P. S. G., and Hamburger, V. Coordinated motor output in the hindlimb of the 7-day chick embryo. *Proceedings of the National Academy of Sciences*, 1975, *72*, 1245–1248.

Bentley, D. R., and Hoy, R. R. Postembryonic development of adult motor patterns in crickets: A neural analysis. *Science*, 1970, *170*, 1409–1411.

Bergstrom, R. M., Hellstrom, P. E., and Stenberg, D. Studies in reflex irradiation in the foetal guinea pig. *Annales Chirurgiae et Gynaecologiae Fenniae*, 1962, *51*, 171–178.

Bernstein, N. *The coordination and regulation of movements.* London: Pergamon Press, 1967.

Berridge, K. C., and Fentress, J. C. Trigeminal deafferentation and control of ingestive and grooming sequences in rats. *Society for Neuroscience* (Abstract), 1984.

Berridge, K. C., and Fentress, J. C. Contextual control of trigeminal sensorimotor function. *Journal of Neuroscience,* in press.

Bizzi, E. Central and peripheral mechanisms in motor control. In G. E. Stelmach and J. Requin (Eds.), *Tutorials in motor behavior.* New York: North Holland, 1980, pp. 131–143.

Bolwig, N. Facial expression in primates with remarks on a parallel development in certain carnivores. *Behaviour,* 1964, *22,* 167–192.

Bosma, J. F. Anatomic and physiologic development of the speech apparatus. In D. B. Tower, (Ed.), *The nervous system.* Vol. 3, *Human communication and its disorders.* New York: Raven Press, 1975, pp. 469–481.

Bower, T. G. R. Repetitive processes in child development. *Scientific American,* 1976, *235*(5), 38–47.

Bower, T. G. R. *Human development.* San Francisco: Freeman, 1979.

Brooks, V. B. *Motor control.* In J. M. Brookhart and V. B. Mountcastle (Eds.), *Handbook of physiology, Section I.* Vol. II. Bethesda, Md.: American Physiological Society, 1981.

Brooks, V. B., and Thach, W. T. Cerebellar control of posture and movement. In J. M. Brookhart and V. B. Mountcastle (Eds.), *Handbook of physiology, Section I.* Vol. II. Bethesda, Md.: American Physiological Society, 1981.

Brown, M. C., and Booth, C. M. Segregation of motor nerves on a segmental basis during synapse elimination in neonatal muscle. *Brain Research,* 1983, *273,* 188–190.

Bruce, I. C., and Tatton, W. G. Sequential output–input maturation of kitten motor cortex. *Experimental Brain Research,* 1980, *39,* 411–419.

Bruner, J. S. The organization of early skilled action. In M. P. M. Richards (Ed.), *The Integration of a Child into a Social World.* London: Cambridge University Press, 1974.

Burghardt, G. M. Ontogeny of communication. In Thomas A. Sebeok (Ed.), *How animals communicate.* Bloomington, Ind.: Indiana University Press, 1977, pp. 71–97.

Burke, R. E. Motor units: Anatomy, physiology, and functional organization. In J. M. Brookhardt and V. B. Mountcastle (Eds.), *Handbook of physiology, Section I.* Vol. II. Bethesda, Md.: American Physiological Society, 1981, pp. 345–422.

Butler, R. G. Effects of context on communicatory exchanges in mice: A syntactic analysis. *Behavioral and Neural Biology,* 1980, *28*(4), 431–441.

Carins, R. B. The ontogeny and phylogeny of social interactions. In M. E. Hahn and E. C. Simmel (Eds.), *Communicative behavior and evolution.* New York: Academic Press, 1976, pp. 115–139.

Carmichael, L. The onset and early development of behavior. In P. Mussen (Ed.), *Carmichael's manual of child psychology.* Vol. 1. New York: Wiley, 1970, pp. 447–563.

Chevalier-Skolnikoff, S. The ontogeny of communication in the stumptail macaque *(Macaca arctoids).* *Contributions to Primatology,* 1974, *2,* 1–174.

Chomsky, N. *Rules and representations.* New York: Columbia University Press, 1980.

Coghill, G. E. *Anatomy and the problem of behavior.* Cambridge, England: Cambridge University Press, 1929.

Collard, E., and Povel, D. J. Theory of serial pattern production: Tree traversals. *Psychological Review,* 1982, *89,* 693–707.

Condon, W. S., and Sander, L. W. Neonate movement is synchronized with adult speech: Interactional participation and language acquisition. *Science,* 1974, *183,* 99–101.

Connolly, K. The nature of motor skill development. *Journal of Human Movement Studies,* 1977, *3,* 128–143.

Cooke, J. Early organization of the central nervous system: Form and pattern. *Current Topics in Developmental Biology,* 1980, *15,* 373–407.

Cooper, W. E. The development of speech timing. In S. J. Segalowitz and F. A. Gruber (Eds.), *Language development and neurological theory.* New York: Academic Press, 1977, pp. 357–373.

Corner, M. Spontaneous motor rhythms in early life—phenomenological and neurophysiological aspects. *Progress in Brain Research,* 1978, *48,* 349–366.

Cowan, W. M. Aspects of neural development. *International Review of Physiology. Neurophysiology III.* 1978, *17,* 149–190.

Cromer, R. F. Developmental strategies for language. In V. Hamilton and M. D. Vernon (Eds.), *The development of cognitive processes.* London: Academic Press, 1976, pp. 305–358.

Cruze, W. W. Maturation and learning in chicks. *Journal of Comparative Psychology,* 1935, *19,* 371–409.

Darwin, C. *The expression of the emotions in man and the animals.* London: Murray, 1872.

Dawkins, R. Hierarchical organization: A candidate principle for ethology. In P. P. G. Bateson and R. A. Hinde (Eds.), *Growing points in ethology.* Cambridge, England: Cambridge University Press, 1976, pp. 7–54.

DeCasper, A. J., and Fifer, W. P. Of human bonding: Newborns prefer their mother's voices. *Science,* 1980, *208,* 1174–76.

DeGroat, W. C., Douglas, J. W., Glass, J., Simonds, W., Weimer, B., and Werner, P. Changes in somato-vesical reflexes during postnatal development in the kitten. *Brain Research,* 1975, *94,* 150–54.

Deliagina, T. G., Feldman, A. G., Gelfand, I. M., and Orlovsky, G. N. On the role of central program and afferent inflow in the control of scratching movements in the cat. *Brain Research,* 1975, *100,* 297–313.

Doty, R. W. The concept of neural centers. In J. C. Fentress (Ed.), *Simpler networks and behavior.* Sunderland, Mass.: Sinauer Associates, Inc., 1976, pp. 251–265.

Drachman, D. B., and Coulombre, A. J. Experimental clubfoot and arthrogryposis multiplex congention. *Lancet,* 1962, *2,* 523–526.

Easton, T. A. On the normal use of reflexes. *American Scientist,* 1972, *60,* 591–599.

Ebbesson, S. O. E. The parcellation theory and its relation to interspecific variability in brain organization, evolutionary and ontogenetic development and neuronal plasticity. *Cell and Tissue Research,* 1980, *213,* 179–212.

Ede, D. A. *An introduction to developmental biology.* New York: Wiley, 1978.

Edelman, G. M. Group selection and phasic reentrant signaling: A theory of higher brain function. In G. M. Edelman and V. B. Mountcastle (Eds.), *The mindful brain.* Cambridge, Mass.: The MIT Press, 1978, pp. 51–100.

Eibl-Eibesfeldt, I. Beobachtungen zur Forpflanzungs biologie und Jugendentwicklund des Eichornchens. *Zeitschrift für Tierpsychologie,* 1951, *8,* 370–400.

Eibl-Eibesfeldt, I. Uber die ontogenetische Entwicklung der Technik des Nusseoffnens vom Eichhornchen (*Sciurus vulgaris* L.). *Zeitschrift für Saugetierkunde,* 1956, *21,* 132–134.

Eibl-Eibesfeldt, I. The interactions of unlearned behaviour patterns and learning in mammals. In D. F. Delafresnaye (Ed.), *Brain mechanisms and learning.* CIOMS Symposium. Oxford: Blackwell, 1961.

Elsner, N. Developmental aspects of insect neuroethology. In K. Immelmann, G. W. Barlow, L. Petrinovich, and M. Main (Eds.), *Behavioral development.* New York: Cambridge University Press, 1981, pp. 474–490.

Elsner, N., and Hirth, C. Short and long term coordination in a stridulating grasshopper. *Naturwissenchaften,* 1978, *63,* 160–161.

Emde, R. N., Gaensbauer, T. J., and Harman, R. J. Emotional expression in infancy: A biobehavioral study. *Psychological issues.* Vol. 10, Monograph 37. New York: International Universities Press, 1976.

Etienne, A. S., Emmanuelli, E., and Zinder, M. Ontogeny of hoarding in the golden hamster: The development of motor patterns and their sequential coordination. *Developmental Psychobiology,* 1982, *15,* 33–45.

Evarts, E. V. The third Stevenson lecture. Changing concepts of the central control of movement. *Canadian Journal of Physiology and Pharmacology,* 1975, *53,* 191–201.

Evarts, E. V. Role of motor cortex in voluntary movements in primates. In J. M. Brookhart and V. B. Mountcastle (Eds.), *Handbook of physiology, Section I.* Bethesda, Md.: American Physiological Society, 1981, pp. 1083–1120.

Fentress, J. C. Observations on the behavioral development of a hand-reared male timber wolf. *American Zoologist,* 1967, *7,* 339–351.

Fentress, J. C. Development and patterning of movement sequences in inbred mice. In J. Kiger (Ed.), *The biology of behavior.* Corvallis: Oregon State University Press, 1972, pp. 83–132.

Fentress, J. C. Specific and non-specific factors in the causation of behavior. In P. P. G. Bateson and P. Klopfer (Eds.), *Perspectives in ethology.* New York: Plenum Press, 1973a, pp. 155–255.

Fentress, J. C. Development of grooming in mice with amputated forelimbs. *Science,* 1973b, *179,* 704–705.

Fentress, J. C. Dynamic boundaries of patterned behavior: Interaction and self-organization. In P. P. G. Bateson and R. A. Hinde (Eds.), *Growing points in ethology.* Cambridge, England: Cambridge University Press, 1976, pp. 135–169.

Fentress, J. C. The tonic hypothesis and the patterning of behavior. *New York Academy of Sciences*, 1977, *290*, 370–395.

Fentress, J. C. Mus musicus. The developmental orchestration of selected movement patterns in mice. In M. Bekoff and G. Burghardt (Eds.), *The development of behavior: Comparative and evolutionary aspects.* New York: Garland, 1978, pp. 321–342.

Fentress, J. C. Sensorimotor development. In R. N. Aslin, J. R. Alberts, and M. R. Petersen (Eds.), *The development of perception: Psychobiological perspectives.* Vol. 1. New York: Academic Press, 1981a, pp. 293–317.

Fentress, J. C. Order in ontogeny: Relational dynamics. In K. Immelmann, G. Barlow, M. Main, and L. Petrinovich (Eds.), *Behavioral development.* New York: Cambridge University Press, 1981b, pp. 338–371.

Fentress, J. C. Ethological models of hierarchy and patterning of species-specific behavior. In E. Satinoff and P. Teitelbaum (Eds.), *Handbook of neurobiology: Motivation.* New York: Plenum Press, 1983a, pp. 185–234.

Fentress, J. C. A view of ontogeny. In J. R. Eisenberg and D. G. Kleiman (Eds.), *Recent advances in the study of mammalian behavior,* Special Publication No. 7, The American Society of Mammalogists, 1983b, pp. 24–64.

Fentress, J. C. Hierarchical motor control. In M. Studdert-Kennedy (Ed.), *Psychobiology of language, Neuroscience Research Program.* Cambridge, Mass.: MIT Press, 1983c, pp. 40–61.

Fentress, J. C. The analysis of behavioral networks. In J. P. Ewert, R. R. Capranica, and D. J. Ingle (Eds.), *Advances in vertebrate neuroethology.* New York: Plenum Press, 1983d, pp. 939–968.

Fentress, J. C. The development of coordination. *Journal of Motor Behavior,* 1984, *16*, 99–134.

Fentress, J. C., and Stilwell, F. P. Grammar of a movement sequence in inbred mice. *Nature,* 1973, *244*, 52–53.

Fentress, J. C., Field, R., and Parr, H. Social dynamics and communication. In H. Markowitz and V. Stevens (Eds.), *Behavior of captive wild animals.* Chicago: Nelson-Hall, 1978, pp. 67–106.

Fentress, J. C., Stanfield, B. B., and Cowan, W. M. Observations on the development of the striatum in mice and rats. *Anatomy and Embryology,* 1981, *163*, 275–298.

Fentress, J. C., Ryon, J., McLeod, P. J., and Havkin, G. Z. A multidimensional approach to agonistic behavior in wolves. In H. Frank (Ed.), *Proceedings of the First International Captive Wolf Conference,* in press.

Ferron, J., and Lefebvre, L. Comparative organization of grooming sequences in adult and young Sciurid rodents, 1983.

Field, R. Vocal behavior of wolves *(Canis lupus):* Variability in structure, context, annual/diurnal patterns, and ontogeny. Ph.D. Thesis, the John Hopkins University, Baltimore, 1978.

Field, R. A perspective on syntactics on wolf vocalizations. In E. Klinghammer (Ed.), *The behavior and ecology of wolves.* New York: Garland, 1979, pp. 182–205.

Forssberg, H., and Wallberg, H. Infant locomotion—a preliminary movement and electromyographic study. In K. Berg and B. O. Eriksson (Eds.), *Children and exercise.* Vol. IX. Baltimore: University Park, 1981, pp. 32–40.

Fox, M. W. *Behavior of wolves, dogs, and related canids.* New York: Harper and Row, 1972.

Fox, M. W., and Cohen, J. A. Canid communication. In T. A. Sebeok (Ed.), *How animals communicate.* Bloomington: Indiana University Press, 1977, pp. 728–748.

Fraley, N. B., and Fernald, R. D. Social control of developmental rate in the African cichlid *Haplochromis burtoni. Zeitschrift für Tierpsychologie,* 1982, *60*, 66–82.

Frégnac, Y., and Imbert, M. Early development of visual cortical cells in normal and dark reared kittens: Relationship between orientation selectivity and ocular dominance. *Journal of Physiology,* 1978, *278*, 27–44.

Galef, B. G., Jr. Development of flavor preference in man and animals: The role of social and non-social factors. In R. N. Aslin, J. R. Alberts, and M. R. Petersen (Eds.), *Development of perception: Psychobiological Perspectives.* New York: Academic Press, 1981, pp. 411–431.

Galef, B. G., Jr., and Henderson, P. W. Mother's milk: A determinant of the feeding preferences of weaning rat pups. *Journal of Comparative and Physiological Psychology,* 1972, *78*, 213–219.

Gallistel, C. R. *The organization of action: A new synthesis.* Hillsdale, N.J.: Lawrence Erlbaum Associates, 1980.

Gatev, V. Role of inhibition in the development of motor coordination in early childhood. *Developmental Medicine and Child Neurology,* 1972, *14*, 336–341.

Gierer, A. Generation of biological patterns and form: Some physical, mathematical, and logical aspects. *Progress in Biophysics and Molecular Biology, 1981, 37,* 1–47.

Glencross, D. J. Levels and strategies of response organization. In G. E. Stelmach and J. Requin (Eds.), *Tutorials in motor behavior.* New York: North Holland, 1980, pp. 551–566.

Golani, I. Homeostatic motor processes in mammalian interactions: A choreography of display. In P. P. G. Bateson and P. H. Klopfer (Eds.), *Perspectives in ethology.* Vol. 2. New York: Plenum Press, 1976, pp. 69–134.

Golani, I. The search for invariants in motor behavior. In K. Immelmann, G. W. Barlow, L. Petrinovitch, and M. Main (Eds.), *Behavioral development.* New York: Cambridge University Press, 1981, pp. 372–392.

Golani, I., and Fentress, J. C. Early ontogeny of face grooming in mice. *Developmental Psychobiology,* in press.

Golani, I., Broncht, G., Moualern, D., and Teitelbaum, P. "Warm-up" along dimensions of movement in the ontogeny of exploratory behaviour in the infant rat and other infant mammals. *Proceedings of the National Academy of Sciences, 1981, 78,* 7226–7229.

Goldman, P. S. Maturation of the mammalian nervous system and the ontogeny of behavior. *Advances in the Study of Behavior, 1976, 7,* 1–90.

Goosen, C., and Kortmulder, K. Relationships between faces and body motor patterns in a group of captive pigtailed macaques *(Macaca nemestrina). Primates, 1979, 20,* 221–236.

Gorska, T., and Czarkowska, J. Motor effects of stimulation of the cerebral cortex in the dog: An ontogenetic study. In N. T. Tankov and D. S. Kosarov (Eds.), *Motor control.* New York: Plenum Press, 1973, pp. 147–166.

Gottlieb, G. Roles of early experience in species-specific perceptual development. *Development of Perception, 1981, 1,* 5–44.

Gould, E. Mechanisms of mammalian auditory communication. In J. F. Eisenberg, and D. G. Kleiman (Eds.), *Advances in the study of mammalian behavior,* Special Publication No. 7. The American Society of Mammalogists, 1983, pp. 265–342.

Green, S., and Marler, P. The analysis of animal communication. In P. Marler and J. G. Vandenbergh (Eds.), *Handbook of behavioral neurobiology.* Vol. 3, *Social behavior and communication.* New York: Plenum Press, 1979, pp. 73–158.

Grillner, S. Control of locomotion in bipeds, tetrapods, and fish. In J. M. Brookhart and V. B. Mountcastle (Eds.), *Handbook of physiology, Section I.* Vol. II. Bethesda, Md.: American Physiological Society, 1981, pp. 1179–1236.

Gruner, J. A., and Altman, J. Swimming in the rat: Analysis of locomotor performance in comparison to stepping. *Experimental Brain Research, 1980, 40,* 374–382.

Gruner, J. A., Altman, J., and Spivack, N. Effects of arrested cerebellar development on locomotion in the rat. Cinematographic and electromyographic analysis. *Experimental Brain Research, 1980, 40,* 361–373.

Guttinger, H. R. Self-differentiation of song organization rules by deaf canaries. *Zeitschrift für Tierpsychologie, 1981, 56,* 323–340.

Halfmann, K., and Elsner, N. Larval stridulation in acridid grasshoppers. *Naturwissenschaften, 1978, 65,* 265.

Hall, W. G., and Bryan, T. E. The ontogeny of feeding in rats: II. Independent ingestive behavior. *Journal of Comparative Physiology and Psychology, 1980, 94,* 746–756.

Hall, W. G., and Williams, C. L. Suckling isn't feeding, or is it? A search for developmental continuities. *Advances in the Study of Behavior, 1983, 13,* 219–254.

Hamburger, V. The developmental history of the motor neuron (The F. O. Schmitt Lecture in Neuroscience, 1976). *Neurosciences Research Program Bulletin, 1977, 15* (Supplement), 1–37.

Hamburger, V., Wenger, E., and Oppenheim, R. Motility in the chick embryo in the absence of sensory input. *Journal of Experimental Zoology, 1966, 162,* 133–160.

Hansen, E. W. The development of maternal and infant behavior in the rhesus monkey. *Behaviour, 1966, 27,* 107–149.

Harris, W. A. Neural activity and development. *Annual Review of Physiology, 1981, 43,* 689–710.

Havkin, G. Z. Symmetry shifts in the development of interactive behaviour of two wolf pups *(Canis lupus).* Unpublished M.A. Thesis, Dalhousie University, Halifax, Nova Scotia, 1977.

Havkin, G. Z. Form and strategy of combative interactions between wolf pups *(Canis lupus).* Ph.D. Thesis, Dalhousie University, Halifax, Nova Scotia, 1981.

Havkin, G. Z., and Fentress, J. C. The form of combative strategy in interactions among wolf pups *(Canis lupus)*. *Zeitschrift für Tierpsychologie*, 1985, *68*, 177–200.

Hebb, D. O. Drives and the CNS (conceptual nervous system). *Psychology Review*, 1955, *62*, 243–254.

Henneman, E., and Mendell, L. M. Functional organization of motoneuron pool and its inputs. In J. M. Brookhart and V. B. Mountcastle (Eds.), *Handbook of physiology, Section I*. Vol. II. Bethesda, Md.: American Physiological Society, 1981, pp. 423–507.

Hinde, R. A. *Animal behavior: A synthesis of ethology and comparative psychology*. 2nd ed. New York: McGraw-Hill, 1970.

Hinde, R. A. *Ethology: Its nature and relations with other sciences*. New York: Oxford University Press, 1982.

Hinde, R. A., and Stevenson, J. G. Sequences of behavior. *Advances in the Study of Behavior*, 1969, *2*, 267–296.

Hines, M. The development and regression of reflexes, postures, and progression in the young macaque. *Contributions to Embryology*, 1942, *30*, 153–209.

Hofer, M. *The roots of human behavior*. San Francisco: Freeman, 1981.

Hofsten, C., von. Development of visually directed reading: The approach phase. *Journal of Human Movement Studies*, 1979, *5*, 160–178.

Hofsten, C., von. Development changes in the organization of prereaching movements. *Developmental Psychology*, 1984, *20*, 378–388.

Hogan, J. A., and Roper, T. J. A comparison of the properties of different reinforcers. In J. S. Rosenblatt, R. A. Hinde, C. Beer, and M. C. Busnel (Eds.), *Advances in the study of behavior*. London: Academic Press, 1978.

Hollyday, M. Motoneuron histogenesis and the development of limb innervation. *Current Topics in Developmental Biology*, 1980, *15*, 181–215.

Houk, J. C., and Rymer, W. Z. Neural control of muscle length and tension. In J. M. Brookhart and V. B. Mountcastle (Eds.), *Handbook of physiology, Section I*. Bethesda, Md.: American Physiological Society, 1981, pp. 257–323.

Hoyle, G. Neuroethology. *The behavioral and brain sciences*, 1984, *7*, 367–381.

Ikeda, K. Genetically patterned neural activity. In J. C. Fentress (Ed.), *Simpler networks and behavior*. Sunderland, Mass. Sináuer Associates, 1976, pp. 140–152.

Immelmann, K., Barlow, G. W., Petrinovich, L., and Main, M. *Behavioral development*. Cambridge: Cambridge University Press, 1981.

Jacobson, M. *Developmental neurobiology*. 2nd ed. New York: Plenum Press, 1978.

Johnston, T. D. Compensation for substrate elasticity in the kinematics of leaping by infant pigtailed macaques *(Macaca nemestrina)*. *Brain Research*, 1980, *184*, 467–480.

Kear, J. Food selection in finches with special reference to interspecific differences. *Proceedings Zoological Society of London*, 1961, *138*, 163–204.

Keele, S. W. Behavioral analysis of movement. In V. Brooks (Ed.), *Handbook of physiology: Motor control*. Washington: American Physiological Society, 1981, pp. 1391–1414.

Kelso, J. A. S. Contrasting perspectives on order and regulation in movement. In J. Long and A. Baddeley (Eds.), *Attention and performance*. Vol. IX. Hillsdale, N.J.: Lawrence Erlbaum Associates, 1981, pp. 437–457.

Kelso, J. A. S., and Tuller, B. A dynamical basis for action systems. In M. S. Gazzaniga (Ed.), *Handbook of cognitive neuroscience*. New York: Plenum Press, 1984, pp. 319–356.

Kent, R. D. Anatomical and neuromuscular maturation of the speech mechanism: Evidence from acoustic studies. *Journal of Speech and Hearing Research*, 1976, *19*, 421–447.

Kent, R. D. Sensorimotor aspects of speech development. In R. N. Aslin, J. R. Alberts, and M. R. Petersen (Eds.), *Development of perception: Psychobiological perspectives*. Vol. 1. New York: Academic Press, 1981, pp. 161–189.

Kent, R. D., and Murray, A. D. Acoustic features of infant vocal utterances at 3, 6, and 9 months. *Journal of the Acoustical Society of America*, 1982, *72*(2), 353–365.

Kortlandt, A. Aspects and prospects of the concept of instinct (vicissitudes of the hierarchy theory). *Archives Neerlandaises De Zoologie*, 1955, *11*, 155–284.

Kristan, W. B., Jr. Generation of rhythmic motor patterns. In H. M. Pinsker and W. D. Willis, Jr. (Eds.), *Information processing in the nervous system*. New York: Raven Press, 1980, pp. 241–261.

Kruijt, J. P. Ontogeny of social behaviour in Burmese red junglefowl *(Gallus gallus spadiceus)* Bonnaterre. *Behaviour Supplement*, 1964, *XII*.

Kugler, P. N., Kelso, J. A. S., and Turvey, M. T. On the control and coordination of naturally developing systems. In J. A. S. Kelso, and J. E. Clark (Eds.), *The development of movement control and coordination.* New York: Wiley, 1982.

Kuhl, P. K., and Meltzoff, A. N. The bimodal perception of speech in infancy. *Science,* 1982, *218,* 1138–1141.

Kutsch, W. The development of the flight pattern in the desert locust, *Schistocerca gregaria. Zeitschrift fuer Vergleichende Physiologie,* 1971, *74,* 156–168.

Kuypers, H. G. J. M. Anatomy of the descending pathways. In V. Brooks (Eds.), *Handbook of physiology: Motor control.* Bethesda, Md.: American Physiological Society, 1981, pp. 595–666.

Labov, W., and Labov, T. The phonetics of cat and mama. *Language,* 1978, *4,* 816–852.

Lagerspetz, K., Margaretha, N., and Strandvik, C. The effects of training in crawling on the motor and mental development of infants. *Scandinavian Journal of Psychology,* 1971, *12,* 192–197.

Landmesser, L. T. The generation of neuromuscular specificity. *Annual Review of Neuroscience,* 1980, *3,* 279–302.

Lang, F., Govind, C. K., and Costello, W. J. Experimental transformation of muscle fiber properties in lobster. *Science,* 1978, *201,* 1037–1039.

Lashley, K. S. The problem of serial order in behavior. In L. A. Jeffress (Ed.), *Cerebral mechanisms in behavior,* New York: Wiley, 1951, pp. 112–136.

Lawrence, D. G., and Hopkins, D. A. The development of motor control in the rhesus monkey: Evidence concerning the role of corticomotoneuronal connections. *Brain Research,* 1976, *99,* 235–254.

Leonard, L. B., Newhoff, M., and Mesalam, L. Individual differences in early child phonology. *Applied Psycholinguistics,* 1980, *1,* 7–30.

Leventhal, A. G., and Hirsch, H. V. B. Cortical effect of early selective exposure to diagonal lines. *Science,* 1975, *190,* 902–904.

Levine, M. S., Hull, C. D., and Buchwald, N. A. Development of motor activity in kittens. *Developmental Psychobiology,* 1980, *13,* 357–371.

Lewis, D. M. The physiology of motor units in mammalian skeletal muscle. In A. L. Towe and E. S. Luschei (Eds.), *Handbook of behavioral neurobiology.* Vol. 5, *Motor Coordination.* New York: Plenum Press, 1981, pp. 1–67.

Leyhausen, P. On the function of the relative hierarchy of moods (as exemplified by the phylogenetic and ontogenetic development of prey-catching in carnivores). In K. Lorenz and P. Leyhausen (Eds.), *Motivation of human and animal behavior.* New York: Van Nostrand Reinhold, 1973, pp. 144–247.

Liberman, A. M., and Studdert-Kennedy, M. Phonetic perception. In R. Held, H. Leibowitz, and H. L. Teuber (Eds.), *Handbook of sensory physiology.* Vol. VIII, *Perception.* Heidelberg: Springer-Verlag, 1978, pp. 143–178.

Lieblich, A., Symmes, D., Newman, J., and Shapiro, M. Development of the isolation peep in laboratory-bred squirrel monkeys. *Animal Behaviour,* 1980, *28,* 1–9.

Lillehei, R. A., and Snowdon, C. T. Individual and situational differences in the vocalization of young stumptail macaques *(Macoca arctoides). Behaviour,* 1978, *65,* 270–281.

Lindberg, D., and Elsner, N. Sensory influence upon grasshopper stridulation. *Naturwissenchaften,* 1977, *64,* 342–343.

Locke, J. L. *Phonological acquisition and change.* New York: Academic Press, 1983.

Loeb, G. Finding common ground between robotics and physiology. *Trends in Neurosciences,* 1983, 203–204.

Lorenz, K. Z. *The foundations of ethology.* New York: Springer-Verlag, 1981.

Machlis, L. An analysis of the temporal patterning of pecking in chicks. *Behaviour,* 1977, *63,* 1–70.

Macken, M. A., and Ferguson, C. A. Phonological universals in language acquisition. *Annals of the New York Academy of Sciences,* 1981, *379,* 110–129.

Marler, P. Animal Communication Signals. *Science,* 1967, *157,* 769–774.

Marler, P. Some ethological implications for neuroethology: The ontogeny of bird song. In J. P. Ewert, R. R. Copranica, and D. J. Ingb (Eds.), *Advances in vertebrate neuroethology.* New York: Plenum Press, 1983, pp. 21–52.

Marler, P., and Peters, S. Birdsong and speech: Evidence for special processing. In P. D. Eimas and J. L. Mille (Eds.), *Perspectives on the study of speech.* Hillsdale, N.J.: Lawrence Erlbaum Associates, 1981, pp. 75–112.

Marler, P., and Peters, S. Structural changes in song ontogeny in the swamp sparrow *(Melospitza georgiona). Auk,* 1982a, *99,* 446–458.

Marler, P., and Peters, S. Developmental overproduction and selective attrition: New processes in the epigenesis of birdsong. *Developmental Psychobiology*, 1982b, *15*, 369–378.

Mason, W. Ontogeny of social behavior. In P. Marler and J. G. Vandenbergh (Eds.), *Handbook of behavioral neurobiology*. Vol. 3, *Social behavior and communication*. New York: Plenum Press, 1979, 1–28.

Maurus, M., and Pruscha, H. Quantitative analysis of behavioral sequences elicited by automated tele-stimulation in squirrel monkeys. *Experimental Brain Research*, 1972, *14*, 372–394.

McGraw, M. B. Neuromuscular development of the human infant as exemplified in the achievement of erect locomotion. *The Journal of Pediatrics*, 1940, *17*, 747–771.

Mellon, D., and Stevens, P. J. Experimental arrest of muscle transformation and asymmetry reversal in alpheid shrimp. *Neurosciences Abstracts*, 1979, *5*, 254.

Meltzoff, A. N., and Moore, M. K. Imitation of facial and manual gestures by human neonates. *Science*, 1977, *198*, 75–78.

Menn, L. Theories of phonological development. *Annals of the New York Academy of Sciences*, 1981, *379*, 130–137.

Menyuk, P., and Menn, L. Early strategies for the perception and production of words and sounds. In P. Fletcher and M. Garman (Eds.), *Language acquisition*. Cambridge, England: Cambridge University Press, 1979, pp. 49–70.

Metafora, S., Felsani, A., Cotrufo, R., Tajana, G. F., Del Rio, A., de Prisco, P. P., Rutigliano, B., and Esposito, V. Neural control of gene expression in the skeletal muscle fibre: Changes in the muscular mRNA population following denervation. *Proceedings of the Royal Society of London, B*, 1980, *209*, 257–273.

Moran, G. The structure of movement in supplanting interactions in the wolf. Unpublished Ph.D. Thesis, Dalhousie University, Halifax, Nova Scotia, 1978.

Moran, G. Long-term patterns of agonistic interactions in a captive group of wolves. *Animal Behaviour*, 1982, *30*, 75–83.

Moran, G., and Fentress, J. C. A search for order in wolf social behavior. In E. Klinghammer (Ed.), *The behavior and ecology of wolves*. New York: Garland Press, 1979, pp. 245–283.

Moran, G., Fentress, J. C., and Golani, I. A description of relational patterns during "ritualized fighting" in wolves. *Animal Behaviour*, 1981, *29*, 1146–1165.

Moran, T. H., Schwartz, G. J., and Blass, E. M. Organized behavioral responses to lateral hypothalamic electrical stimulation in infant rats. *The Journal of Neuroscience*, 1983, *3*, 10–19.

Moss, S. C., and Hogg, J. The development and integration of fine motor sequences in 12- to 18-month-old children: A test of the modular theory of motor skill acquisition. *Genetic Psychology Monographs*, 1983, *107*, 145–187.

Narayanan, C. H., Fox, M. W., and Hamburger, V. Prenatal development of spontaneous and evoked activity in the rat *(Rattus norvegicus albinus)*. *Behaviour*, 1971, *40*, 100–134.

Navarrete, R., and Vrbová, G. Changes in activity patterns in slow and fast muscles during postnatal development. *Developmental Brain Research*, 1983, *8*, 11–19.

Newman, J. D., and Symmes, D. Inheritance and experience in the acquisition of primate acoustic behavior. In C. Snowdon, C. Brown, and M. Petersen (Eds.), *Primate communication*. New York: Cambridge University Press, 1982, pp. 259–278.

Nice, M. M. Studies in the life history of the song sparrow. *Transactions of the Linnean Society of New York*, 1943, *6*, 1–328.

Nicolai, J., Das lernprogramm in der gesangsausbildung der strohwitwe *Tetraenura fischeri* Reichenow. *Zeitschrift für Tierpsychologie*, 1973, *32*, 113–138.

Northup, L. R. Temporal patterning of grooming in three lines of mice. Some factors influencing control levels of a complex behaviour. *Behaviour*, 1977, *61*, 1–25.

Nottebohm, F. Brain pathways for vocal learning in birds: A review of the first 10 years. *Progress in Psychobiology and Physiological Psychology*, 1980, *9*, 85–124.

Okado, N., Kakimi, S., and Kojima, T. Synaptogenesis in the cervical cord of the human embryo: Sequence of synapse formation in a spinal reflex pathway. *Journal of Comparative Neurology*, 1979, *184*, 491–518.

Oller, B. K. The emergence of the sounds of speech in infancy. In G. H. Yeni-Komshian, J. F. Kavanagh, and C. A. Ferguson (Eds.), *Child phonology*. Vol. 1, *Production*. New York: Academic Press, 1980, pp. 93–112.

Oppenheim, R. W. Metamorphosis and adaptation in the behavior of developing organisms. *Developmental Psychobiology*, 1980, *13*, 353–356.

Oppenheim, R. W. Ontogenetic adaptations and retrogressive processes in the development of the nervous system and behaviour: A neuroembryological perspective. In K. J. Connolly and H. F. R. Prechtl (Eds.), *Maturation and development: Biological and psychological perspectives.* Philadelphia: Lippincott, 1981, 73–109.

Oppenheim, R. W. The neuroembryological study of behavior: Progress, problems and perspectives. In R. K. Hunt (Ed.), *Current topics in developmental biology.* Vol. 17, *Neural development.* New York: Academic Press, 1982.

Oppenheim, R. W., and Reitzel, J. Ontogeny of behavioral sensitivity to strychine in the chick embryo: Evidence for the early onset of CNS inhibition. *Brain Behavior and Evolution,* 1975, *11,* 130–159.

Peiper, A. *Cerebral function in infancy.* New York: Consultants Bureau, 1973.

Peters, A. M. Language learning strategies: Does the whole equal the sum of the parts? *Language,* 1977, *53*(3), 560–573.

Petersen, M. R. Perception of acoustic communication signals by animals: Developmental perspectives and implications. *Development of Perception,* 1981, *1,* 67–109.

Pfeifer, S. Role of the nursing order in social development of mountain lion kittens. *Developmental Psychobiology,* 1980, *13,* 47–53.

Pittman, R., and Oppenheim, R. W. Cell death of motoneurons in the chick embryo spinal cord. IV. Evidence that a functional neuromuscular interaction is involved in the regulation of naturally occurring cell death and the stabilization of synapses. *Journal of Comparative Neurology,* 1979, *187,* 425–446.

Prechtl, H. F. R. Problems of behavioural studies in the newborn infant. In D. H. Lehrman, J. S. Rosenblatt, R. A. Hinde, and E. Shaw (Eds.), *Advances in the study of behaviour.* New York: Academic Press, 1965.

Prechtl, H. F. R. The study of neural development as a perspective of clinical problems. In K. J. Connolly and H. F. R. Prechtl (Eds.), *Maturation and development: Biological and psychological perspectives.* Philadelphia: Lippincott, 1981, pp. 198–215.

Preyer, W. *Specielle Physiologie des Embryo.* Leipzig: Grieben, 1885.

Prosser, C. L. Correlation between development of behaviour and neuromuscular differentiation in embryos of *Eisenia foetida. The Journal of Comparative Neurology,* 1933, *58,* 603–641.

Provine, R. Development of function in nerve nets. In J. C. Fentress (Ed.), *Simpler Networks and Behavior.* Sunderland, Mass.: Sinauer Associates, 1976, pp. 203–220.

Provine, R. R. Behavioural neuroembryology: Motor prespectives. In W. T. Greenough and J. Juraska (Eds.), *Developmental psycho/neurobiology.* New York: Academic Press, 1983.

Purves, D., and Lichtman, J. W. Elimination of synapses in the developing nervous system. *Science,* 1980, *210,* 153–157.

Reed, E. S. An outline of a theory of action systems. *Journal of Motor Behaviour,* 1982, *14,* 98–134.

Reite, M., and Short, R. A biobehavioural developmental profile (BDP) for the pigtail monkey. *Developmental Psychobiology,* 1980, *13,* 243–285.

Richmond, G., and Sachs, B. D. Grooming in Norway rats: The development and adult expression of a complex motor pattern. *Behaviour,* 1980, *75,* 82–96.

Rogers, C. M. Implications of a primate early rearing experiment for the concept of culture. In E. W. Menzel, Jr. (Ed.), *Precultural primate behaviour.* New York: S. Karger, 1973, 185–191.

Rosenblatt, J. S. Stages in the early behavioural development of altricial young of selected species of non-primate mammals. In P. P. G. Bateson and R. A. Hinde (Eds.), *Growing points in ethology.* Cambridge, England: Cambridge University Press, 1976, pp. 345–383.

Rosenblatt, J. S. The sensorimotor and motivational basis of early behavioural development of selected altricial mammals. In. N. E. Spear and B. A. Cambell (Eds.), *Ontogeny of learning and memory.* New York: Lawrence Erlbaum, 1980, 1–38.

Sackett, G. P., Sameroff, A. J., Cairns, R. B., and Suomi, S. J. Continuity in behavioural development: Theoretical and empirical issues. In K. Immelmann, G. W. Barlow, L. Petrinovich, and M. Main (Eds.), *Behavioural development.* New York: Cambridge University Press, 1981, pp. 23–57.

Saito, K. Development of reflexes in the rat fetus studied *in vitro. Journal of Physiology,* 1979, *294,* 581–594.

Sedlacek, J. The development of supraspinal control of spontaneous motility in chick embryos. *Progress in Brain Research,* 1978, *48,* 367–384.

Selverston, A. I. Are central pattern generators understandable? *The Behavioural and Brain Sciences,* 1980, *3,* 535–571.

Seyfarth, R. M., and Cheney, D. L. The ontogeny of vervet monkey alarm calling behaviour: A preliminary report. *Zeitschrift für Tierpsychologie*, 1980, *54*, 37–56.

Seyfarth, R. M., and Cheney, D. L. How monkeys see the world: A review of recent research on East African vervet monkeys. In C. T. Snowdon, C. H. Brown, and M. R. Petersen (Eds.), *Primate communication*. Cambridge, England: Cambridge University Press, 1982, pp. 239–252.

Seyfarth, R. M., Cheney, D. L., and Marler, P. Monkey responses to three different alarm calls: Evidence to predator classification and semantic communication. *Science*, 1980a, *210*, 801–803.

Seyfarth, R. M., Cheney, D. L., and Marler, P. Vervet monkey alarm calls: Semantic communication on a free-ranging primate. *Animal Behaviour*, 1980b, *28*, 1070–1094.

Shalter, M. D., Fentress, J. C., and Young, G. W. Determinants of response of wolf pups to auditory signals. *Behaviour*, 1977, *60*, 98–114.

Sherrington, C. S. *The integrative action of the nervous system.* New Haven: Yale University Press, 1906.

Simpson, M. J. A. Social display and the recognition of individuals. In P. P. G. Bateson and P. H. Klopfer (Eds.), *Perspectives in ethology.* New York: Plenum Press, 1973, pp. 225–279.

Smiley, C., and Wilbanks, W. A. Effects of noise on early development in the rat. *Bulletin of the Psychonomic Society*, 1962, *19*, 181–183.

Smith, W. J. Message, meaning, and context in ethology. *American Naturalist*, 1965, *99*, 405–409.

Soffe, S. R., Clarke, J. D. W., and Roberts, A. Swimming and other centrally generated motor patterns in newt embryos. *Journal of Comparative Physiology A*, 1983, *152*, 535–544.

Stark, R. E. Stages of speech development in the first year of life. In G. H. Yeni-Kormshian, J. F. Kavanagh, and C. A. Ferguson (Eds.), *Child phonology*, Vol. 1. New York: Academic Press, 1980, pp. 73–92.

Stelzner, D. J. The normal postnatal development of synaptic end-feet in the lumbosacral spinal cord and of responses in the hind limbs of the albino rat. *Experimental Neurology*, 1971, *31*, 337–357.

Stent, G. S. Strength and weakness of the genetic approach to the development of the nervous system. *Annual Review of Neuroscience*, 1981, *4*, 163–194.

Streter, F. A., Luff, A. R., and Gergely, J. Effect of cross-reinnervation on physiological parameters and on properties of myosin and sacroplasmic reticulum of fast and slow muscles of the rabbit. *Journal of General Physiology*, 1975, *66*, 811–821.

Studdert-Kennedy, M. The beginnings of speech. In K. Immelmann, G. W. Barlow, L. Petrinovich, and M. Main (Eds.), *Behavioral development.* New York: Cambridge University Press, 1981, pp. 533–561.

Super, C. Environmental effects on motor development: The case of African infant precocity. *Developmental Medicine and Child Neurology*, 1976, *18*, 561–567.

Szechtman, H., and Hall, W. G. Ontogeny of oral behaviour induced by tail pinch and electrical stimulation of the tail of rats. *Journal of Comparative and Physiological Psychology*, 1980, *94*, 436–445.

Szekely, G. Developmental aspects of locomotion. In R. M. Herman, S. Grillner, P. S. G. Stein, and D. G. Stuart (Eds.), *Neural control of locomotion.* New York: Plenum Press, 1976, pp. 735–757.

Tamasy, V., Koranyi, L., and Phelps, C. P. The role of dopaminergic and serotonergic mechanisms in the development of swimming ability of young rats. *Developmental Neuroscience*, 1981, *4*, 389–400.

Tanner, J. M. Physical growth. In P. H. Mussen (Ed.), *Carmichael's manual of child psychology* (Third Edition). New York: Wiley, 1970, pp. 77–155.

Teitelbaum, P. Levels of integration of the operant. In W. K. Honig and J. E. R. Staddon (Eds.), *Handbook of operant behavior.* Englewood Cliffs, N.J.: Prentice-Hall, 1977, pp. 7–27.

Teitelbaum, P. Disconnection and antagonistic interaction of movement subsystems in motivated behavior. In A. R. Morrison, and P. Strick (Eds.), *Changing concepts of the nervous system: Proceedings of the first institute of neurological sciences symposium in neurobiology.* New York: Academic Press, 1982, pp. 467–498.

Teitelbaum, P., Schallert, T., De Ryck, M., Whishaw, I. W., and Golani, I. Motor subsystems in motivated behavior. In R. F. Thompson, L. H. Hicks, and V. B. Shyrkou (Eds.), *Neural mechanisms of goal-directed behavior and learning.* New York: Academic Press, 1980, pp. 127–143.

Terzuolo, C. A., and Viviani, P. Determinant and characteristics of motor patterns used for typing. *Neuroscience*, 1980, *5*, 1085–1103.

Thelen, E. Kicking, rocking, and waving: Contextual analysis of rhythmical stereotypies in normal human infants. *Animal Behaviour*, 1981, *29*, 3–11.

Thelen, E. Rhythmical behavior in infancy: An ethological prespective. *Developmental Psychology*, 1981, *17*, 237–257.

Thelen, E. Learning to walk is still an "old" problem: A reply to Zelazo (1983). *Journal of Motor Behavior*, 1983, *15*, 139–161.

Thelen, E., and Fisher, D. M. From spontaneous to instrumental behavior: Kinematic analysis of movement changes during very early learning. *Child Development*, 1983, *54*, 129–140.

Thelen, E., Ridley-Johnson, R., and Fisher, D. M. Shifting patterns of bilateral coordination and lateral dominance in the leg movements of young infants. *Developmental Psychobiology*, 1983, *16*, 29–46.

Tinbergen, N. On aims and methods of ethology. *Zeitschrift für Tierpsychologie*, 1963, *20*, 410–433.

Tooker, C. P., and Miller, R. J. The ontogeny of agonistic behaviour in the blue gourami, *Trichogaster trichopterus* (Pisces, Anabantoidei). *Animal Behaviour*, 1980, *28*(4), 973–988.

Tran Thanh Van, K. Control of morphogenesis by inherent and exogenously applied factors in thin cell layers. *International Review of Cytology, Supplement 11a*, 1980, 175–194.

Tuller, B., Kelso, A. S., and Harris, K. S. On the kinematics of articulatory control as a function of stress and rate. *Status Report on Speech Research SR-71/72*. New Haven, Conn.: Haskins Laboratories, 1982, pp. 81–88.

Turkewitz, G., and Kenny, A. Limitations on input as a basis for neural organization and perceptual development: A preliminary theoretical statement. *Developmental Psychobiology*, 1982, *5*, 357–368.

Turkewitz, G., Lewkowicz, D. J. and Gardner, J. M. Determinants of infant perception. *Advances in the Study of Behavior*, 1983, *13*, 39–62.

Twitchell, T. E. Reflex mechanisms and the development of prehension. In K. Connolly (Ed.), *Mechanisms of motor skill development*. London: Academic Press, 1970, pp. 25–38.

Vanderwolf, C. H. The role of the cerebral cortex and ascending activating systems in the control of behavior. In E. Satinoff and P. Teitelbaum (Eds.), *Handbook of behavioral neurobiology*. Vol. 6, *Motivation*. New York: Plenum Press, 1983, pp. 67–104.

Vince, M. A. Postnatal effects of prenatal sound stimulation in the guinea pig. *Animal Behaviour*, 1979, *27*, 908–918.

Viviani, P., and Terzuolo, C. Space-time invariance in learned motor skills. In G. E. Stelmach and J. Requin (Eds.), *Tutorials in motor behavior*. Amsterdam: North-Holland, 1980, pp. 525–533.

Walker, B. E., and Quarles, J. Palate development in mouse foetuses after tongue removal. *Archives of Oral Biology*, 1976, *21*, 405–412.

Watson, J. S. Contingency experience in behavioral development. In K. Immelmann, G. W. Barlow, L. Petrinovich, and M. Main (Eds.), *Behavioral development*. New York: Cambridge University Press, 1981, pp. 83–89.

Weber, Th. Stabilisierung des Flugrhythmus durch Erfahrung bei *Gryllus campestris* L. *Naturwissenschaften*, 1972, *59*:366.

Weber, Th. Elektrophysiologische Untersuchungen zur Entwicklung und zum Verlauf von Verhaltensweisen bei *Gryllus campestris* L. Dissertation, University of Cologne, 1974.

Weismer, G., and Fennell, A. Studies of phrase-level speech timing. Invited Paper, 105th meeting of the Acoustical Society of America, University of Wisconsin, Madison, 1983, 84–142.

Weismer, G., and Ingrisano, D. Phrase-level timing patterns in English: Effects of emphatic stress location and speaking rate. *Journal of Speech and Hearing Research*, 1979, *22*, 516–533.

West, M. J., King, D. H., and Eastzer, D. H. The cowbird: Reflections on development from an unlikely source. *American Scientist*, 1981, *69*, 56–66.

Whiting, H. T. A. Dimensions of control in motor learning. In G. E. Stelmach and J. Requin (Eds.), *Tutorials in motor behavior*. New York: North-Holland, 1980, pp. 537–550.

Wiesel, T. N. Postnatal development of the visual cortex and the influence of environment. *Nature*, 1982, *299*, 583–591.

Wilson, D. M. The central nervous control of flight in a locust. *Experimental Biology*, 1961, *38*, 471–490.

Windle, W. F. *Physiology of the fetus*. Philadelphia: Saunders, 1940.

Windle, W. F. Genesis of somatic motor function in mammalian embryos: A synthesizing article. *Physiological Zoology*, 1944, *17*, 247–260.

Wolff, P. H. The causes, controls, and organization of behavior in the neonate. *Psychological issues*, Vol. 5, No. 1, Monograph 17. New York: International Universities Press, 1966.

Wolff, J. R. Some morphogenetic aspects of the development of the central nervous system. In K.

Immelmann, G. W. Barlow, L. Petrinovich, and M. Main (Eds.), *Behavioral development.* New York: Cambridge University Press, 1981, pp. 164–190.

Wolgin, D. L. Motivation, activation, and behavioral integration. In R. L. Isaacson and N. E. Spear (Eds.), *The expression of knowledge: Neurobehavioral transformations of information into action.* New York: Plenum Press, 1982, pp. 243–290.

Wolgin, D. L., Hein, A., and Teitelbaum, P. Recovery of forelimb placing after lateral hypothalamic lesions in the cat: Parallels and contrasts with development. *Journal of Comparative and Physiological Psychology,* 1980, *94,* 795–807.

Woolridge, M. W. A quantitative analysis of short-term rhythmical behaviour in rodents. D. Phil. Thesis, Oxford University, 1975.

Wyman, R. J. A simpler network for the study of neurogenetics. In J. C. Fentress (Ed.), *Simpler networks and behavior.* Sunderland, Mass.: Sinauer Associates, 1976, pp. 153–166.

Zelazo, P. R. The development of walking: New findings and old assumptions. *Journal of Motor Behavior,* 1983, *15,* 99–137.

Environmental and Neural Determinants of Behavior in Development

Timothy H. Moran

I. Introduction

Developmental characteristics of age-typical behaviors have been investigated extensively in a number of altricial species during the past decade. These investigations have outlined (1) the proximal determinants of a behavior during a particular developmental period, (2) how these controls change as development progresses and perceptual domains are expanded, and (3) the bases of these changes.

The framework in which developmental analyses were often cast held that the behavioral capabilities of the neonate followed an ontogenetic course similar to that of sensorimotor development (Teitelbaum, 1971; Teitelbaum, *et al.,* 1969; Campbell and Coulter, 1976). However, as will be shown, limited sensorimotor capabilities often belie more sophisticated motivational and associative complexity. Studied in settings consonant with their natural environments, infants often express otherwise hidden behavioral capabilities at birth or shortly thereafter. The experimental conditions necessary for the expression of these behavioral capabilities are the very ones present in the nest setting: high ambient temperature and other forms of external activation.

Timothy H. Moran   Department of Psychiatry and Behavioral Sciences, Johns Hopkins University School of Medicine, Baltimore, Maryland 21205.

When these conditions are provided, rat pups, even on the day of birth, demonstrate independent ingestion of nutrients (Hall, 1979a,b) and fluids (Bruno, 1981; Wirth and Epstein, 1976); perform tasks that are instrumental to receiving food (Johanson and Hall, 1979) or electrical stimulation of the medial forebrain bundle (Moran *et al.*, 1981), and acquire new associative relationships that allow previously neutral stimuli to achieve behavioral significance (Johanson and Hall, 1982; Pedersen *et al.*, 1982). In response to diverse forms of stimulation, pups demonstrate components of adultlike complex motor patterns (Williams, 1979; Moran *et al.*, 1983a). These findings concerning proximal causation and variability in infant behavior provide a sufficient basis for examining the roles of environmental and neural factors in infant behavior. This analysis will attempt to identify principles of behavioral development in select species, incorporate sensorimotor and perceptual development into a broader motivational scheme, assess changes in performance capabilities and determine the role of experience in the manifestation of behavior. Together these results suggest that the organization underlying certain behaviors is remarkably complex at birth or shortly thereafter and that domains of behavioral competence expand to incorporate sensorimotor advances and behavioral experiences.

The following sections will focus on four series of experiments that identify factors underlying behavioral competence in infant rats. Suckling and huddling will be discussed as examples of complex behaviors, occurring throughout early development, that exhibit variability and undergo a series of transitions in underlying controls. Independent ingestion and activation-induced behaviors will be presented as examples of precocial behavioral complexity in neonates requiring a special set of experimental conditions to be elicited. By focusing on the details of these particular behavioral capabilities, general attributes of neonatal behaviors and their underlying mechanism will be generated.

II. SUCKLING AS AN ORGANIZED BEHAVIOR

Suckling is a complex behavior that is restricted to an early developmental period. Although it is the only behavior available to altricial mammals to satisfy nutritional, hydrational, and mineral needs, the performance of the suckling act, even in the absence of ingestion, appears to be intrinsically rewarding. In this section suckling will be analyzed as (1) a behavior that occurs in response to a complex array of sensory cues, (2) a behavior for meeting nutritional requirements, and (3) a reinforcing act. Together, these results demonstrate that suckling, although stereotyped in its motor expression, exhibits flexibility in initiating stimuli, is under a variety of internal controls, and supports the acquisition of a range of approach behaviors that utilize different motor patterns in their expression. Furthermore, through suckling and its surround circumstances, the pup is able to expand its range of incentives.

A. EXTERNAL CONTROLS

101

ENVIRONMENTAL
AND NEURAL
DETERMINANTS OF
BEHAVIOR IN
DEVELOPMENT

The first nipple attachment in albino rats occurs only if parturient mothers' saliva or amniotic fluid coats the nipple (Teicher and Blass, 1977). Since the rat mother eats the placenta during parturition and actively licks the pups and then her nipple line during the interval between pup deliveries (Roth and Rosenblatt, 1967), a behavioral mechanism exists for the mother to deposit saliva and amniotic fluid on her nipples. A number of authors (Teicher and Blass, 1976; Blass *et al.*, 1977; Bruno *et al.*, 1980; Hofer *et al.*, 1975) have demonstrated that in order for suckling to occur in albino rats 2–30 days of age, the dam's nipples must be coated with pup saliva. These results pose the problems of how amniotic fluid and later pup saliva gain control over nipple attachment. The first problem is not trivial. Amniotic fluid is dynamic, reflecting the mother's hormonal status at parturition (Saunders and Rhodes, 1973) and, as a plasma filtrate, the mother's diet. Moreover, mammalian fetuses swallow and excrete amniotic fluid during the last trimester of gestation, further altering its composition (Lev and Orlic, 1972).

Pedersen and Blass (1982) provided an experimental basis for the initial nipple attachment by approximating the pre- and postnatal events that precede the first nipple attachment. First, they adulterated the amniotic fluid *in utero* by the addition of a novel olfactory stimulus (citral). Second, they activated the pups postnatally, in the presence of citral, with a soft brush, mimicking the mother's licking behavior. Pups that had experienced citral both pre- and postnatally did not attach to normal nipples but readily attached to washed nipples scented with citral. This finding could be obtained only when citral was experienced by pups both prenatally and postnatally. Either experience alone was not sufficient. Thus a specific prenatal manipulation in conjunction with a specific postnatal one severely biased the rats away from amniotic fluid, the stimulus that normally elicits the first nipple attachment. It biased the rats toward the stimulus that was part of the rats external milieu during the end of the gestational period and immediately surrounding birth. This experimental design closely mimicked natural events in two important ways. First, the fetus was provided with an aromatic substance during the final 2 days of gestation. Second, the neonate was stimulated in the presence of that substance during the first hour following parturition. Providing the pup with activation allowed the odor that was present in the amniotic fluid to direct the behavior of the pup in its first suckling act.

These observations led Pedersen *et al.* (1982) to address how pup saliva could replace amniotic fluid as the necessary olfactory cue for nipple attachment in the days following birth by further exploring postnatal events that could confer behavioral significance upon neutral stimuli. Since suckling is normally preceded by the maternal rat handling and licking her pups upon entering the nest, activating them prior to a suckling bout (MacFarlane and Blass, 1980, unpublished observation) and activation allowed the cue for the initial nipple attachment to be altered, the role of activation in determining the significance of the olfactory environment for suckling maintenance was examined. Two routes of activation were employed. Pups were either stroked with a soft artist's brush as in the previous experiment or activated pharmacologically, via an intraperitoneal injection of amphetamine (0.5 mg/kg). Pups received these treatments either at room or nest temperature. When stimulation occurred in the presence of an odor, the pups would generally attach to washed nipples scented with that particular odor. Pups did not attach to washed nipples in the presence of a volatile if activation occurred in unscented conditions and did not attach to washed nipples in the absence of the "conditioned" stimuli, nor did they attach to nipples scented with a different odor. The effect occurred under the specific pairing of a neutral stimulus with activation. Independent manipulation of the amount of stroking, amphetamine, temperature, and odor concentration yielded the finding that there was an optimal level of stimulation that induced normal attachment. Combining too many of these variables at high levels blocked the effect.

The sensory flexibility exhibited in these experiments indicates that although suckling is sufficiently stereotyped in its motor expression in the days immediately following birth as to have been mistakenly considered reflexive (Peiper, 1963; Teitelbaum, 1977), the probability of its being elicited by a particular stimulus is related to the animal's previous experience with that stimulus.

Even though experience with the novel odor was not directly paired with suckling, the odor does come to elicit suckling. This result can be viewed in two ways. Activation in the presence of olfactory stimulation appears to be a reliable antecedent to suckling in the nest situation. Prior to a suckling bout, the dam licks and handles her pups (MacFarlane and Blass, 1980, unpublished observation). This pairing of activation with suckling, as a natural occurrence in the pups experience, may lead directly to the novel odor, having been paired with activation, eliciting suckling. Alternatively, this demonstration may represent an interaction of experience with a behavior of high probability. When presented with the odor that had been paired with activation, a conditioned activation may occur, generally increasing behavior. Since suckling is such a highly probable behavior for the neonate, activation results in nipple attachment in this setting. Whatever the mechanism, these experiments demonstrate that suckling can be elicited by stimuli that were previously neutral, and these stimuli, by being paired with activation, take on functional relevance for the pup.

Flexibility is also demonstrated under conditions in which pups have been deprived of olfactory or tactile input. The act of nipple attachment depends upon olfactory cues to bring the pup into contact with the nipple (MacFarlane *et al.*, 1983). Once the nipple is contacted, the tactile properties of the nipple are crucial for attachment to take place (Hofer *et al.*, 1981). Depriving pups of either form of information disrupts the suckling act (Singh *et al.*, 1976; Teicher *et al.*, 1978; Hofer *et al.*, 1981). Yet, although either olfactory bulbectomy or infraorbital nerve section, producing anesthesia of the snout area, may be so disruptive to suckling as to result in relentless weight loss and eventual death, pups 2 days of age or younger can recover from either procedure (Teicher *et. al.*, 1978; Hofer *et al.*, 1981). Pups can compensate, presumably, by relying on alternative cues to gain access to the suckling situation.

103

ENVIRONMENTAL
AND NEURAL
DETERMINANTS OF
BEHAVIOR IN
DEVELOPMENT

These results, demonstrating the ability of infant pups both to acquire new eliciting olfactory cues and to compensate for loss of major sources of sensory information in the suckling situation, present an example of developmental plasticity. The behavioral patterns underlying the suckling act are available to the animal at birth, but the particular stimulus configuration that elicits these patterns appears not to be completely specified. Newborn pups initially respond to the familiar odor of amniotic fluid as an eliciting stimulus for nipple attachment (Teicher and Blass, 1977), but this olfactorant does not remain in the environment. A constituent of the pups' own saliva comes to elicit the behavior patterns of suckling as development progresses (Teicher and Blass, 1976). Both of these cues can be altered, however, as has been demonstrated (Pedersen and Blass, 1982; Pedersen *et al.*, 1983). Furthermore, removing either olfactory or tactile inputs does not totally disrupt the behavior if done at an early age (Teicher *et al.*, 1978; Hofer *et al.*, 1981). This plasticity allows the neonate to respond to possible changes in the dam's olfactory or tactile configurations that may occur in response to diet, nest location and composition, and stage of cycle.

There are limits, however. As discussed by Rosenblatt (1983), young animals are responsive to stimulation in a manner by which low-intensity stimulation elicits approach responses and high-intensity stimulation elicits withdrawal responses (Shneirla, 1965). For example, kittens will suckle at a brooder box shaped like the nursing mother, covered with soft tactile cues (Rosenblatt *et al.*, 1962) but not at a brooder box with sharp angles and discontinuous surfaces (Kovach and Kling, 1967). They are able to accommodate to minor changes in the situation where low-level stimulation is maintained. In rat pups, too high a level of stimulation paired with a novel odor will block the ability of the pup to associate that odor with suckling (Pedersen *et al.*, 1983). Furthermore, the range of acceptable tactile cues for suckling to occur in the rat pup appears to be narrower than in the kitten, as attempts at constructing an artificial nipple to which the pup will attach have been generally unsuccessful (Blass, unpublished observation).

B. Internal Controls of Suckling

While suckling can be elicited under certain conditions by a variety of environmental stimuli, there has been disagreement as to the nature of its underlying internal controls. Suckling performs a number of functions for the pup. It is the behavior through which the pup satisfies its nutritional and hydrational needs as well as being a primary source of social contact between the pup and the dam. The latter, in fact, appears to be the main motivational determinant of suckling. The evidence against nipple attachment being under the control of the internal stimuli that determine food and water consumption in adults is impressive. The latency with which pups attach to the nipples of an anesthetized dam is not markedly affected by maternal separation until about 2 weeks of age (Hall *et al.*, 1975, 1977).

Likewise, nipple attachment is unaffected by a variety of manipulations that markedly inhibit adult feeding. Neither gastric loading (Hall and Rosenblatt, 1977; Williams *et al.*, 1980), systemic injections of amphetamine (Raskin and Campbell, 1981), the intestinal peptide cholecystokinin (Blass *et al.*, 1979), nor acute dehydration (Bruno *et al.*, 1982; Cramer *et al.*, 1984) interfere with nipple attachment. Moreover, once a pup is attached to the nipple, only catastrophic events lead the pup to release the nipple. Specifically, when obtaining milk through an intraoral cannula (Hall and Rosenblatt, 1977) or when withdrawing milk from milk-replete females (Cramer and Blass, 1983), rats release the nipple only after their stomachs have become so engorged that milk refluxes into their nares and lungs, compromising breathing (Hall and Rosenblatt, 1977). When milk is delivered directly into the stomach via an indwelling gastric cannula, rats remain attached even though they have become grossly distended to the point of stomach rupture. If removed from the nipple, they reattach in a matter of seconds (Hall and Blass, unpublished observation, 1975). Suckling, therefore, appears to be immune to the immediate events that normally terminate feeding. Nipple attachment in pups 10 days of age or less is not responsive to a variety of changes in the internal milieu and is not dependent on separation from the dam for its appearance. In fact, these very young rats do not leave the nipple even after 12 hr without receiving milk delivery (Cramer *et al.*, 1980). In this regard suckling does not appear to demonstrate one characteristic of adult-motivated behaviors: varying threshold (Tinbergen, 1969). Nipple attachment appears to occur with the same intensity and latency despite the animal's recent attachment history and is not sensitive to postabsorptive consequences.

Suckling has nonnutritional benefits, however, that extend to thermal support (Alberts, 1978), stress reduction as measured by heart rate (Hofer and Weiner, 1975), and corticosterone reduction (Levine, 1966) and, in some species, transport (Brewster and Leon, 1980). With suckling serving such a multiplicity of functions for the neonatal pup, attachment at any time may be determined by a variety of factors.

While attachment does not exhibit varying threshold in neonatal pups, milk intake on the nipple well may. As discussed previously, pups continue to ingest a constant supply of milk while attached to the nipple (Hall and Rosenblatt, 1977), but in these paradigms the pup must detach from the nipple in order to stop the experimentally controlled milk delivery from a tongue cannula. The natural supply of milk from the mother is intermittent rather than constant. The volume of milk ingested by day-10 pups placed with an awake lactating mother for a test of set duration increases following gastric preloads (Houpt and Epstein, 1973; Houpt and Houpt, 1975). In fact, as demonstrated by Cramer and Blass (1982), deprivation from suckling results in increased milk intake in pups as young as day 5 under the proper testing situation. Cramer and Blass (1983) have differentiated between the total volume of intake a pup is willing to ingest and the rate at which ingestion takes place. In pups younger than 15 days, the rate of ingestion is sensitive to the nutritional status of the pups, while the total volume they are willing to take is not.

105

ENVIRONMENTAL
AND NEURAL
DETERMINANTS OF
BEHAVIOR IN
DEVELOPMENT

The probability of pups withdrawing milk from the nipple at any given opportunity appears to be under the control of internal signals. Lorenz *et al.* (1982), examining milk intake and attachment in the nest situation, have found a similar result. Gastrointestinal preloads reduced intake in 1- and 10-day-old pups but did not affect attachment. At day 20, both intake and attachment were reduced by preloading the pups. This differentiation resolves the differences between the results of Hall and Rosenblatt (1977) and those of Houpt and Epstein (1973) and point to ingestion while suckling and to attachment as different events for the pup: one being under internal controls of nutritive state and the other representing the interplay of other benefits for the developing pup. However, even in weaning-age pups, deprivation is defined at least in part by suckling abstinence aside from its nutritive consequences. As demonstrated by Cramer and Blass (1984), depriving pups of only the nutritive component of suckling was not equivalent to deprivation in isolation. Pups kept with dams whose nipples had been ligated, allowing suckling but not ingestion, could not be differentiated from nondeprived pups until 20 days of age in the amount of milk ingested in a subsequent test. Suckling itself during the deprivation period was sufficient to reduce subsequent intake. This difference is further highlighted in the following section.

C. SUCKLING AND REINFORCEMENT

Behavioral variability is expressed by the pups in their ability to acquire new behavioral patterns to attach to a nipple. Kenny and Blass (1977) demonstrated that pups as young as 7 days of age developed a spatial discrimination in a "Y" maze when allowed to suckle a nonlactating dam. Pups demonstrated a preference for the nonlactating dam to rooting into the gauze-covered ventrum of the dam, although the latter was preferred over entering an empty goal box. The opportunity to suckle a nonlactating dam was also sufficient to sustain behavior in a runway (Amsel *et al.*, 1976; Amsel *et al.*, 1977). In a choice test, pups 10 to 12 days of age did not demonstrate any preference for nutritive versus nonnutritive suckling. Only at 17 days of age did a clear preference for nutritive suckling develop (Kenny *et al.*, 1979). This result indicates that a preference for nutritive suckling develops at the same time that nipple attachment begins to come under the internal controls related to ingestion. Prior to that time suckling is continued even in the face of potentially negative consequences, and nonnutritive suckling is a powerful reinforcing act. However, using running speed as a different measure of avidity for a reinforcer, 11-day-old pups do run faster for nutritive than for nonnutritive suckling (Letz *et al.*, 1978). The onset of preference for nutritive suckling varies according to the measure employed but appears to coincide with the ability of pups to begin to interact independently with their environment.

The preference of the pup for suckling versus independent ingestion follows a different developmental course. Pups choose access to suckling over feeding through 24 days of age even when food deprived for 24 hr (Stoloff and Blass, 1983). Similarly, dehydrated pups continue to prefer maternal contact rather than drinking in isolation (Bruno *et al.,* 1983). Even though independent feeding and drinking have become the primary modes of nutrition and hydration for these pups, suckling and the chance to establish contact with the dam still support the acquisition of an operant and are preferred over independent ingestion.

Through suckling the pup may increase its range of reinforcing stimuli. Brake (1981) has demonstrated that suckling infant rats 11 to 14 days of age exhibit a preference for a novel olfactory stimulus that has been paired with suckling. The olfactory preference can be acquired in the absence of milk delivery but is stronger when paired directly with the delivery of milk while suckling. This offers the neonate a mechanism for acquiring preferences in relation to suckling, expanding their domains through experiences while in the nest. The pairing of environmental cues with the reinforcement surrounding the suckling situation may instill incentive value on those cues enabling the pup to learn about its environment through maternal contact. This idea has been discussed by Rosenblatt (1983) and will be treated in greater detail later in this chapter.

In summary, suckling is a complex behavioral phenomenon that undergoes a number of developmental transitions. The cue for nipple attachment changes during the early neonatal period and can be brought under experimental control by linking a novel olfactory stimulus to either manual or pharmacological stimulation at a high ambient temperature. These conditions mimic those existing in the nest setting and allow new stimuli to gain behavioral significance for the pup. The internal controls of suckling are multiple and extend beyond its nutritional consequences. Young pups remain attached to a nipple even when intake has reached its biometric limit. As pups approach weaning, the internal controls affecting adult ingestion begin to influence both nipple attachment and intake. The act of suckling is reinforcing to pups, allowing the pup both to acquire new approach patterns to gain access to nipples and to expand its range of reinforcing stimuli through the association of environmental cues with suckling.

III. Huddling

Huddling is a prominent feature of behavior in neonates. Under seminatural conditions the mother's visits to the nest are intermittent (MacFarlane and Blass, 1983) and huddling significantly improves the capacity for pups for thermoregulation. Alberts and colleagues (Alberts, 1978a,b; Alberts and Brunjes, 1978; Brunjes and Alberts, 1979) have identified the changing tactile, thermal, and olfactory controls underlying the expression of this behavior during different developmental periods.

Huddling attenuates the rate of temperature loss in neonatal pups by allowing an individual to become part of an aggregate, thereby reducing the individual's exposed surface area and altering its surface-to-mass ratio. In a huddling situation, pups better maintain body temperature and need to expend less energy to do so (Alberts, 1978b). Alberts addressed the issue of whether the thermoregulatory

107

ENVIRONMENTAL
AND NEURAL
DETERMINANTS OF
BEHAVIOR IN
DEVELOPMENT

effects of huddling were merely the passive outcome of a natural tendency toward aggregation in pups or whether the dynamics of the huddling situation represented a regulatory behavior. By mapping the exposed area of the huddle, Alberts demonstrated that as the ambient temperature fell, the total exposed surface area of the huddle decreased. In contrast, as ambient temperature rose, huddle surface area increased as the extent of contact between the individual pups diminished. This regulation of huddle size was evident in pups as young as 5 days of age and continued through day 20. Furthermore, the huddle is dynamic with the position of individual pups relative to the rest of the group constantly changing. This dynamic shifting of huddle was a result both of cooler pups from the outside exposed surface burrowing down toward the center of the clump, displacing other pups toward the periphery, and of warmer pups maximizing their own exposed body surface by actively moving toward the periphery. Thus the huddle as an active regulatory unit represents the outcome of the behavior of its individual members in maximizing their own temperature regulation and minimizing metabolic cost.

The sensory controls underlying huddling change with development. Although rat pups and adults gain thermoregulatory benefits from huddling, a variety of cues can direct and maintain huddling behavior. Alberts (1978a) demonstrated that heat cues are crucial in the 5-day-old pup for huddling to occur, and in the absence of heat emanating from an immobile furry source no huddling would occur. Olfactory cues become important at 10 days of age, as anosmia disrupted huddling in pups 10–20 days of age. An inability to locate a stimulus pup does not completely explain this deficit, as pups reliably approached and contracted anesthetized stimulus pups but did maintain contact through huddling. Overall, the range of cues for huddling increases with age. Olfactory input and tactile cues, related to the "furriness" of a stimulus, become increasingly important during the second week and photic cues become meaningful at the time of eye opening.

While a variety of cues come to elicit huddling, two-choice preference tests revealed developmental differences in the attractiveness of huddling stimuli (Alberts and Brunjes, 1978). At 5 days of age a warm furry tube was preferred as a huddling object over an anesthetized littermate. This preference was reversed in the day-10 to day-20 pups, who reliably preferred contact with an anesthetized littermate. Five- and 10-day-old pups did not show any preference for an anesthetized littermate over an anesthetized gerbil. Only at 15 days of age did this preference develop. The emergence of this preference for a conspecific has an olfactory basis as anosmia blocks its formation. Alberts and Brunjes (1978) interpret these transitions as a postnatal ontogeny of contact behavior from physiological to filial huddling. Physiological huddling is viewed as a homeostatic response of the neonate to thermoregulatory immaturity. Fillial huddling is directed by nonhomeostatic cues that are representative of social affiliation. This transition represents a developmental progression in the controls underlying a behavior that is carried from an early developmental epoch into adulthood.

Brunjes and Alberts (1979) examined the developmental origins of olfactory-guided huddling. Pups were reared in an artificial odor from day 1 and tested for preference in a huddling situation with an artificially scented littermate or a conspecific. Five- and 10-day-old pups did not display any preference. Beginning at 15 and 20 days of age, however, pups reared in an artificial odor demonstrated a clear preference for huddling with pups scented with that odor over normally reared conspecifics. This preference was specific to the rearing odor and not generalized to any novel odor. In further studies, Alberts (1981) demonstrated that simple experience with the odor was sufficient to instill this preference. Active interaction with the dam or nest situation was not necessary. This contrasts with instilling a novel odor for suckling elicitation (Pedersen *et al.*, 1981), where activation in the presence of the odor appears to be a necessary condition. However, a preference for a nipple scent can be instilled as early as the time of birth with a single pairing of a novel odor with activation. Olfaction does not become a primary cue for huddling until day 15, at which time simple experience with an odor appears to be sufficient to convey a functional association to that odor in the huddling situation.

Although exposure to an odor is sufficient to instill a preference for that odor in the huddling situation, Alberts (1981) has demonstrated that the relative degree of odor preference depends upon the setting in which that odor was experienced. Mere exposure was sufficient to instill a preference but did not lead to the same degree of attraction derived from olfactory experience combined with maternal care. In two-choice preference tests, pups chose the odor that had been associated with active maternal care over odors to which they had merely been exposed. Furthermore, unlike the results of Kenny and Blass (1977), the nutritive rewards of suckling made no difference in the strength of this preference, as an odor paired with suckling was not preferred over an odor paired with a maternal but nonlactating dam.

Rosenblatt (1983) has suggested that a demonstration of this type represents the use of olfaction by the developing animal to recognize a source of stimulation that is rewarding. In this view the development of such an olfactory preference as can be demonstrated in huddling is similar to that seen in home orientation and nest recognition. Experience in the presence of an odor can convey either positive or negative meaning to that odor. Thus rat pups distinguish their own nests from others and begin to orient on an olfactory basis (Sczerzenie and Hsiao, 1977; Altman and Sudershan, 1975). The specific olfactory configuration of the nest environment leads the pup to recognize its own nest, and the experiences of the pup within that nest situation convey incentive value to that olfactory configuration. Similarly, the home orientation of the kitten has an olfactory basis whose incentive appears to be an outcome of contact with the olfactory configuration of the mother (Rosenblatt *et al.*, 1969; Rosenblatt, 1983). However, in these cases, the specific nest experience that conveys incentive to the olfactory configuration has not been isolated as it has with suckling and huddling.

In summary, huddling, like suckling, represents a behavioral capability of the neonate that does not require an extraordinary system of environmental supports.

Pups huddle in response to a variety of stimuli, the variety of which can increase with age based on experience. Thermal support and contact appear to represent incentives that the altricial pup acquires through huddling. The association of an olfactory configuration with these goal objects within the nest situation instills that olfactory configuration with incentive value and this olfactory preference guides later filial huddling and may allow the pup to increase its range of incentives.

109

ENVIRONMENTAL
AND NEURAL
DETERMINANTS OF
BEHAVIOR IN
DEVELOPMENT

IV. INDEPENDENT INGESTION

Independent ingestion does not normally appear in rat pups until after the second postnatal week (Bolles and Woods, 1964; Babicky *et al.*, 1973; Galef, 1979), when intake from suckling is supplemented by the ingestion of bits of food in and around the nest site. The neonatal rats' abilities to ingest water and nutrients freely at earlier ages have only recently been appreciated. This section will outline these capacities, focusing on the capabilities of very young pups, increasing behavioral complexity with development and the role of external activation as a necessary eliciting factor.

A. WATER INTAKE

The thirst mechanism may actually be present during gestation. In the final week of gestation, fetal rabbits swallow larger amounts of amniotic fluid following systemic dehydration by hypertonic mannitol (Bruns *et al.*, 1963). Wirth and Epstein (1976) demonstrated that the development of responsiveness to thirst challenges is sequential. Two-day-old rat pups, in response to cellular dehydration, swallow more fluid than their littermate controls when water is infused directly into their mouths. At 4 days of age, pups drink to hypovolemia and at 6 days of age to isoproterenol-induced beta activation. Central (Misantone *et al.*, 1980) and peripheral (Wirth, 1984) angiotensin administration first causes water intake at day 2, and its dipsogenic potency appears to rise sharply at day 5 or 6.

Bruno (1981) has extended these findings by outlining a developmental progression in drinking. During the first postnatal week, rat pups subjected to intracellular dehydration swallow fluid infused directly into the mouth and initiate drinking by vigorously licking and swallowing fluid spread in a thin film across the floor of a test chamber. At these ages, intake is partially sensitive to the type of fluid presented: Milk and water are readily accepted, whereas hypertonic saline is not. Pups drink, however, only when in direct contact with the fluid. They do not approach a fluid source until day 10, when dehydrated pups locate a fluid source at a distance or drink from a cup.

The development of responsiveness to thirst challenges is precocial, since it is present before the pup ingests water as an isolated substance. Ingestion occurs both at room temperature (Wirth and Epstein, 1976) and at high ambient temperature. However, the neonatal pup requires direct contact with the fluid source for ingestion to occur. As pups develop, they are able to detect and approach a fluid source from a distance.

B. NUTRIENT INTAKE

Hall and his colleagues (Hall, 1979a,b; Johanson and Hall, 1980; Hall and Bryan, 1980), in a series of experiments, have carefully documented neonatal ingestive behavior and have examined the developmental aspects of independent nutrient ingestion. They demonstrated that even 1-day-old rat pups ingest milk that is directly infused into the anterior or posterior portion of the mouth (Hall, 1979). The amount of milk ingested with anterior oral infusions was directly related to the length of deprivation from the mother and from suckling. In consuming the milk, pups became highly activated and emitted a variety of behavioral responses, which included mouthing and licking and other responses that resembled adult ingestion.

Hall and Bryan (1980) outlined the progression of the pup's ingestive behavior from the limited terminal act to an integration of approach patterns with that act. As early as 3 days of age, rat pups ingested large quantities of liquid or semisolid food spread over the floor of a test chamber. As with adults, intake was affected by the deprivational status of the pup through all age groups tested. With increasing age, however, pups became better able to locate and ingest food from a specific location within the testing arena. Although younger pups consumed some food from a localized source, they appeared unable to maintain orientation to the food source. A gradual decline in ingestive behaviors occurred as the termination of feeding approached, a pattern similar to the type of satiety sequence evidenced by adults (Antin et al., 1975).

There are, however, important differences between the independent ingestion of pups and adult feeding. Milk is ingested in response to both dehydration and deprivation in younger pups. It is not clear whether milk intake in these settings represents hunger or thirst or even if the two states are differentiated in controlling intake (Bruno et al., 1982). Another critical difference is expressed in the level of behavioral activation that accompanies intake in younger pups. In response to oral milk infusions, or when in direct contact with a fluid nutrient spread over the floor of the test chamber, pups become highly activated, exhibiting mouthing, licking, probing, increased locomotion rolling and full body stretches which resemble the responses of pups to maternal milk letdowns (Hall, 1979a; Hall and Bryan, 1980; Moran et al., 1983b). As intake increases in response to deprivation, so also does the level of activation. This aspect begins to disappear in pups older than 6 days of age.

Both intake and activational responses depend upon testing conditions. Behavioral activation and intake were obtained only in pups tested at high ambient temperature. At room temperature, young pups showed very little ingestion and no behavioral activation. Johanson and Hall (1980) examined the role of temperature as a facilitator for independent ingestion in neonatal rats. Even in the absence of a nutrient source, pups displayed high levels of oral behaviors and activation when placed in an extremely warm testing environment (37°C). Up until 6 days of age, the higher the temperature, the more the behavioral activation. Older pups were not obviously activated by high temperature. Conversely, when oral infusions were given at room temperature, pups displayed no activation and allowed the milk simply to run out of their mouths. Johanson and Hall demonstrated that the critical feature was environmental temperature rather than the core temperature of the pups. Feeding and activation in deprived pups depended upon exteroceptive stimulation arising from warm ambience. Pups, therefore, require activation from an external source to display adultlike ingestive responses.

Johanson and Hall (1979) also demonstrated that ingestion in 1-day-old pups can support the acquisition of an operant response. By exploiting the propensity of pups to probe upwards along the walls of a small test chamber, they showed that 1-day-old pups deflected a response paddle to receive oral infusions of milk. Pups learned the relationship between their behavior and its consequence when the opportunity to ingest was the reinforcer. Pups older than 6 days of age, for whom oral infusions are not activating, did not perform the operant to receive oral infusions of milk (Hall, personal communication). Furthermore, dehydrated pups, which do not show activational responses to ingestion, do not learn an operant to receive fluids (Bruno and Hall, personal communication). Seemingly, activational responses not only are necessary for the ingestion of milk following deprivation in neonatal pups but also contribute to milk's reinforcing potential. These results indicate that altricial rats can learn an operant on the day after birth and that the feeding system, under the special circumstances of testing in the heat, is functional in both its terminal and its appetitive aspects, even at this early age. Whether activation permits the expression of reinforcement mechanisms or is reinforcing in itself is not clear from these data.

Just as stimuli associated with suckling can acquire reinforcing properties, pairing a novel odor with an oral infusion of milk can confer reinforcing potential on that odor. Johanson and Hall (1982) have demonstrated a conditioned orientation to an odor paired with an oral milk infusion in 5-day-old rat pups. Thus both suckling and independent ingestion in a warm environment provide the pup with opportunities for expanding their range of reinforcing stimuli. Again the role of activation in this phenomenon appears to be critical.

It is evident from these studies that suckling is not the only ingestive behavior available to the newborn. Components of drinking systems are available around birth. As well, when deprived pups are activated by warmth, certain components of adult feeding systems are expressed.

It is unlikely that these ingestive responses would have any survival value for the neonatal pups. In the natural environment, pups would be unlikely to be severely deprived or have the opportunity to conserve milk outside of the suckling situation. Hall (1979) has suggested that the existence of ingestive responses in the neonate might be usefully conceptualized in terms of Anokhim's (1964) idea of "systemogenesis," in which components of a behavior develop prior to the require-

111

ENVIRONMENTAL
AND NEURAL
DETERMINANTS OF
BEHAVIOR IN
DEVELOPMENT

ment for their utilization. Independent ingestion in neonatal rats, thus, presents a capability for a behavior long before it is utilized by the pup.

The factors that control independent ingestion appear to differ markedly from those controlling suckling. Ingestion is demonstrably responsive to internal stimuli yet requires a special set of circumstances for its manifestation. The organizational complexity underlying independent ingestion in neonates is far from complete in the immediate postnatal period. Initially, pups need to be in contact with the nutrient source for ingestion to occur. Intake is sensitive to deprivational status and a pattern of satiation does occur, but initially, intake does not appear to be specific to hunger or thirst. Only with increasing development do the terminal ingestive acts become integrated. New approach patterns, to gain access to oral infusions, can be acquired by young pups, but only later do pups seem able to orient to or maintain contact with an isolated fluid source. This latter aspect of the development of independent ingestion stands in contrast to the ability of young pups to locate and contact the dam. Variability in initiating stimuli is not apparent, although such capabilities have not been fully explored. External activation arising from the sensory inputs from a warm ambience is critical for the behavior to be elicited in neonatal pups. With maturation, differences diminish. Older pups exhibit greater behavioral integration, more flexibility in behavioral patterns, and the need for exogenous activation wanes.

V. Brain Stimulation and Behavioral Organization

Intrahypothalamic electrical stimulation induces a variety of organized behavior patterns in adult animals (Margulles and Olds, 1962; Hoebel and Teitelbaum, 1962; Caggiula, 1970; Mogenson and Stevenson, 1966; MacDonald and Flynn, 1966; Wise, 1969; Valenstein et al., 1970). In adult animals integration of the terminal consummatory act into a behavioral sequence is a feature of stimulation-induced behaviors. Direct medial forebrain bundle electrical stimulation also induces behavior in the neonate. Isolated portions of diverse terminal acts that in the adult are components of organized behavioral sequences are elicited by medial forebrain bundle (MFB) electrical stimulation in pups (Moran and Blass, 1982; Moran et al., 1983a,b). Increasing complexity of organization is evident with increasing development in the expression and interaction of these behaviors in response to stimulation. Medial forebrain bundle electrical stimulation in neonatal pups provides a methodology for outlining the development of behavioral organization underlying these terminal sequences.

Pups 3 days of age and older are able to acquire operant responses to obtain electrical stimulation to the MFB at the level of the lateral hypothalamus (Moran et al., 1981) and other central areas (Lithgow and Barr, 1984). Utilizing a modification of the operant procedure developed by Johanson and Hall (1979), we demonstrated that pups would acquire a two-paddle discrimination and would respond significantly more frequently than yoked controls over an 18-hour test session. These results indicate that neural development in 3-day-old rat pups is sufficiently integrated to support the acquisition of appetitive learning for central reinforcement (i.e., reinforcement without a peripheral referent). Thus neural substrates in the area of the MFB and other forebrain structures are involved in the central mediation of reinforcement in neonates. These same areas support electrical self-stimulation in adults (Gallistel, 1973).

In examining the ability of 3-day-old rat pups to acquire an operant response to receive electrical stimulation, we were struck by the behavioral activation accompanying stimulation and how it resembled that seen during and following oral infusions of milk (Hall, 1979). Behavioral activation elicited by stimulation, in fact, reliably predicted both electrode placement in the area of the MFB and pups' ability to acquire self-stimulation successfully (Moran *et al.*, 1981). We have explored this behavioral activation by examining the behavioral responses of 3-, 6-, 10-, and 15-day-old pups receiving stimulation to the MFB in a number of situations.

The stimulation used in these experiments was direct current pulse trains of a 500-msec duration. Each train consisted of 30 pulses of a 2-msec duration. These were the same stimulation parameters that supported self-stimulation in the prior experiments. This type of stimulation is considerably less than that normally utilized to initiate stimulation-bound behavior in adult rats. In fact, trains of this duration, even when administered every second, do not support stimulation-induced ingestion in adult animals (Ball, 1969).

In their initial experiment, Moran *et al.* (1983a) observed the responses of pups receiving stimulation at frequencies of one train per 30, 20, or 10 sec. In response to these stimulation trains, pups 3, 6, and 10 days of age displayed a reliable and orderly series of behavioral responses. Some of the prominent behaviors observed during stimulation are shown in Figure 1. Following infrequent stimulation, mouthing and licking were the predominant behaviors elicited. As stimulation continued and increased in frequency of occurrence, probing, gaping (a response similar to that seen in nipple grasping), and even organized behaviors resembling the stretch responses of pups receiving maternal milk deliveries and lordosis responses were reliably elicited. The occurrrence of all behaviors increased with increased frequency of stimulation. As pups became more activated by the stimulation, their behavioral output increased.

Behavior became more organized with age. The duration of stretch and lordosis responses increased and became temporally paired with the time of stimulation. As show in Figure 2, in 3-day-old pups, both stretch and lordosis responses had a high probability of being immediately preceded by oral (ingestive) responses of mouthing and licking. This, however, changed with age. At day 6 and at day 10 the probability of a stretch response occurring within 10 sec of an oral response remained high. However, the probability of a lordosis response being paired with these oral behaviors was significantly decreased. In fact, ear wiggling, another component of adult female sexual behavior, preceded lordosis responses in 50% of the

113

ENVIRONMENTAL
AND NEURAL
DETERMINANTS OF
BEHAVIOR IN
DEVELOPMENT

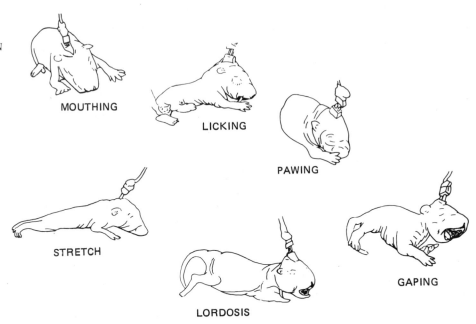

Fig. 1. Characteristic behavioral responses to medial forebrain bundle (MFB) electrical stimulation depicted in a 3-day-old rat pup. (From Moran *et al.*, 1983.)

day-10 pups. The increased duration of these terminal behaviors, their segregation with age, and the additional complexity of ear wiggling in day-10 pups indicate increased organization underlying behavior in these older pups.

Behaviors observed during central stimulation are similar in important ways to those seen during other forms of activation in pups. Pups activated by oral infu-

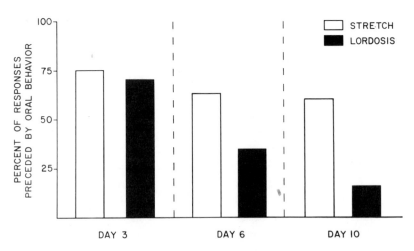

Fig. 2. Mean ± S.E. duration (seconds) of stretch and lordosis responses in 3-, 6-, and 10-day-old pups in response to stimulation. (From Moran *et al.*, 1983.)

sions of milk and stroking emit a series of oral and general activation behaviors that resemble responses to electrical stimulation (Hall, 1979a,b; Pederson *et al.*, 1982). Williams (1979) has demonstrated that lordosis and ear wiggling can be reliably elicited in pups as young as 3 days of age following steroid pretreatment and manual stimulation. In response to MFB stimulation, however, a variety of behavioral responses are elicited and a developmental progression is evident. This contrasts with the responses obtained during other forms of activation. As discussed, pups are no longer activated by oral infusions after 6 days of age and the behavioral responses to other forms of activation appear to be somewhat situation specific.

In order to determine the degree to which the particular form of responses to MFB stimulation could be determined by the testing environment, Moran *et al.* (1983b) stimulated pups in the presence of milk utilizing the procedure of Hall and Bryan (1980), in which stimulated and control pups were placed on a milk-soaked towel and tested for 30 min. MFB stimulation resulted in significantly increased milk intake over that of nonstimulated littermate controls. Pups receiving stimulation showed significantly more behavioral responses than nonstimulated controls. They remained behaviorally active throughout the 30-min test session and did not exhibit the satiety pattern seen in control pups. Of more interest is the effect of milk on the classes of stimulation-induced behaviors emitted by these pups. At day 3, milk availability increased the frequency of all behavioral responses to stimulation, as if increasing the overall level of activation. Milk or stimulation alone were not as activating as the two in combination. At day 6, and even more so on day 10, milk differentially affected the pattern of behavioral responses. As shown in Figure 3, activation by milk and stimulation combined synergistically to increase mouthing as the primary behavior leading to ingestion, but behaviors not compatible with ingestion were inhibited. Although lordosis occurred in response

115

ENVIRONMENTAL
AND NEURAL
DETERMINANTS OF
BEHAVIOR IN
DEVELOPMENT

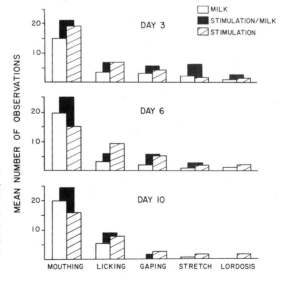

Fig. 3. Mean number of occurrences of individual behaviors during five 1-min sampling intervals for 3-, 6-, and 10-day-old pups. The three groups represented are control pups in the presence of milk, stimulated pups in the presence of milk, and stimulated pups tested on a bare felt surface in the absence of environmental supports. (Moran *et al.*, 1983.)

to stimulation in a situation without obvious environmental supports (i.e., milk), the presence of milk and its ingestion channeled behavior away from lordosis responses. Behaviors incompatible with ingestion did not occur.

Day-15 pups did not respond to stimulation of this type either in the absence of environmental supports or in the presence of milk. It is not that stimulation of this type is completely ineffective in pups of this age. Day-15 pups do acquire operant responses to receive stimulation (Moran and Schwartz, unpublished observation; Lithgow and Barr, 1984). Stimulation at these parameters simply is not sufficient to activate behavioral sequences at day 15.

These studies raise a number of points concerning behavioral organization in developing rats. Infants are similar to adults in that direct MFB electrical stimulation seems to be rewarding. Second, there is good agreement between a central locus and behavioral reward. Third, a number of diverse behavioral patterns can be elicited by brain stimulation.

In addition to these similarities, there are important differences that focus on the nature of organization underlying stimulation-induced behavioral responses at different ages. Hypothalamic stimulation in adults evokes many fully integrated behavior patterns that are not only stimulation bound but "stimulus" bound as well (Valenstein et al., 1970). These behaviors include feeding (Margullis and Olds, 1962; Hoebel and Teitlebaum, 1962), satiety sequences (Wyrwicka and Dobrecka, 1960), sexual behavior (Caggiula, 1970), aggression (MacDonnell and Flynn, 1966), and drinking (Mogenson and Stevenson, 1966). The particular behavior elicited in adults depends to a great extent upon the presence of the appropriate environmental stimuli. In the absence of such stimuli, fully integrated behaviors are not elicited by electrical stimulation. Diverse integrated responses are seen in 3- to 10-day-old rat pups in the absence of environmentally supporting stimuli. This is a striking difference and suggests a relative weakness of inhibitory controls over the behavioral expression of central organization of pups aroused by stimulation of the MFB. In adults one may postulate that the various behavioral patterns that are available under stimulation are normally inhibited and that this inhibition is released by specific environmental stimuli. The environmental stimuli, in the presence of central stimulation, therefore, make available species-specific motor patterns. In pups, however, diverse terminal behaviors appear to be more readily available for expression and do not require a specific environmental stimulus for their elicitation. This difference begins to disappear in 6- and 10-day-old pups. The presence of milk in the environment does bias behavior in the direction of particular terminal behaviors, but environmental stimuli still are not necessary for behavior to occur in infants aroused by central stimulation.

A second difference between the responses of neonates and adults to MFB stimulation is that day-3 pups often display sudden shifts from oral behaviors to what in adults are components of sexual behavior. This difference also diminishes developmentally. By day 10, organization appears hierarchical in structure; once a certain behavior pattern is initiated, terminal acts that define other states are inhibited and behavior is selective. Fentress (1976) has postulated "boundaries"

117

ENVIRONMENTAL
AND NEURAL
DETERMINANTS OF
BEHAVIOR IN
DEVELOPMENT

between motivation sytems. In this conception, as behavior progresses and becomes more directed, the probability of another behavior intervening in the adult behavior pattern diminishes. Boundaries are strengthened at high levels of behavioral activation and behavior becomes more channeled. This does not seem to be true in young pups who readily shift between terminal acts. Activation of one behavioral response does not preclude what in the adult would be an incompatible behavior.

Shifting from one behavioral pattern to another in younger pups may reflect a lack of intrinsic inhibitory systems in highly activated younger pups. Blocking the development of cortical inhibitory mechanism by prenatal treatment with the mitotic inhibitor methazoxymethanol (MAM) timed to produce a specific hypoplasia within the cortex prevents the increased level of behavioral organization usually demonstrated in day-10 pups (Moran *et al.,* 1983). As shown in Figure 4, MAM-treated day-10 pups readily demonstrate these behavioral shifts.

It is of interest that as weaning approaches and pups begin to interact independently with their environment, stimulation of this type no longer elicits the patterns of behavior described previously. At the point where pups no longer require exogenous stimulation for a number of behaviors and are approaching the sensorimotor capabilities of the adult (Gottlieb, 1971; Rose, 1968; Crowley and Hepp-Raymond, 1966), inhibitory control over behavior elicited under stimulation appears to increase. Again, this appears to be due, in part, to cortical development, as MAM-treated day-15 pups continue to respond to these stimulation parameters even though their eyes and ears have opened.

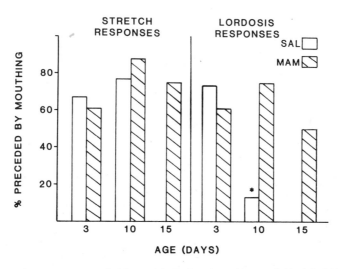

Fig. 4. Percent of responses preceded by oral behaviors (mouthing or licking) in MAM- and saline-treated pups. The number of stretch responses preceded by oral behaviors remains high in both groups. The number of lordosis responses preceded by oral behaviors significantly decreases at day 10 in saline-treated pups but remains high at days 10 and 15 in MAM pups.

VI. Discussion

TIMOTHY H. MORAN

This analysis has presented experimental data illustrating two distinct classes of neonatal behavior; one consists of behaviors that occur naturally and are vital for survival. These include suckling and huddling. The other class consists of behavioral capabilities that can be elicited under a restricted set of experimental conditions. The functional relevance of behaviors in this class is not clear.

Altricial pups are tied to the nest during the first 2 weeks of life, dependent upon the dam as a source of nutrients, contact, warmth, and stimulation. During this time, although the range of their behavioral repertoire is limited, complexity is evident. Behaviors such as suckling and huddling exhibit flexibility in initiating stimuli allowing the pup to interact with a changing environment. The conditions that exist within the nest of high ambient temperature and periodic maternal activation are the very ones that enable rat pups to acquire new functional relationships. The pups' developmental interactions within this context may allow them to acquire information about their surrounding environments and possibly enhance future functional interaction.

In contrast with these rather stereotyped functional behaviors that can occur within a variety of settings, neonatal pups also have the capacity to emit a wider range of behaviors under proper testing conditions. Two interrelated issues are raised by this contrast. The first deals with the requirement of external stimulation for the manifestation of some selected terminal behavior in neonates and how this requirement is alleviated as development progresses. The second focuses on the pattern of increased behavioral integration and eventual independence with development.

The distinction has been made between general and specific activational systems in studies of CNS organization underlying adult motivated behaviors. A simplified view holds that general activating systems appear to consist of the reticular activating formation (French *et al.*, 1953, Jaspers *et al.*, 1958; Lindsley, 1961) and, as more recently identified, the catecholamine pathways of the basal forebrain (Ungerstadt, 1971; Stricker and Zigmond, 1976). Destruction of these catecholamine systems results in global deficits manifested in the sensorimotor, appetitive, and consummatory aspects of all motivated behaviors. In animals with damage to these systems, severe specific internal challenges (e.g., dehydration) cannot overcome the behavioral inertia even when the challenge is life threatening (Teitelbaum and Stellar, 1954). Rats in the first stage of recovery from lateral hypothalamic lesions (Teitelbaum and Epstein, 1962) or dopamine depletions (Stricker and Zigmond, 1976) neither eat nor drink and will die unless maintained by artificial feeding and hydration. This cataleptic indifference may be reversed, however, by hyperstimulation of a general nature (Wolgin and Teitelbaum, 1975; Marshall *et al.*, 1976; Teitelbaum and Wolgin, 1978). When such rats are roughly handled, immersed in an ice bath, or given large doses of amphetamine, they locomote normally, groom, eat, and drink for short periods. Within minutes of the termination of stimulation, they sag into catalepsy once again.

119

ENVIRONMENTAL
AND NEURAL
DETERMINANTS OF
BEHAVIOR IN
DEVELOPMENT

Although the contribution of general activation in the behavior of intact animals cannot be directly or reliably measured, Stricker and Zigmond (1976) have called attention to particular interactions between general and specific activational systems based on the behavioral deficits of catecholamine-depleted animals. They suggest that the behavior of dopamine-depleted animals reflects the absence of a relevant general arousal component. In this case, dopaminergic pathways are viewed as providing a release from a tonic inhibition (possibly cortical) on behavior. Thus in the absence of catecholamine input, specific stimulation arising from need states cannot overcome this tonic inhibition. Accordingly, dopamine-depleted animals engage in ingestive behavior following intense stimulation (e.g., immersion in an ice bath) not because hunger and thirst stimuli are specifically heightened, but because this stimulation produces increased transmission in residual dopaminergic pathways in the presence of receptor supersensitivity resulting from the original depletion. Behavior is permitted as the residual transmission overcomes the endogenous tonic inhibition for a brief period. This allows specific internal and external stimuli access to normally functioning motor systems, resulting in specific terminal behavioral patterns.

In neurologically intact animals, high levels of general activation can combine synergistically with low levels of environmental stimulation to produce behavior. In the absence of any obvious internal bias or need state, activation by either hypothalamic electrical stimulation (Valenstein *et al.*, 1970) or tail pinch (Antelman *et al.*, 1976) yields behavior that appears to be selected and directed by environmental stimuli. The consummatory behaviors elicited by brain stimulation are interchangeable to a large extent, even at a given electrode site (Valenstein *et al.*, 1970, for review). The specific visual, gustatory, or olfactory inputs produce sufficient specific stimulation so that, when combined with suprathreshold general activation, behavior occurs. Likewise, tail-pinch elicits a variety of species-typical behaviors, determined by the available consummatory objects (Antelman *et al.*, 1976; Barfield and Sachs, 1968; Marques *et al.*, 1976). The behavioral effects of both forms of stimulation depend on intact dopamine transmission (Antelman and Szechtman, 1975; Phillips and Fibiger, 1973).

It seems reasonable to speculate that the neural substrate underlying activation as permissive for behavioral complexity in neonatal pups may correspond to that involved in general activation of adults. The development of catecholamine pathways bears a major postnatal component (Loizou, 1972; Coyle and Henry, 1973), and only recently has the functional development of these pathways been addressed. Studies by Cheronis *et al.* (1979) have indicated that the nigrostriatal dopamine pathways do not support impulse traffic under basal conditions when pups are tested at room temperature until after 6 days of age. This conclusion is based on the following evidence. Following axotomy there is no buildup of metabolites around the severed pathway, a result that would occur if ongoing activity was taking place within this system. As well, no decrease of transmitter level in the striatum following administration of synthesis blockers was evident in these neonatal pups as would be found in the adult. However, following direct pharmacological stimulation, impulse traffic in this pathway can be demonstrated. Moreover, when these studies were repeated at nest temperature (32–33°C), ongoing activity characteristic of adult impulse flow is readily measured (Horowitz *et al.,* 1982). Also, as shown in Figure 5, in response to stroking, another activating stimulus in pups, the level of DOPAC, a primary dopamine metabolite, is significantly increased in relation to dopamine concentration within the striatum. Thus the pathways are capable of function in the neonate but require the same forms of stimulation as those necessary for demonstrating behavioral capabilities in these newborn pups. Specifically, either pharmacological (Cheronis *et al.,* 1979), thermal (Horowitz *et al.,* 1982) or physical (Saad *et al.,* 1984) stimulation induces both impulse flow within the nigrostriatal dopamine pathway and activates complex behaviors in neonatal pups.

Following dopamine depletion in adult animals, organized behavior can occur with high levels of activation, as receptors are supersensitive to a stimulated impulse flow. A similar situation may exist in the young pup. A possible model provided by the dopamine-depleted animal suggests that at early ages impulse flow produced by activation interacts with receptor fields that are not regularly activated. A supersensitivity may result from a release against a background of relative quiescence. As development progresses and catecholamine pathways become more sensitive to environmental stimuli, the ongoing behavior of the pup may stimulate a tonic ongoing level of activity in these pathways, removing the supersensitivity and the high level of activation to stimuli such as oral infusions of milk and MFB stimulation. In the maturing art not dependent on external stimulation, a tonic ongoing release of transmitter frees the animal from the constraint of external activation. This developmental progression may in part explain the lack of responsiveness of the day-15 pup to MFB stimulation. Younger pups are being stimulated

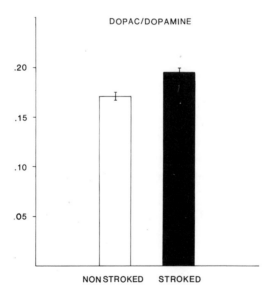

Fig. 5. Striatal DOPAC–dopamine concentratons in stroked and nonstroked 3-day-old rat pups. Stroking with a soft brush for 5 min increases dopamine turnover.

against a background of relative inactivity within catecholamine pathways. Stimulation in this setting results in a high level of behavioral activation. The same stimulation and putative release of transmitter against a background of ongoing activation may no longer be behaviorally meaningful for the pup.

121

ENVIRONMENTAL
AND NEURAL
DETERMINANTS OF
BEHAVIOR IN
DEVELOPMENT

As the pup develops and the need for external activation wanes, behavior becomes better integrated. Terminal behaviors become increasingly differentiated based on environmental stimuli and ongoing behavior (Moran *et al.,* 1983a,b). Appetitive components become linked to terminal behaviors and pups are able to sustain contact with a goal object (Bruno, 1981; Hall and Bryan, 1980). The internal controls underlying adult behaviors emerge as behavior becomes responsive to ingestive consequences (Hall and Rosenblatt, 1977; Kenny *et al.,* 1979; Bruno *et al.,* 1983).

The day-3 rat appears to have neither spontaneous excitatory or inhibitory systems when behaviorally activated by heat, stimulation, or other arousing stimuli. Behavioral responses are produced but they may not be appropriate to the setting, and "boundaries" between diverse behavioral systems appear to be lacking. With development, behavior becomes linked to the environment and takes on increasing complexity. This increasing complexity is due in part to the maturational aspect of perceptual motor development. As the motoric competence of the pup increases, it is better able to locomote toward and maintain contact with environmental stimuli. In addition, the sensory modalities of vision (Gottlieb, 1971; Ross, 1968) and audition (Crowley and Hepp-Raymond, 1966) are reaching maturity, providing the pup with an additional influx of sensory information and allowing the pup to locate objects within its environment. A variety of other changes also occurs with development. Neonatal pups do not have the ability to thermoregulate independently, and hypothermia can produce immobility (Wishaw *et al.,* 1979) and eventual anesthesia. The neonate is dependent upon the dam and littermates as sources for behavioral thermoregulation (Alberts, 1978). With development the pup acquires physiological mechanisms of thermoregulation (Hull, 1973; Fowler and Kellog, 1975). As discussed previously, aspects of neural function appear to be temperature dependent.

This period of increased behavioral independence corresponds to rapid cortical development as the cortex doubles in size from day 8 to day 15 and there is a three-fold increase in the levels of glutamate decarboxylase, a marker for cortical intrinsic GABAergic neurons (Johnston and Coyle, 1980). Cortical inhibitory mechanisms, absent in younger animals, are emerging, as evidenced by the onset of behavioral changes to cortical spreading depression begin to appear (Hicks and D'Amato, 1975). During this period the amount of sensory information that pups receive also increases. The emergence of these inhibitory mechanisms combined with the functional development of catecholamine pathways, as already discussed, may enable the pup to direct behavior appropriately to the demands of the environment, taking into account environmental cues and the animal's own ongoing activity.

Such a formulation allows a number of hypotheses about the behavior of the neonate both within the nest setting and as behavioral independence is gained. Initially, activation appears necessary for pups to attain new information and expand their range of meaningful environmental cues. Activation within the proper test setting also appears to release behavior patterns such as ingestion or sexual responses, which eventually take on functional significance for the animal. Hofer et al. (1984) have recently demonstrated that pups sleep almost all the time within the first few days of life, even while suckling. They awaken during or following a maternal milk letdown and stretch response. In other words, the only time the infant is awake and highly aroused is within the context of suckling and maternal contact. It is within this seting that new associative relationships can be formed as demonstrated by Brake (1981), and these are just the conditions used by Pedersen and Blass (1983) and Pedersen et al. (1983) to alter the olfactory cue for suckling. It would appear from our results (Saad et al., 1984) that these instances are accompanied by a high level of dopamine utilization, mediating these associations and possibly conferring upon them reinforcing potential for the neonate. These initial demonstrations of learning and reinforcement may then depend upon the warm ambience and activation, necessitating an intact dopamine system.

Yet normal rat pups are able to attach to anesthetized mothers at room temperature, and neonatal 6-OHDA or lateral hypothalamic lesions resulting in large dopamine depletions are not catastrophic during the first postnatal week (Bruno et al., 1984; Almli et al., 1979). Following such lesions, pups continue to suckle and grow, although at a slower rate (Bruno et al., 1984). Thus suckling maintenance does not appear to require activation or an intact dopamine system. We do not know whether pups treated with 6-OHDA can be activated by the types of stimuli that allow the intact pup to acquire associative relationships. The model proposed here would predict that these pups would initially be unable to acquire relationships that appear to be activation dependent.

A point that requires more attention is to determine what specific information is derived from the pup's early experience in the nest that is utilized by the animal as it reaches behavioral independence. As weaning approaches, pups begin to ingest bits of food around the nest site and the initial forays from the nest occur. Even these first attempts at behavioral independence are facilitated by the presence of the dam. Pups ingest more water when they can contact the dam than when she is absent from the testing environment (Bruno et al., 1983) and the initial foray from the nest is made by following the mother (Alberts and Leimbach, 1980). Dietary choice is also in part determined by nest experience as weaning age pups demonstrate a preference for foods that were part of the maternal diet (Galef, 1979). Eventually the pup gains independence as ingestive behavior occurs in the absence of the dam or littermates.

Recent work by Fillion and Blass (Fillion, 1985; Fillion and Blass, 1985) has demonstrated a link between nest experience and later sexual behavior in male rats. They outlined a developmental progression in the olfactory cue eliciting a component of attachment behavior, probing. Neonatal pups probe to a variety of odors, but by day 10 the odor of the estrous female elicits higher levels of probing than the odor of diestrous females, ovariectomized females, or male rats. By introducing an artificial odor into the early developmental period, they demonstrated that this transition has significance for adult sexual behavior. Pups exposed to citral both pre- and postnatally mated more readily with citral-scented females than normal females in their first mating experience. These results suggest that the normal developmental transition underlying probing may be tied to the experiences of the pups during the postpartum estrous of the dam and may have lasting functional implications. Demonstrations of this type point to the acquisition of information within the nest situation that allow the pup to interact with an expanding environment at weaning and in later life.

123

ENVIRONMENTAL
AND NEURAL
DETERMINANTS OF
BEHAVIOR IN
DEVELOPMENT

VII. Concluding Remarks

The work discussed in this chapter has focused on specific experimental paradigms that highlight two trends in the study of behavioral development. The first examines natural patterns of behavior and identifies the conditions that allow developmental transitions to occur. The second identifies behavioral capabilities before they would normally be expressed and attempts to understand the mechanisms that allow their expression.

The neonate is capable of behavioral complexity, which is reflected in such normally occurring behaviors as suckling and huddling and can be elicited under a variety of experimental conditions, as has been discussed for independent ingestion and in response to central electrical stimulation. The work presented here has attempted to tie together these phenomena through a view of neural development interacting with changing environment demands. The newborn requires external stimulation in order both to activate neural pathways and to display a variety of behavioral capabilities. The conditions that exist within the nest of warm ambience and periodic activation by the dam are just the ones that allow the pup to display behavioral complexity in experimental situations, and the mechanisms underlying these displays may mediate functional transitions within and surrounding the nest situation.

References

Alberts, J. R. Huddling by rat pups: Multisensory controls of contact behavior. *Journal of Comparative and Physiological Psychology*, 1978a, *92*, 222–230.

Alberts, J. R. Huddling by rat pups: Group behavioral mechanisms of temperature regulation and energy conservation. *Journal of Comparative and Physiological Psychology*, 1978b, *92*, 213–245.

Alberts, J. R. Ontogeny of olfaction: Reciprocal roles of sensation and behavior in the development of perception. In R. N. Aslin, J. R. Alberts, and M. R. Petersen (Eds.), *Development of perception: Psychobiological perspectives*. Vol. 1. New York: Academic Press, 1981, pp. 322–352.

Alberts, J. R., and Brunjes, P. C. Ontogeny of thermal and olfactory determinants of huddling in the rat. *Journal of Comparative and Physiological Psychology*, 1978, *92*, 897–906.

Alberts, J. R., and Liembach, M. The first foray: Maternal influence on nest egression in the weanling rat. *Developmental Psychobiology*, 1980, *13*, 417–429.

Almli, C. R., Hill, D. L., McMullen, N. T., and Fisher, R. S. Newborn rats: Lateral hypothalamic damage and consumatory-sensorimotor ontogeny. *Physiology and Behavior*, 1979, *22*, 767–773.

Altman, J., and Sudarshan, K. Postnatal development of locomotion in the laboratory rat. *Animal Behavior*, 1975, *23*, 896–920.

Amsel, A., Burdette, D. R. and Letz, R. Appetitive learning, patterned alternation and extinction in 10 day old rats with non-lactating suckling as a reward. *Nature*, 1976, *262*, 816–818.

Amsel, A., Letz, R., and Burdette, D. R. Appetitive learning and extinction in 11 day old rat pups. Effects of various reinforcement conditions. *Journal of Comparative and Physiological Psychology*, 1977, *91*, 1156–1167.

Amsel, A., and Stanton, M. The ontogeny and phylogeny of the paradoxical reward effects. In J. R. Rosenblatt, R. A. Hinde, C. Beer, and M. Busnel (Eds.), *Advances in the study of behavior*. New York: Academic Press, 1980.

Antelman, S. M., Rowland, N. E., and Fisher, A. E. Stimulation bound ingestive behavior: A view from the tail. *Physiology and Behavior*, 1976, *17*, 743–748.

Antelman, S. M., and Szechtman, H. Tail pinch induces eating in sated rats which appears to depend on nigrostriatal dopamine. *Science*, 1975, *189*, 731–733.

Antin, J., Gibbs, J., Holt, J., Young, R. C., and Smith, G. P. Cholecystokinin elicits the complete behavioral sequence of satiety in rats. *Journal of Comparative and Physiological Psychology*, 1975, *89*, 784–790.

Babicky, A., Ostadaloca, J., Parizek, J., Kolar, R., and Bibr, B. Onset and development of the physiological weaning period of infant rats reared in nests of different sizes. *Physiologia Bohemoslovaca*, 1973, *22*, 449–456.

Ball, G. A. Seperation of electrical stimulation and electrically induced eating in the hypothalamus. *Communications in Behavioral Biology*, 1969, *A3*, 5–10.

Barfield, R. J., and Sachs, B. D. Sexual behavior: Stimulation by painful electric shock to the skin in male rats. *Science*, 1986, *161*, 392–395.

Blass, E. M., Beardsley, W., and Hall, W. G. Age-dependent inhibition of suckling by cholecystokinin. *American Journal of Physiology*, 1980, *236*, E567–E570.

Blass, E. M., and Pedersen, P. E. Surgical manipulation of the uterine environment in rat fetuses. *Physiology and Behavior*, 1981, *25*, 993–995.

Blass, E. M., Teicher, M. H., Cramer, C. P., Bruno, J. P., and Hall, W. G. Olfactory, thermal and tactile controls of suckling in preauditory and previsual rats. *Journal of Comparative and Physiological Psychology*, 1977, *91*, 1248–1260.

Bolles, R. C., and Woods, P. J. The ontogeny of behavior in albino rats. *Animal Behavior*, 1964, *12*, 427–441.

Brake, S. Suckling infant rats learn a preference for a novel olfactory stimulus paired with milk delivery. *Science*, 1981, *211*, 506–508.

Brewster, J. A., and Leon, M. Relocation of the site of mother young contact: Maternal transport behavior in Norway rats. *Journal of Comparative and Physiological Psychology*, 1980, *94*, 69–79.

Brunjes, P. C., and Alberts, J. R. Olfactory stimulation induces filial preferences for huddling in rat pups. *Journal of Comparative and Physiological Psychology*, 1979, *93*, 548–555.

Bruno, J. P. The development of drinking behavior in preweanling rats. *Journal of Comparative and Physiological Psychology*, 1981, *95*, 1016–1027.

Bruno, J. P., Blass, E. M., and Amin, F. Determinants of suckling vs. drinking in weanling albino rats: Influence of hydrational state and maternal contact. *Developmental Psychobiology*, 1983, *16*, 177–184.

Bruno, J. P., Craigmyle, L., and Blass, E. M. Dehydration inhibits suckling behavior in infant rats. *Journal of Comparative and Physiological Psychology*, 1982, *96*, 405–415.

Bruno, J. P., Snyder, A. M., and Stricker, E. M. Effect of dopamine depleting brain lesions on suckling and weaning in rats. *Behavioral Neuroscience*, 1984, *98*, 156–161.

Bruno, J. P., Teicher, M. H., and Blass, E. M. Sensory determinants of suckling behavior in weanling rats. *Journal of Comparative and Physiological Psychology*, 1980, *94*, 115–127.

Bruns, P. D., Drase, V. E., and Battaglia, F. C. The placental transfer of water from the fetus to mother following IV infusions of hypertonic mannitol to the maternal rabbit. *American Journal of Obstetrics and Gynecology*, 1968, *86*, 160.

Caggiula, A. R. Analysis of the copulation reward properties of posterior hypothalamic stimulation in male rats. *Journal of Comparative and Physiological Psychology*, 1970, *70*, 399–412.

125

ENVIRONMENTAL
AND NEURAL
DETERMINANTS OF
BEHAVIOR IN
DEVELOPMENT

Campbell, B. A., and Coulter, X. The ontogenesis of learning and memory. In M. R. Rosenzwig and E. L. Bennett (Eds.), *Neural mechanisms of learning and memory*. Cambridge, Mass.: MIT Press, 1976.

Cheronis, J. C., Erinoff, L., Heller, A., and Hoffman, P. C. Pharmacological analysis of the functional ontogeny of the nigrostriatal dopaminergic neurons. *Brain Research*, 1979, *169*, 545, 560.

Coyle, J. T., and Henry, D. Catecholamines in fetal and newborn rat brains. *Journal of Neurochemistry*, 1973, *21*, 61–67.

Cramer, C. P., and Blass, E. M. Rate vs. volume in the milk intake of suckling rats. In D. Novin and B. Hoebel (Eds.), *Neural bases of feeding and reward*. Brunswick, Maine: Haer Press, 1982, pp. 67–74.

Cramer, C. P., and Blass, E. M. Mechanisms of control of milk intake in suckling rats. *American Journal of Physiology*, 1983, *245*, R154–R159.

Cramer, C. P., and Blass, E. M. Nutritive and non-nutritive determinants of milk intake of suckling rats. *Behavioral Neuroscience*, 1985, *99*, 578–582.

Cramer, C. P., Blass, E. M., and Hall, W. G. The ontogeny of nipple shifting behavior in albino rats: Mechanisms of control and possible significance. *Developmental Psychobiology*, 1980, *13*, 165–180.

Cramer, C. P., Pfister, J. F., and Blass, E. M. Transitions in dehydration-induced inhibition of milk intake in suckling rats. *Physiology and Behavior*, 1984, *32*, 691–694.

Crowley, D. E., and Hepp-Raymond, M. C. Development of cochlear function of the ear in the infant rat. *Journal of Comparative and Physiological Psychology*, 1966, *62*, 427–430.

Fentress, J. C. Dynamic boundaries of patterned behavior: Interaction and self organization. In P. P. G. Bateson and R. A. Hinde (Eds.), *Growing points in ethology*. London: Cambridge University Press, 1976, pp. 135–170.

Fillion, T. Suckling and the development of male sexual responsivity to estrous chemostimuli in the rat. Baltimore, Maryland: John Hopkins University, 1985, Dissertation.

Fillion, T., and Blass, E. M. Infantile behavioral reactivity to estrous chemostimuli in norway rats. *Animal Behaviour*, 1985 (in press).

Fowler, S. J., and Kellog, G. Ontogeny of thermoregulatory mechanisms in the rat. *Journal of Comparative and Physiological Psychology*, 1975, *89*, 738–746.

French, J. D., Verzeano, M., and Magnoun, H. W. A neural basis for the anesthetic state. *Archives of Neurology and Psychiatry*, 1953, *69*, 519–529.

Galef, B. G. The ecology of weaning: Parasitism and the achievement of independency in altricial animals. In P. H. Klopfer and D. J. Gubernick (Eds.), *Parental care*. New York: Plenum Press, 1979.

Ballistel, C. R. Self-stimulation: The neurophysiology of reward and motivation. In J. A. Deutsch (Ed.), *The physiological basis of memory*. New York: Academic Press, 1973.

Gottlieb, G. Ontogenesis of sensory function in birds and mammals. In E. Tobach, L. Aaronson, and E. Shaw (Eds.), *The Biopsychology of development*. New York: Academic Press, 1971.

Hall, W. G. Feeding and behavioral activation in the infant rat. *Science*, 1979a, *205*, 206–209.

Hall, W. G. The ontogeny of feeding in rats: I. Ingestive and behavioral responses to oral infusions. *Journal of Comparative and Physiological Psychology*, 1979b, *93*, 977–1000.

Hall, W. G., and Bryan, T. E. The ontogeny of feeding in rats: II. Independent ingestive behavior. *Journal of Comparative and Physiological Psychology*, 1980, *94*, 746–756.

Hall, W. G., Cramer, C. P. and Blass, E. M. Developmental changes in suckling in rat pups. *Nature*, 1975, *258*, 318–320.

Hall, W. G., Cramer, C. P. and Blass, E. M. Ontogeny of suckling in rats: Transitions toward adult ingestion. *Journal of Comparative and Physiological Psychology*, 1977, *91*, 1141–1155.

Hall, W. G., and Rosenblatt, J. S. Suckling behavior and intake control in the developing rat pup. *Journal of Comparative and Physiological Psychology*, 1977, *91*, 1232–1247.

Hicks, S. P., and D'Amato, C. J. Motor sensory cortex, cortical spinal systems and developing locomotion and placing in rats. *American Journal of Anatomy*, 1975, *143*, 1–42.

Hoebel, B. G., and Teitelbaum, P. Hypothalamic control of feeding and self-stimulation. *Science*, 1962, *135*, 375–376.

Hofer, M. A., Fisher, A., and Shair, H. Effects of intraorbital nerve section on survival growth and suckling behavior of developing rats. *Journal of Comparative and Physiological Psychology*, 1981, *95*, 123–132.

Hofer, M. A., Shair, H., and Singh, P. Evidence that maternal ventral skin substances promote suckling in infant rats. *Physiology and Behavior*, 1976, *17*, 131–136.

Hofer, M. A., and Weiner, H. Physiological mechanisms for cardiac control by nutritional intake after early maternal separation in the young rat. *Psychosomatic Medicine*, 1975, *37*, 8–24.

Hogan, J. A., and Roper, T. J. A comparison of the properties of different reinforcer. In J. S. Rosen-blatt, R. A. Hinds, C. Beer, and M. C. Busnel (Eds.), *Advances in the study of behavior.* New York: Academic Press, 1978.

Horowitz, J., Heller, A., and Hoffman, P. C. The effect of development of thermoregulatory function on the biochemical assessment of the ontogeny of neonatal dopaminergic neuronal activity. *Brain Research*, 1982, *255*, 245–252.

Houpt, K. A., and Epstein, A. N. The ontogeny of food intake in the rat: G.I. fill and glucoprivation. *American Journal of Physiology*, 1973, *225*, 58–60.

Houpt, K. A., and Houpt, T. R. Effects of gastric loads and food deprivation on subsequent food intake in suckling rats. *Journal of Comparative and Physiological Psychology*, 1975, *88*, 764–772.

Hull, D. Thermoregulation in young mammals. In C. G. Whitlow (Ed.), *Comparative physiology of thermoregulation.* New York: Academic Press, 1973.

Jasper, H. H., Proctor, L. D., Kaighton, R. S., Noshway, W. C., and Costello, R. T. *Reticular formation of the brain,* Boston: Little, Brown, 1958.

Johanson, I. B., and Hall, W. G. Appetitive learning in one day old rat pups. *Science*, 1979, *205*, 419–421.

Johanson, I. B., and Hall, W. G. The ontogeny of feeding in rats: III. Thermal determinants of early ingestive responding. *Journal of Comparative and Physiological Psychology*, 1980, *94*, 977–992.

Johanson, I. B., and Hall, W. A. Appetitive conditioning in neonatal rats: Conditioned orientation to a novel odor. *Developmental Psychobiology*, 1982, *15*, 379–397.

Johnston, M. V., and Coyle, J. T. Ontogeny of neurochemical markers for noradrenergic, GABAergic and cholinergic neurons in neocortex lesioned with MAM. *Journal of Neurochemistry*, 1980, *34*, 1429–1441.

Kenny, J. T., and Blass, E. M. Suckling as an incentive to instrumental learning. *Science*, 1977, *196*, 898, 899.

Kenny, J. T., Stoloff, M. L., Bruno, J. P., and Blass, E. M. The ontogeny of preferences for nutritive over non-nutritive suckling in albino rats. *Journal of Comparative and Physiological Psychology*, 1979, *85*, 363–377.

Kovach, J. K., and Kling, A. Mechanisms of neonate sucking behavior in the kitten. *Animal Behavior*, 1967, *15*, 91–101.

Letz, R., Burdette, D. R., Gregg, B., Kittrell, M. E., and Amsel, A. Evidence for a transitional period for development of persistence in infant rats. *Journal of Comparative and Physiological Psychology*, 1978, *92*, 856–866.

Lev, R., and Orlic, D. Protein absorption by the intestines of the fetal rat in utero. *Science*, 1972, *177*, 522–524.

Levine, S. Infantile stimulation and adaptation to stress. In S. Levine (Ed.), *Endocrines and the central nervous system.* 1966, Baltimore: Williams & Wilkins, pp. 280–291.

Lindsley, D. B. The reticular activating system and perceptual organization. In D. E. Sheer, (Ed.), *Electrical stimulation of the brain.* Hogg Foundation for Mental Health, University of Texas, Austin, 1961.

Lithgow, T., and Barr, G. A. Self stimulation in 7 and 10 day old rats. *Behavioral Neuroscience*, 1984, *98*, 479–486.

Lorenz, D. N., Ellis, S. B. and Epstein, A. N. Differential effects of upper gastrointestinal fill on milk ingestion and nipple attachment in the suckling rat. *Developmental Psychobiology*, 1982, *15*, 309–330.

Loizou, L. A. The postnatal ontogeny of monoamine containing neurons in the CNS of albino rats. *Brain Research*, 1972, *40*, 395–418.

MacDonnell, M. F., and Flynn, J. F. Control of sensory fields by stimulation of the hypothalamus. *Science*, 1966, *152*, 1406–1408.

MacFarlane, B. A., Pedersen, P. E., Cornell, C. E., and Blass, E. M. Sensory control of suckling associated behaviors in the domestic Norway rat, *Rattus Norvegicus. Animal Behavior*, 1983, *31*, 462–471.

Margules, D. L., and Olds, J. Identical feeding and rewarding systems in the lateral hypothalamus of rats. *Science*, 1962, *135*, 374–375.

Marques, D. M., Fisher, A. E., Okrutny, M. S., and Rowland, N. E. Tail pinch induced fluid ingestion: Interactions of taste and deprivation. *Physiology and Behavior*, 1979, *22*, 37–41.

Marshall, J. F., Levitan, D., and Stricker, E. M. Activation induced restoration of sensorimotor functions in rats with dopamine depleting brain lesions. *Journal of Comparative and Physiological Psychology*, 1976, *90*, 536–546.

Misantone, L. T., Ellis, S., and Epstein, A. N. Development of angiotensin-induced drinking in the rat. *Brain Research,* 1980, *186,* 195–202.

Mogenson, G. J., and Stevenson, J. A. F. Drinking and self stimulation with electrical stimulation of the lateral hypothalamus. *Physiology and Behavior,* 1966, *1,* 251–254.

Moran, T. H., and Blass, E. M. Organized response patterns and self stimulation induced by intrahypothalamic electrical stimulation in 3 day old rats. In D. Novin and B. G. Hoebel (Eds.), *Neural bases of feeding and reward.* Brunswick, Maine: Haer Press, 1982, pp. 59–74.

Moran, T. H., Lew, M. F., and Blass, E. M. Intracranial self stimulation in 3 day old rats. *Science,* 1981, *214,* 1366–1368.

Moran, T. H., Sanberg, P. R., and Coyle, J. T. Stimulation induced behavior in normal and methazoxymethanol treated rats. *Neuroscience Abstract,* 1983, *9,* 978.

Moran, T. H., Schwartz, G. N., and Blass, E. M. Organized behavioral responses elicited by lateral hypothalamic electrical stimulation in neonatal rats. *Journal of Neuroscience,* 1983, *3,* 10–19.

Moran, T. H., Schwartz, G. N., and Blass, E. M. Stimulation induced ingestion in neonatal rats. *Developmental Brain Research,* 1983, *7,* 197–204.

Pedersen, P. E., and Blass, E. M. Prenatal and postnatal determinants of the first suckling episode in albino rats. *Developmental Psychobiology,* 1982, *15,* 349–355.

Pedersen, P. E., Williams, C. L., and Blass, E. M. Classical conditioning of suckling behavior in three day old rats. *Journal of Experimental Psychology,* 1983, *8,* 329–341.

Phillips, A. G., and Fibiger, H. C. Deficits in stimulation induced feeding after intraventricular administration of 6-hydroxydopamine in rats. *Behavioral Biology,* 1975, *9,* 749–754.

Peiper, A. *Cerebral function in infancy and childhood,* New York: Consultants Bureau, 1963.

Raskin, L. A., and Campbell, B. A. Ontogeny of amphetamine anorexia: A behavioral analysis. *Journal of Comparative and Physiological Psychology,* 1981, *95,* 425–435.

Rose, G. H. Development of visually evoked electrocortical response in the rat. *Developmental Psychobiology,* 1968, *1,* 35–40.

Rosenblatt, J. S. Olfaction mediates developmental transitions in the altricial newborns of selected species of mammals. *Developmental Psychobiology,* 1983, *16,* 347–375.

Rosenblatt, J. S., Turkewitz, G., and Schneirla, T. C. Development of suckling and related behavior in newborn kittens. In E. L. Bliss (Ed.), *Roots of behavior.* New York: Hoebez, 1962.

Rosenblatt, J. S., Burkewitz, G., and Schneirla, T. C. Development of home orientation in newly born kittens. *Transactions of the New York Academy of Sciences,* 1969, *31,* 231–250.

Roth, L. L., and Rosenblatt, J. S. Changes in self-licking during pregnancy in the rat. *Journal of Comparative and Physiological Psychology,* 1967, *63,* 397–400.

Rowland, N. E., and Antelman, S. M. Stress induced hyperphagia and obesity in rats: A possible model for understanding human obesity. *Science,* 1976, *191,* 310–312.

Rudy, J. W., and Cheatle, M. D. Odor aversion learning by neonatal rats. *Science,* 1977, *198,* 845–846.

Rudy, J. W., and Cheatle, M. D. Ontogeny of associative learning; acquisition of odor aversion by neonatal rats. In N. E. Spear and B. A. Campbell (Eds.), *Ontogeny of memory and learning.* Hillsdale, New Jersey: Lawrence Erlbaum Associates, 1979.

Saad, K. M., Blass, E. M., Robinson, R. G., and Moran, T. H. Behavioral activation in neonatal rat pups affects nigrostriatal DA function. *EPA Abstract,* 1984, *55,* 146.

Saunders, P., and Rhodes, P. The origin and circulation of amniotic fluid. In D. V. I. Fairweather and T. K. A. B. Eskes (Eds.), *Amniotic fluid.* Amsterdam: Experta Medica, 1973.

Schneirla, T. C. Aspects of stimulation and organization in approach/withdrawal processes underlying vertebrate behavioral development. In D. S. Lehrman, R. A. Hinde, and E. Shaw (Eds.), *Advances in the study of behavior.* Vol. 1. New York: Academic Press, 1965.

Sczerzenie, V., and Hsiao, S. Development of locomotion toward home nesting material in neonatal rats. Developmental Psychobiology, 1977, *10,* 315–321.

Singh, P. J., Tucker, A. M., and Hofer, M. A. Effects of nasal $ZnSO_4$ irrigation and olfactory bulbectomy on rat pups. *Physiology and Behavior,* 1976, *17,* 373–382.

Stoloff, M. L., and Blass, E. M. Changes in appetitive behavior in weanling age rats: Transitions from suckling to feeding behavior. *Developmental Psychobiology,* 1983, *16,* 439–453.

Stricker, E. M., and Zigmond, M. J. Recovery of function after damage to central catecholamine containing neurons: A neurochemical model for the lateral hypothalamic syndrome. In J. M. Sprague and A. N. Epstein (Eds.), *Progress in psychobiology and physiological psychology.* New York: Academic Press, 1976.

127

ENVIRONMENTAL
AND NEURAL
DETERMINANTS OF
BEHAVIOR IN
DEVELOPMENT

Teicher, M. H., and Blass, E. M. Suckling in newborn rats: Eliminated by nipple lavage, reinstated by pup saliva. *Science,* 1976, *193,* 422–425.

Teicher, M. H., and Blass, E. M. First suckling response of the newborn: The roles of olfaction and amniotic fluid. *Science,* 1977, *198,* 635–636.

Teicher, M. H., Flaum, L. E., Williams, M., Eckert, S. J., and Lumia, A. R. Survival, growth and suckling behavior in neonatally bulbectomized rats. *Physiology and Behavior,* 1978, *21,* 553–561.

Teitelbaum, P. The encephalization of hunger, In J. M. Sprague and A. N. Epstein (Eds.), *Progress in physiological psychology.* (Vol. 4) New York: Academic Press, 1971.

Teitelbaum, P. Levels of integration of the operant. In W. K. Honig and J. E. R. Staddon (Eds.), *Handbook of operant behavior.* Englewood Cliffs, N.J.: Prentice-Hall, 1977.

Teitelbaum, P., and Epstein, A. N. The lateral hypothalamic syndrome: Recovery of feeding and drinking after lateral hypothalamic lesions. *Psychological Review,* 1962, *69,* 74–90.

Teitelbaum, P., and Stellar, E. Recovery from the failure to eat produced by hypothalamic lesions. *Science,* 1954, *120,* 894–895.

Teitelbaum, P., and Wolgin, D. L. Neurotransmitters and the regulation of food intake. In W. H. Gispen, Tj. B. Van Wimersma Griedanus, B. Bohus, and D. de Wied (Eds.), *Hormones, homeostasis and brain: Progress in brain research.* Vol. 24. Amsterdam: Elsevier, 1975.

Teitelbaum, P., Cheng, M. F., and Rozin, P. Development of feeding parallels its recovery after hypothalamic damage. *Journal of Comparative and Physiological Psychology,* 1969, *67,* 430–441.

Tinbergen, N. *The study of instinct,* London: Oxford University Press, 1951.

Valenstein, E. S., Cox, V. C., and Kakolewski, J. W. Reexamination of the role of the hypothalamus in motivation. *Psychological Review,* 1970, *77,* 16–31.

Whishaw, I. Q., Schallert, T., and Kolb, B. The thermal control of immobility in developing infant rats: Is the neocortex involved? *Physiology and Behavior,* 1979, *23,* 757–762.

Williams, C. L. The ontogeny of steriod facilitated lordosis and ear wiggling in infant rats. *Developmental Psychobiology Meeting* (Abstract), Atlanta, Georgia, November 1979.

Williams, C. L., Hall, W. G., and Rosenblatt, J. S. Changing oral cues in suckling of weaning age rats: Possible contributions to weaning. *Journal of Comparative and Physiological Psychology,* 1980, *94,* 472–483.

Wirth, J. B. Ontogeny of angiotensin induced thirst in infant rats, submitted for publication, 1984.

Wirth, J. B., and Epstein, A. M. Ontogeny of thirst in the infant rat. *American Journal of Physiology,* 1976, *320,* 188–198.

Wolgin, D. L., and Teitelbaum, P. Role of activation and sensory stimuli in recovery from lateral hypothalamic damage in the cat. *Journal of Comparative and Physiological Psychology,* 1978, *92,* 474–500.

Wyrwicka, W., and Dobrecka, C. Relationship between feeding and satiation centers in the hypothalamus. *Science,* 1960, *132,* 805–806.

The Ontogeny of Vocal Learning in Songbirds

Sarah W. Bottjer and Arthur P. Arnold

I. Introduction

Songbirds (order: Passeriformes; suborder: Oscines) are one of the few groups of animals that makes extensive use of vocal communication and learns the sounds used for such communication during the course of development. Careful experimental study of song behavior itself, starting with the classic work of Thorpe (1958), Konishi (1965), and Marler (1970), has resulted in the accrual of an extensive body of literature detailing behavioral aspects of song learning. More recently, examination of the neuroanatomical basis of song learning, starting with the ground-breaking work of Nottebohm (Nottebohm, 1971; Nottebohm *et al.,* 1976), has greatly advanced our knowledge of the neural and hormonal correlates of song learning and behavior.

It is now well established that song is a complex learned behavior that is controlled by a highly localized, interconnected *system* of neurons. We also know that song behavior and its associated neural network are sexually dimorphic in many species and that gonadal hormones act directly on neurons in some nuclei of the song system. We feel this system offers unprecedented opportunities for studying neural, hormonal, and behavioral aspects of vocal learning and for applying the findings from this system to problems of vocal learning in humans. In what follows we selectively review the literature on song development in birds, point out the

Sarah W. Bottjer and Arthur P. Arnold   Department of Psychology and Brain Research Institute, University of California, Los Angeles, California 90024.

areas and questions we feel are most important, and speculate on possible mechanisms of song learning.

II. Background

A. Behavioral Data

1. Acoustic Experience. Early experiments designed to assess the role of acoustic experience in avian vocal development revealed that song was a learned behavior. For example, Thorpe (1958) and Marler (1970) reported that young male songbirds raised in acoustic isolation never develop normal vocal behavior, although they may produce a song structure that includes some species-typical elements (Figure 1). This result has been obtained for all species of songbirds studied (for general reviews see Arnold, 1982; Konishi and Nottebohm, 1969; Kroodsma, 1978, 1982; Marler, 1976; Marler and Mundinger, 1971; Nottebohm, 1970, 1975, 1980b). Furthermore, birds that are isolated but "tutored" with recordings of con-

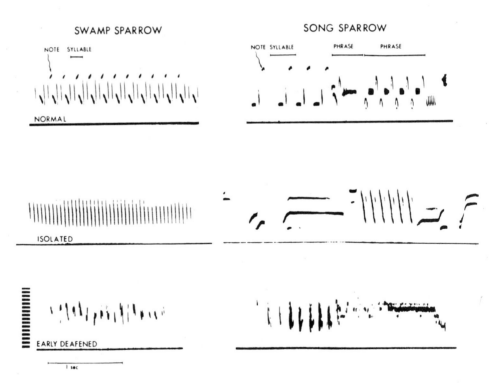

Fig. 1. Sonagrams for two different species demonstrating the differences between normal, isolate-reared, and early-deafened birds. Acoustically isolated birds develop an abnormal song, but preserve some species-typical elements of the song pattern. Early-deafened birds produce a highly abnormal song, but may also retain some species-typical song structure. Abscissa = time in seconds; ordinate = frequency in 500-Hz steps. (Courtesy of Peter Marler.)

specific vocalizations during a particular "critical" period of development do learn to produce normal song. In the case of juvenile white-crowned sparrows *(Zonotrichia leucophrys)* reared in the laboratory, exposure to recordings of adult males must occur between 8 and 56 days of age in order for birds to produce a good copy of the tutor song later on (Marler, 1970). Thorpe (1958) found that chaffinches *(Fringilla coelebs)* that were captured in the fall of their first year and isolated thereafter produced songs that were superior to those of birds isolated shortly after hatching but that were still inferior to songs of normal adult males. According to Thorpe, the fine detail of song is not acquired until the bird's second spring, when it begins to sing and establish a territory, at which time the bird may learn to imitate some characteristics of the songs of birds in adjacent territories. Thus chaffinches can apparently learn new song patterns at least until they are approximately 1 year old.

Exposing isolation-reared birds to the song of the species after the so-called "critical" or sensitive period does not result in normal song development (see pp. 152–154 for discussion of sensitive periods). White-crowned sparrows exposed extensively to song after they are 100 days old retain the features of abnormal song exhibited by birds that have never heard conspecific song (Marler, 1970). Interestingly, the acoustic stimulation received during the sensitive period seems to be stored as a memory trace (or "template") and only becomes apparent at a later stage of development when the juvenile male itself begins to sing—white-crowned sparrows that hear conspecific song early in life and are isolated thereafter nevertheless reproduce the original song model some months later (Marler, 1970).

Songbirds exhibit a distinct stage of vocalizing prior to the emergence of the final song pattern. Thus in many species of songbirds, random "warbling" sounds are produced during the first summer. These birds fall silent over the winter but begin singing again as day length increases the following spring. During this period of sustained singing a closer and closer approximation of the final song pattern is gradually produced (see Marler and Peters, 1982a,c). Marler and Peters (1981a, 1982a) tutored juvenile male swamp sparrows *(Melospiza georgiana)* with conspecific song and carefully monitored their vocal behavior from the time the birds were 50 days old to approximately 1 year. They found that these birds did not begin to produce any sounds resembling the tutor syllables until approximately 240 days after they had last heard them. Thus swamp sparrows are capable of memorizing song syllables, which they imitate faithfully 8 months later without benefit of intervening exposure or rehearsal.

2. EARLY PERCEPTUAL SELECTIVITY. Young birds of some species seem to have a predilection for the sounds of their own species' song. Young male white-crowned sparrows, if exposed to recorded songs of adult males of different species, learn only the song of their own species (Marler, 1970). It is possible for young birds to learn the song of another species under some circumstances: Young zebra finches *(Poephila guttata)* learn the song of a Bengalese finch *(Lonchura striata)* when the latter serves as a foster father, even when they can see and hear adult male zebra finches (Immelmann, 1969), suggesting that social bonding may play an

important role. However, zebra finches raised by (nonsinging) females but allowed to hear and see males of several other species of grass finches, including zebra finches, later produce a song consisting only of conspecific elements (Immelmann, 1969).

This selective, species-typical learning was demonstrated more dramatically in swamp sparrows by Marler and Peters (1977), who showed that swamp sparrows tutored with synthetic songs composed of notes produced by either swamp sparrows or song sparrows *(Melospiza melodia)* copied only the sounds of their own species (see also Marler and Peters, 1981b). In fact, swamp sparrows copied only conspecific notes regardless of whether the overall song pattern matched that of their own species or was instead similar to the pattern of song sparrows. Thus selectivity of young male swamp sparrows is apparently based on individual component sounds (notes) rather than on the overall temporal structure of the song. In contrast, young song sparrows seem to base their selection of species-specific song elements primarily on the overall syllables from tutor songs having the temporal and organizational features of conspecific song (Marler and Peters, 1981b).

Direct evidence for early perceptual selectivity was provided by Dooling and Searcy's (1980) demonstration that young (ca. 20 days) swamp sparrows exhibit a significantly greater cardiac-orienting response to conspecific vocalizations than to song sparrow vocalizations upon first exposure to these sounds. These authors pointed out that this finding argues for innate perceptual biases as a basis for selective learning in the swamp sparrow. Dooling and Searcy's experiment differed from other experiments in that they did not infer perceptual selectivity from subsequent vocal behavior. In fact, their subjects were tested before they had produced *any* song-related vocalizations, indicating that the greater response obtained to conspecific song was purely sensory.

3. AUDITORY FEEDBACK. As mentioned earlier, the initial vocalizations of songbirds bear no resemblance to their final, stereotyped song pattern. The first sounds to be produced lack any stereotypy, even at the level of individual syllables. In addition, incipient song has a greater length and frequency range than adult song and tends to be low in volume (see Marler and Peters, 1982c). This phase of vocalizing is termed *subsong* and gradually gives way to "plastic song," during which the morphology of individual syllables gradually becomes stereotyped, although they are still produced in a variable order. Finally, a "crystallized" song pattern emerges that, in many species, does not change throughout the life of the individual. Plastic song is characterized by a gradual attrition in the number of syllable types produced, as crystallization begins to occur. Marler and Peters (1982b) found that hand-reared swamp sparrows produced many syllable types early in plastic song, only a few of which were retained in the final, crystallized song pattern. This progressive development of vocal production has frequently given rise to the notion that song development entails an auditory–motor phase of learning, during which birds learn to match the sounds of the species' song with the motor (articulatory) patterns necessary to produce those sounds.

Konishi (1965) provided support for this idea when he showed that white-crowned sparrows that were exposed to conspecific song in the field but deafened at approximately 100 days of age produced a completely disrupted song pattern, lacking most of the basic features common to the normal song of the species. The abnormal vocalizations of these birds were similar to acoustically isolated white-crowns that were deafened at 40 days. Thus exposure to conspecific song prior to deafening had no mitigating effects: Even though the template was already established, birds that were deafened produced grossly abnormal song. Although the song patterns of early-deafened birds are more abnormal than those of isolate-reared birds, the former may preserve some vestiges of the characteristics of normal conspecific song (Marler and Sherman, 1983; Figure 1).

Nottebohm (1968) extended the results of Konishi by demonstrating that delaying deafening until progressively later stages after the onset of song in chaffinches results in correspondingly better approximations of normal song in adult birds. A somewhat different result is obtained in zebra finches, which reach maturity at approximately 90–100 days. Price (1979) found that deafening birds as old as 83 days that were already producing a normal, stable song pattern produced as much disruption as deafening at much younger ages. (We have replicated this result in our own lab; Bottjer, unpublished data.) Because all of Price's birds were recorded between 4 and 5 months of age, it is not known whether the song patterns of birds deafened around 80 days displayed a more gradual course of deterioration than did those of early-deafened birds.

Thus young birds must be able to hear their own incipient vocalizations in order to be able to compare their motor output with their (already established) auditory memory of song. In contrast, deafening adult birds after they have already acquired a stable song pattern (i.e., following "crystallization" of the song pattern) does not disrupt performance of that song in most species. Konishi (1965) found that an adult white-crowned sparrow maintained all the fine details of its song pattern up to 14 months after being deafened. Nottebohm (1968) reported that the song patterns of chaffinches that were deafened after they had produced full song for one or more breeding seasons were indistinguishable from intact adult males. Price (1979) found that songs of three out of four zebra finches deafened after they were 10 months old showed only minor changes 6 months after the operation. We have also found minor changes in the songs of some, but not all, adult-deafened zebra finches; the majority of birds exhibit no immediate changes (Bottjer and Arnold, 1984a), but we have noticed a tendency for minor changes to develop over several months in some additional birds.

Not all species develop resistance to the effects of deafening. Nottebohm *et al.*, (1976) found that deafening produced gradual deterioration in the song repertoire of an adult male canary until its postoperative song resembled that of a young-deafened bird. Dittus and Lemon (1970) reported that the song pattern of male cardinals *(Richmondena cardinalis)* deteriorated following deafening, with disruption being most pronounced in some birds 1 year following the operation. It is interesting that deafening disrupts adult song production in the canary, since canaries differ from other species we have discussed in that they are able to change their repertoire and add new song elements in successive breeding seasons (Nottebohm and Nottebohm, 1978). (It is not known whether cardinals retain song plasticity in later years.)

B. THEORETICAL FRAMEWORK

1. PERCEPTUAL LEARNING. The evidence for early perceptual sensitivity to species-specific sounds suggests that the auditory system of passerine birds is somehow predisposed to respond selectively to certain acoustic features or patterns. Because exposure to these specific acoustic patterns is necessary for song to develop, theoretical conceptions of the mechanisms underlying song learning have emphasized the formation of an auditory "template" (see e.g., Marler, 1970). The original template hypothesis encompassed the idea of some innate selectivity for species-specific sounds as well as its elaboration into a more complex schema based on experience with conspecific song. As Marler and Mundinger (1971) expressed it, "The same template can serve both as a kind of filter for focusing attention on sounds that match its crude specification, and as the vehicle for retaining information about the more detailed characteristics of sounds."

Thus a phase of perceptual learning that involves the development and/or maintenance of an auditory memory, based on initial sensitivity as well as continued exposure to certain acoustic features, seems to be a prerequisite for the appearance of normal species-typical vocalizations. The fully formed template presumably consists of some specific neural representation of the exact characteristics of the species-typical sounds heard during development.

2. AUDITORY-MOTOR LEARNING: THE MATCHING PROCESS. The failure of deafened birds that have already established an auditory representation of song to reproduce that song vocally suggests that normal vocal development also involves a phase of motor learning during which auditory feedback is used to match vocal output to the previously established auditory template. During plastic song, birds gradually produce a better and better match of the sounds they have heard earlier in development. The initial efferent commands giving rise to vocalizations produce highly variable sounds bearing no obvious relation to the sounds encoded in the template. Auditory feedback of these sounds is evidently used to realize this fact and perhaps to assist in the programming of the appropriate efferent commands; removal of auditory feedback renders the template unusable.

Thus the earlier phase of perceptual learning is followed by a stage of sensorimotor integration. Because deafening adult birds does not disrupt singing behavior in most species, it has been hypothesized that this period of motor learning involves either (1) transfer of control from auditory feedback to proprioceptive feedback or (2) the establishment of a central motor program that controls vocal output independently of peripheral feedback (Konishi, 1965).

1. PERIPHERAL NERVES AND HYPOGLOSSAL DOMINANCE. Nottebohm and his co-workers (1971; Nottebohm and Nottebohm, 1976; Nottebohm *et al.*, 1979) have shown that the left and right halves of the vocal organ of songbirds (the syrinx) are separately innervated by the left and right tracheosyringeal branches of the hypoglossal nerve with each half independently controlling different, isolable song elements. Neural control of song in many passerine species (e.g., chaffinches, canaries, white-crowned sparrows) is extremely lateralized, as evidenced by the fact that each half of the syrinx does not exert an equivalent degree of control over singing behavior: Severing the left hypoglossal nerve severely disrupts fully developed song, whereas severing the right nerve produces only a mild impairment of the song pattern. In addition, although young male chaffinches are able to develop normal song with only right or left innervation of the syrinx, adult birds cannot reintroduce song elements that are lost following denervation of one or the other side of the syrinx (Nottebohm, 1971; cf. Nottebohm *et al.*, 1979).

2. CENTRAL PATHWAYS FOR VOCAL CONTROL. Nottebohm *et al.* (1976) reported the discovery of highly localized central nervous pathways controlling song in canaries (see Figure 2). Those authors demonstrated that destruction of discrete portions of the brain resulted in severe deficits in the singing performance of adult

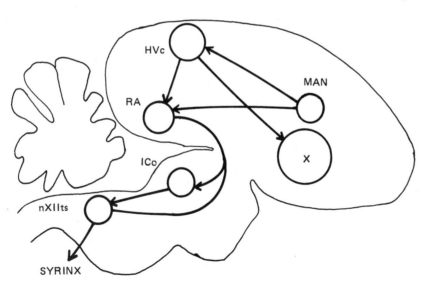

Fig. 2. Schematic diagram of the neural network which controls song (sagittal view). HVc, RA, and nXIIts form part of a direct efferent system linking the telencephalon with the vocal organ (syrinx) of songbirds. Based on work of Nottebohm *et al.* (1976, 1982) and Kelley and Nottebohm (1979). Abbreviations: HVc, hyperstriatum ventrale, pars caudale; RA, robust nucleus of the archistriatum; nXIIts, tracheosyringeal portion of the hypoglossal nucleus; MAN, magnocellular nucleus of the anterior neostriatum; X, Area X; ICo, nucleus intercollicularis of the midbrain.

males. For example, bilateral lesion of a telencephalic nucleus, hyperstriatum ventrale pars caudale (HVc), completely eliminated song immediately after the operation, and no recovery of singing behavior was observed over a period of months. Interestingly, birds would adopt the posture characteristic of singing behavior, even though no sound was forthcoming! Bilateral lesion of nucleus robustus archistriatalis (RA) disrupted song significantly but did not cause birds to become aphonic. Furthermore, unilaterally lesioning left HVc produced much greater effects on the song structure than did lesioning right HVc. Neuroanatomical tracing studies revealed that HVc projects monosynaptically to RA and that both RA and HVc are connected (directly and indirectly, respectively) with the cranial motor neurons that innervate the syrinx (the posterior portion of the hypoglossal nucleus: nXIIts), as shown in Figure 2.

3. OTHER PROJECTIONS OF THE SONG-CONTROL SYSTEM. RA also projects to nucleus intercollicularis (ICo), a region known to be a very-low-threshold site for eliciting vocalizations via electrical stimulation (see e.g., Brown, 1971), although unilateral lesions of ICo do not seem to affect song (Nottebohm *et al.,* 1976). In turn, the dorsomedial nucleus of ICo projects to nXIIts (Gurney and Konishi, 1980). Additional neuroanatomical investigations by Kelley and Nottebohm (1979) and Nottebohm *et al.* (1982) revealed that a group of cells adjacent to the telencephalic auditory projection area, nucleus interfacialis of the neostriatum (NIf), projects onto HVc. A nucleus of the dorsal thalamus, nucleus uvaeformis (Uva), projects directly to HVc as well as to NIf.

Two nuclei in the anterior forebrain are known to connect monosynaptically with HVc and/or RA: Area X in the lobus parolfactorius receives a large projection from HVc (Nottebohm *et al.,* 1976), and the magnocellular nucleus of the anterior neostriatum (MAN) projects directly to both HVc and RA (Nottebohm *et al.,* 1982; Figure 2). These nuclei were assumed to be somehow involved in song because of their connections with HVc and RA. However, unilateral lesions in Area X of adult male canaries produced no effect on stable song patterns (Nottebohm *et al.,* 1976). Recently we have discovered that MAN seems to be critically involved in song learning by juvenile zebra finches but not in maintenance of song by adult birds (see p. 142).

4. LATERALIZATION. As mentioned, section of the left hypoglossal nerve or lesion of left HVc each exerts more detrimental effects on song production in adult birds than does the corresponding operation on the right side. These results have been interpreted by Nottebohm (see e.g., 1979) as evidence for neural lateralization of song control. According to this hypothesis, the left and right sides of the syrinx are independently controlled, and each side acts as a separate sound source (see Greenewalt, 1968). The left side of the syrinx seems to produce the vast majority of elements of a song and, because the central connections of the vocal-control system seem to be overwhelmingly ipsilateral, the system is lateralized throughout. That is, left HVc projects to left RA, which projects to left nXIIts such that lesion of left HVc disrupts song elements produced by the left side of the syrinx.

However, McCasland and Konishi (McCasland, 1983) have recently obtained results that seem difficult to reconcile with the neural lateralization hypothesis. First, they recorded multiunit activity in HVc of awake, singing birds and found a song-correlated pattern of activity that was indistinguishable in left and right HVc. Even more problematic was their discovery that plugging one bronchus, thereby cutting off airflow through that side of the syrinx, did *not* mimic effects produced by nerve-section experiments: Birds were able to produce "easily recognized facsimiles" of most preoperative song syllables regardless of whether the left or right bronchus was plugged. (However, McCasland and Konishi did not determine the combined effect of plugging one bronchus and simultaneously cutting the nerve on the same side.) Although it is difficult to know exactly how to interpret the results of these experiments, they are clearly not compatible with the present version of the neural lateralization hypothesis. However, because lesions of left HVc disrupt not only song elements, but also phrase structure (the latter remaining unmodified following lesion of right HVc), we feel current evidence suggests that there is some neural asymmetry in the song control system (see Arnold and Bottjer, 1985).

D. Sexual Dimorphisms and Brain–Behavior Correlations

1. Sex Differences in Adult Birds. In most species of songbirds song is a complex reproductive vocalization typical of males. Females sing less or not at all. Because the initial neuroanatomical investigations of Nottebohm and his colleagues had revealed a surprising degree of localization in terms of an entire neural system for song control, it was natural to wonder whether the brains of females would reflect the fact that they do not sing. Nottebohm and Arnold (1976) provided evidence for a dramatic sex difference in the volume of vocal-control nuclei: In both zebra finches and canaries, the volume of HVc, RA, Area X, and the hypoglossal nucleus were up to five times larger in males than in females, while nuclei not involved with song control showed no sex differences (Figure 3). A comparison of these two species revealed a greater dimorphism in zebra finches than in canaries; this finding correlates with the fact that adult female canaries implanted with testosterone begin to sing, whereas female zebra finches never sing in adulthood, even when they are treated with testosterone. Sexual dimorphisms have also been described at the cellular level in RA: Cell size and number are greater in adult male zebra finches than in females and the dendritic field of neurons is greater in male than in female canaries (DeVoogd and Nottebohm, 1981a; Gurney, 1981). In the white-browed robin chat *(Cossypha heuglini),* a species in which females participate in vocal duets with males, there is less sexual dimorphism in the volume of vocal control regions than in canaries and zebra finches (Brenowitz *et al.,* in press).

Arnold *et al.* (1976) injected adult male zebra finches with radioactive-labeled testosterone. Examination of these brains revealed that neurons in HVc, MAN, ICo, and nXIIts accumulated testosterone or its metabolites. Furthermore, Arnold

SARAH W. BOTTJER
AND ARTHUR P.
ARNOLD

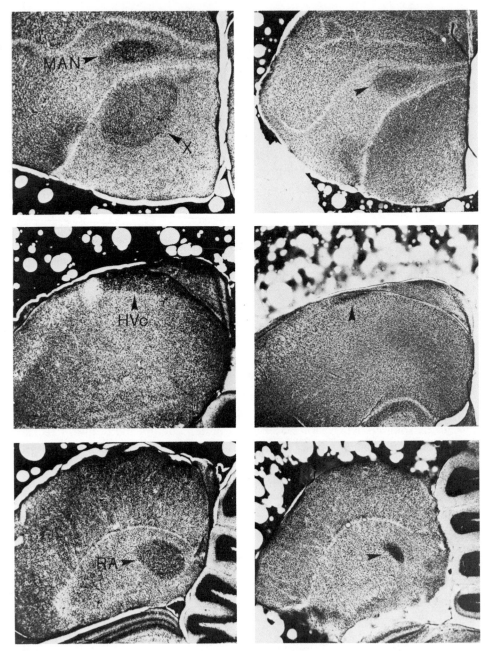

Fig. 3. Photomicrographs showing brain regions in male (left) and female (right) zebra finches. Cresyl violet-stained sections 50µm thick were used. Top: MAN and Area X are clearly seen in males; the boundaries of MAN are more diffuse in females, since the MAN cells are smaller. Area X cannot be seen in females. The middle and bottom photomicrographs provide comparisons of HVc and RA, respectively. The magnification is identical in each case.

and Saltiel (1979) and Arnold (1980) demonstrated that there is also a striking sex difference in terms of such steroid accumulation: HVc and MAN of male zebra finches have a much higher concentration of androgen-accumulating cells than do those of females. In HVc, for example, the percentage of labeled cells was approximately three times higher in males than in females.

2. DEVELOPMENT OF SEX DIFFERENCES IN THE SONG SYSTEM. Gurney has studied the influence of steroid hormones on the ontogeny of sex differences in the volume of song control nuclei of the zebra finch (Gurney and Konishi, 1980; Gurney, 1981; 1982; Konishi and Gurney, 1982). Female chicks that were implanted with estradiol on the first day of life had significantly larger HVcs and RAs as adults than normal intact females, although the volume of each nucleus was still smaller than those of adult males. Administration of estradiol to adult females had no effect on the size of HVc or RA, indicating a sensitive period for masculinization of the song control system. However, administration of testosterone or dihydro-testosterone to adult females that had received neonatal exposure to estradiol induced a still greater increase in the volume of HVc and RA, suggesting that early exposure to estrogen caused RA and HVc to develop a sensitivity to androgen. According to Gurney, these latter females also began to sing and to develop a stereotyped vocal pattern, the quality of which was sometimes as good as a normal male song.

3. BRAIN–BEHAVIOR RELATIONSHIPS. The sexual dimorphisms that have been discovered in songbirds highlight another striking aspect of the vocal control system—the correlation between singing behavior and its localized neural substrate. We have already seen that both singing behavior and the neural network controlling song production are sexually dimorphic and that hormone-induced morphological masculinization of the song nuclei of female zebra finches coincides with the ability to sing. In addition, Nottebohm *et al.* (1981) demonstrated a correlation between the size of the song repertoire in adult male canaries and the volume of HVc and RA: The more syllables a bird was able to produce, the larger the size of HVc and RA. This result marks the first discovery of a direct relationship between variation in the volume of a localized neural substrate and complexity (or amount) of the specific learned behavior that it controls. A correlation between "brain space" and song behavior was also revealed by the finding that administration of testosterone to *adult* female canaries causes them to sing and results in a significant increase in the volume of HVc and RA (Nottebohm, 1980a). At least part of the increased size of RA is attributable to dendritic growth of RA neurons (DeVoogd and Nottebohm, 1981a).

Finally, Nottebohm (1981) reported a seasonal change in the volume of HVc and RA in adult male canaries. For example, HVc is twice as large in the spring, when androgen levels have risen and males have established a new song repertoire, than in the fall, when androgen levels are low and birds are not singing. Nottebohm hypothesized that these cyclic fluctuations are related to the ability to acquire new motor patterns that underlie the new song syllables that male canaries are able to produce as adults (see p. 154).

SARAH W. BOTTJER
AND ARTHUR P.
ARNOLD

A. The Auditory Template

Songbirds respond selectively to species-typical sounds and, with sufficient exposure to their own species' song, can reproduce that pattern weeks or months later. These behavioral findings gave rise to the theoretical notion of an auditory template, but it is not known whether there exists a discrete area in the brain to which an auditory filter and/or memory of song are localized. Katz and Gurney (1981) found that single neurons within HVc of zebra finches responded to noise bursts; all HVc cells that gave an auditory response were shown to project to Area X (as opposed to RA), and their axons gave off collaterals that arborized within HVc. McCasland and Konishi (1981) reported much greater multiunit activity in HVc in response to elements of a song when it was played forward as opposed to backward. They also found that auditory neurons in HVc of Wasserschlager canaries responded much more to the song of another Wasserschlager individual than to that of another canary strain or to the songs of other species.

Margoliash (1983) investigated the response of single auditory units in HVc of white-crowned sparrows to playback of tone or noise bursts, different dialects of white-crowned sparrow song, and computer-generated stimuli modeled after specific parameters of song phrases. He discovered a small number of units that were song specific (i.e., that responded only to certain components of a song when it was played in the forward direction). Tone or noise bursts, as well as reversed playback of the song, were ineffective stimuli. Tests with computer-synthesized stimuli revealed that two song elements in sequence (e.g., whistle–whistle) were often extremely effective for excitation of a single cell. A most interesting aspect of Margoliash's work was that the song-specific units of a given bird were maximally responsive to the song of that individual. Particularly striking was the fact that some song-specific units responded to only a few songs out of a large number of similar songs of the same dialect. In addition, single cells of birds that sang abnormal (isolate) songs responded only to each bird's own song and not to playback of normal song.

Leppelsack and Vogt (1976) recorded single units in the telencephalic auditory projection area (Field L) of starlings *(Sturnus vulgaris)* in response to playback of starling calls and songs. They found many units that were highly selective for a restricted subset of sounds, responding to fewer than 10 of the 80 natural sounds that were presented. Kirsch *et al.* (1980) recorded single auditory units in a restricted area of the frontal neostriatum (Field GA) in starlings. In contrast to the properties of units in Field L, Field GA cells exhibited broad turning curves. Field GA is located anterolateral to MAN and Area X, but it is not known whether there are any synaptic connections between these nuclei. Kelley and Nottebohm (1979) reported the existence of direct connections between cells in and around Field L of canaries and regions adjacent to HVc and RA. Fibers from a nucleus that lies immediately anterolateral to Field L (nucleus interfacialis or NIf) project directly into HVc (Kelley and Nottebohm, 1979; Nottebohm *et al.*, 1982).

These results raise the possibility that Field L, Field GA, NIf, and/or auditory neurons in HVc contain a representation of the song heard by juvenile birds that is critical for normal vocal development. Further work along these lines may reveal other areas within or connected to the song system that contain auditory neurons selective for song elements. These experiments constitute an important first step in searching for the neural representation of the auditory template. However, the isolation of single cells in adult birds that are selectively responsive to species-specific song elements reveals little about the specific role such neurons may play in vocal learning. It would be particularly interesting if developmental studies revealed a change in the response properties of such neurons correlated with different phases of song learning (e.g., before and after tutoring) and if lesioning such neurons after template formation (i.e., after tutoring) prevented emergence of normal song. Such experiments will greatly increase our knowledge of the neural correlates of the auditory template.

B. Auditory–Motor Integration: The Matching Process

1. Auditory and Nonauditory Contributions to Song Development. Even if we assume a discrete neural locus for encoding the auditory memory of song, we know that an auditory template is not sufficient for vocal learning to proceed normally, because auditory feedback of self-produced vocalizations is necessary after the template is established. This result suggests that such auditory feedback is compared to the template and that this information is used to modify the motor systems that give rise to vocalizations.

McCasland and Konishi (1981) reported a paradoxical relation between singing behavior and the response of auditory neurons in HVc of adult birds—the normal pattern of motor activity seen in HVc during singing behavior was accompanied by a profound and long-lasting inhibition of auditory responses to recordings of song. If motor inhibition of auditory responses in HVc also occurs in early development, such auditory responses would presumably make little contribution to song learning. However, a much different picture of the relationship between "auditory" and "motor" neurons of HVc may be revealed in young birds, as McCasland and Konishi suggest.

A variety of nonauditory sources of feedback may also be important during the matching phase of vocal learning, such as feedback from the syrinx, larynx, beak region, and respiratory system. Bottjer and Arnold (1982) reported the existence of sensory afferent fibers originating in the syrinx of zebra finches and terminating in the medulla. The nature of the receptors that are associated with these fibers as well as the exact termination site in the brain is not yet known; however, investigation of the peripheral course of these syringeal afferent fibers revealed that they traveled separately from motor axons to the syrinx at a point outside the brain. We utilized this information in a preliminary attempt to determine whether these fibers transmit signals from the syrinx that could contribute importantly to song development (Arnold and Bottjer, 1983). These syringeal afferents were bilaterally sectioned in young male zebra finches around the time that subsong is just beginning (ca. 25–35 days of age); additional juvenile males were subjected to control surgery. The outcome of this experiment was equivocal: Song development in many, but not all, of the experimental birds was severely disrupted, and song development in some (a minority) of the control birds was also abnormal. Although the results were highly suggestive of a significant role for feedback along syringeal afferent fibers in the development of learned song behavior, further investigation is necessary to confirm this result. It may be that the absence of any substantial disruption in some experimental birds was due to rapid regrowth of the sectioned fibers. With this idea in mind, our next objective is to map out and lesion the central termination site of these fibers.

2. THE ROLE OF MAN IN SONG LEARNING. Nottebohm and his colleagues demonstrated that lesions of HVc or RA disrupt production of crystallized song patterns in adult birds (Nottebohm *et al.*, 1976; see Section II. C. 2). However, the ability to learn song in zebra finches is restricted in time to a specific period of development. We therefore wondered if there existed a discrete brain region that could be related specifically to song development in juvenile male zebra finches. As noted previously, there are two nuclei in the anterior forebrain, MAN and Area X, which have monosynaptic connections with HVc and/or RA (see Figure 2). Nottebohm *et al.* (1976) had also shown that unilateral lesion of Area X has no effect on song in adult canaries, despite the fact that it receives a direct projection from HVc. (No data were available concerning the behavioral effects of lesions in MAN.) It was therefore tempting to speculate that MAN and/or Area X are importantly involved in encoding some aspects of the experience that are necessary for normal song development in juvenile birds. We began to test these ideas by lesioning MAN in juvenile zebra finches during the early stages of subsong. Bilateral lesion of MAN in male birds approximately 35–50 days old produced drastic disruption of song behavior (Bottjer *et al.*, 1984). The "songs" of these birds were extremely simple, usually consisting of one or two highly abnormal syllables. The sounds produced by these birds were often produced at very low volume and were very scratchy or noisy in character (see Figure 4). Juvenile birds that received lesions that completely missed MAN exhibited normal song development (Figure 4). Starting at 50–60 days of age and thereafter lesions in MAN had little or no effect on song development (Figure 5).[1]

The time course of the effectiveness of MAN lesions parallels the development of the motor pattern of song. That is, birds that were producing highly variable vocalizations showed immediate disruption of song behavior in response to lesions of MAN. In contrast, young birds that had just learned to produce a fairly stereotyped song pattern showed no immediate effect of MAN lesions, although their song patterns tended to become simpler and/or display slight abnormalities over

---

[1]Because of the close physical proximity of MAN and X, it seemed possible that lesions in MAN exerted their effect by damaging axons that project from HVc to X. Neuroanatomical mapping of the HVc to X projection did not support this idea. In the zebra finch, axons leave HVc and travel along the ventricle, entering Area X from its medial side without traversing MAN (Bottjer *et al.*, in preparation).

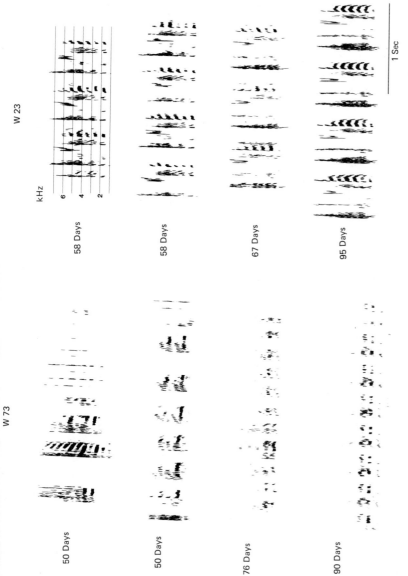

Fig. 4. Sonagrams showing abnormal song development in a bird that received a complete bilateral lesion of MAN at 35 days (W73, left) and normal song development in a bird that received a control lesion at 38 days (W23, right). The sounds produced by W73 lacked the frequency modulations characteristic of normal zebra finch song and were produced in extremely long bouts of singing that lacked normal phrasing. The vertical lines on the right of the sonagrams labeled 50 days were audible only as faint clicks. The other note produced by this bird shown at 50, 76, and 90 days resembled a hoarse, croaking sound. The song pattern of W23 consisted of short stereotyped phrases including 6 notes; individual notes were highly frequency-modulated and/or contained evenly spaced harmonics.

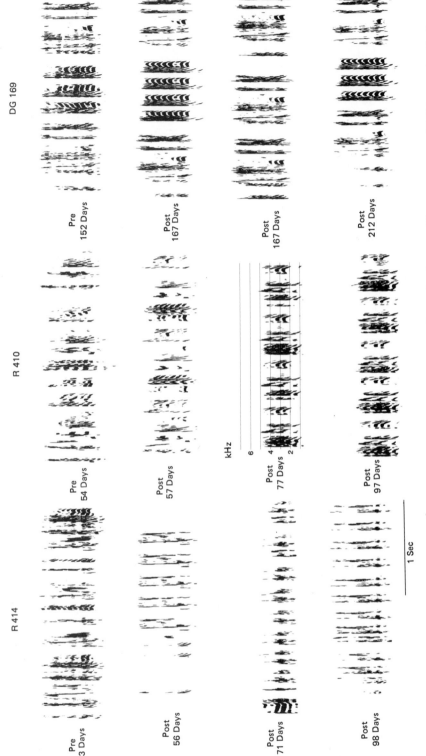

Fig. 5. Sonagrams demonstrating the effect of bilateral MAN lesions at different stages of song development. The song pattern of R414 was not well advanced at 53 days of age. Although individual notes were recognizable, the overall song pattern lacked any stereotypy. This bird was lesioned immediately after this recording session; his postoperative song exhibited the same abnormalities as that of W73 (see Figure 4). The song pattern of R410 was quite well advanced at 54 days of age—comparison of several sonagrams from this recording session indicated a high degree of stereotypy in terms of the overall song pattern. Bilateral lesion of MAN had no immediate effect on the song pattern of this bird—in fact, his first postoperative songs actually showed some improvement (R410, 57 days). However, by 77 days the song of this bird had regressed somewhat and showed no improvement by the time he was fully adult (97 days). DG169 was an adult who had been producing a highly stereotyped song pattern for some time. Bilateral lesion of MAN had no effect on the song pattern of this bird during 7 weeks of postoperative recording.

time. Adult birds that had been producing crystallized song for some time were completely unaffected by MAN lesions at postoperative intervals of up to 7 weeks (Figure 5).

These results are very exciting for a variety of reasons. First, they demonstrate the existence of a neural region that is important for song learning, as opposed to maintaining production of an already-learned song. Second, this is the only instance, to our knowledge, of a situation in which lesion of a specific brain region is effective only during part of the development of a learned behavior and may therefore offer unique opportunities for studying neuronal correlates of learning. Finally, a high proportion of MAN cells in adult male zebra finches accumulate testosterone or its metabolites (Arnold *et al.*, 1976). This finding suggests that hormonal action on MAN cells may be importantly involved with song learning.

Thus far we can only speculate about the specific mechanisms by which MAN lesions disrupt song acquisition. Lesions in MAN may exert their effects purely by interfering with normal neural development in HVc and RA. Because lesions of MAN decrease afferents to HVc and RA, neurons in these latter regions may undergo abnormal development. As a preliminary test of this idea we measured the total volume of RA in birds that had received complete MAN lesions as juveniles and those with lesions that had missed all or most of MAN. Although the volume of RA was slightly smaller in birds whose MAN was completely destroyed, this difference was not significant (Bottjer *et al.*, 1984). In addition, we have measured the soma size of individual RA neurons in these two groups of birds (Bottjer and Miesner, unpublished data). The average size of RA somas in birds with complete MAN lesions was 132.39 μm^2 (range = 112.05–168.37 μm^2), whereas the size of RA somas in birds with incomplete lesions or no damage to MAN was 153.10 μm^2 (range = 132.11–169.33 μm^2). This difference is also not significant but indicates a slight decrease in the size of RA neurons following a complete lesion of MAN. However, it seems unlikely that a difference of this magnitude could account for the drastic changes in vocal behavior that are induced by MAN lesions.

An alternative idea is that functions important for vocal learning may be intrinsic to MAN during a restricted period of development. Of course, we can only hypothesize at this time as to the exact nature of such functions. Because the time course of the effectiveness of lesions parallels the development of song as a motor pattern, we feel it is likely that MAN may be importantly involved in the formation of an auditory–motor transcription. It seems less likely that MAN acts as a conduit for auditory information relevant to song-related feedback, since we know that deafening profoundly disrupts song through approximately 80 days of age (see p. 133) when the motor pattern is fixed and MAN lesions are no longer effective. This pattern of results may indicate that functions important for motor learning of song are being carried out in MAN during a restricted period of development. Since MAN is clearly not on the main motor output path in adulthood (i.e., lesions in adult birds do not affect song), vocal development must either become independent of these functions or be "transferred" somehow to another area of the brain.

Electrophysiological investigations and neuroanatomical tracing studies will help to reveal the exact functions of MAN. For example, if proprioceptive feedback proves to be important for song learning, we may find that MAN receives transsynaptic input from the syrinx or from other peripheral organs. We have not yet assessed the effect of bilateral lesion of Area X during song development, but it is possible that Area X may also be intimately involved in song learning. If so, it will be interesting to observe the type of deficit in song behavior caused by Area X lesions and to determine if the time course of such deficits is the same as or different from those engendered by MAN lesions.

3. DEVELOPMENT OF THE SONG SYSTEM IN ZEBRA FINCHES. We also became interested in examining the normal ontogeny of the song system in juvenile zebra finches. We reasoned that a comparison of normative events in the development of the song system with changes in song behavior of juvenile male birds might serve to generate specific hypotheses concerning some of the neural events underlying song learning. Bottjer *et al.* (1985) began to examine the development of the song system by measuring the total volume of various song control nuclei in male zebra finches at different stages of song development. Birds were divided into three age groups with means of 12, 25, and 53 days. These ages correspond to the time (1) prior to production of any song, (2) when song sounds are first produced (initial subsong), and (3) when the final song pattern begins to stabilize. The mean volume of each of the brain regions measured and the ratios of 25- to 12-day volumes and 53- to 25-day volumes are shown in Table 1. The volumes of HVc, RA, and Area X all increased markedly during the time periods examined, whereas the size of brain regions not involved with song control (Rt, Tel) increased much less or not at all. Interestingly, increases in the volume of HVc preceded those in RA and Area X. HVc increased much more steeply in size than RA between 12 and 25 days,

TABLE 1. VOLUMES (mm^3) $\pm$ S.D. OF VARIOUS BRAIN REGIONS IN 12-, 25-, AND 53-DAY MALE ZEBRA FINCHES[a]

Region	Age				
	12	25	53		
HVc	0.165 ± 0.057	0.408 ± 0.085	0.521 ± 0.124	2.47	1.27
RA	0.155 ± 0.022	0.226 ± 0.028	0.402 ± 0.101	1.46	1.78
X	—[b]	1.536 ± 0.331	2.262 ± 0.475	—	1.47
MAN	—[b]	0.565 ± 0.153	0.247 ± 0.050	—	0.44
Tel[c]	0.910 ± 0.055	1.151 ± 0.085	1.119 ± 0.134	1.26	0.97
Rt[c]	1.098 ± 0.133	1.166 ± 0.111	0.980 ± 0.166	1.06	0.84

[a]The two columns on the right represent the corresponding ratios for each area between 25- and 12-day birds, and 53- and 25-day birds, respectively.
[b]MAN and Area X were too ill-defined to be traced in 12-day males (see text).
[c]Rt (nucleus rotundus) is a thalamic nucleus involved with processing of visual information; Tel represents the average cross-sectional area of the telencephalon at the level of the anterior commissure. Both these areas were measured as controls to determine whether changes in the volume of brain regions were specific to the song system.

while RA increased in size more sharply than HVc between 25 and 53 days. Area X could not be seen in three of five 12-day birds and was well defined enough to be traced in only one of these latter two birds. By 25 days Area X had attained a large volume, which continued to increase between 25 and 53 days. In adult males, HVc contains a higher proportion of androgen-accumulating cells than RA, and androgen-labeled cells are not seen in Area X (Arnold, 1980). This pattern of findings may indicate that hormones act directly in HVc to trigger growth and that HVc then induces secondary growth in RA and Area X via some trophic, transsynaptic influence.

The most surprising outcome of this experiment was that, in contrast to other song nuclei, the volume of MAN *decreased* sharply between 25 and 53 days. MAN could be seen in all 12-day birds, but its borders were too diffuse to be traced reliably; we estimated its volume to lie between .20 and .38 mm^3 at this age. By 25 days the volume of MAN was quite large, but it then decreased steeply in size between 25 and 53 days (Table 1). A graphic comparison of the volume changes in HVc, RA, and MAN is shown in Figure 6.

In order to determine whether a decrease in number of neurons could account for the decreased volume of MAN, we measured the somal area and density of neurons in MAN of each of the 25- and 53-day males. There were no differences between these two groups in the distribution of soma size or in neuronal density (Bottjer *et al.*, 1985). Thus the total number of neurons in MAN must decrease sharply during this period of song development; our data indicate a loss of slightly more than 50% of the neurons from MAN between 25 and 53 days.

In the case of zebra finches, therefore, brain nuclei that are known to be directly involved with song production by adult birds, such as HVc and RA, are increasing dramatically in size prior to and during the initial period of song production. By 53 days the size of all song control nuclei appear to be at or very close

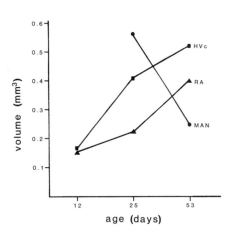

Fig. 6. Total volume of HVc, RA, and MAN in juvenile male zebra finches at 12, 25, and 53 days of age. See text for additional details.

to their adult values. We do not yet know whether these increases in gross nuclear volume are due to increases in neuronal soma size or dendritic arborization, decreases in cell density, and/or increases in cell number. At the same time that these areas are increasing in size, the volume of MAN is decreasing because of a massive loss of neurons. We consider this finding to be particularly exciting for two major reasons. First, MAN is quite large when birds are first learning to produce song and decreases markedly in size by the time that the motor pattern of song begins to stabilize. In addition, the decrease in volume of MAN also correlates with the time period when lesions of MAN disrupt song development; we have already mentioned that lesions become ineffective between 50 and 60 days as soon as the motor pattern of song becomes stereotyped.

We do not know the mechanism of cell loss from MAN, but we consider two alternatives here that seem to have quite different implications for neural mechanisms of vocal learning. One possibility is that there is extensive neuronal death in MAN. If so, it may be that neurons are lost from MAN as the motor pattern of song develops; perhaps neurons involved with control of "erroneous" sounds that are not included in the final song pattern are not retained in MAN. We do not know whether the number of syllables produced by juvenile zebra finches decrease as song development proceeds (i.e., whether some syllables are eliminated during the course of song learning). However, Marler and Peters (1982b,c) have demonstrated that swamp sparrows produce many more syllables during subsong than are actually retained in the crystallized song (see Section II. A. 3). This overproduction of song sounds is followed by a marked attrition in the number of syllables produced as crystallization occurs. A similar type of attrition may occur in zebra finches as song development proceeds and MAN decreases in neuron number; cell loss then presumably would be curtailed as the motor pattern of song becomes stereotyped. Thereafter, MAN lesions are ineffective in disrupting song development, suggesting that the function subserved by MAN is no longer important for song learning or that functions important for song learning are shifted to other areas of the brain (such as HVc, RA, and/or Area X?).

An alternative explanation of the loss of cells from MAN is that some neurons redifferentiate and migrate outside the borders of MAN. For example, some neurons within MAN may change their morphology, their histochemical characteristics, and their function. In this instance one can imagine that such cells may be somehow "selected" because they are involved with control of sounds that *are* retained in the final song pattern. Such cells would then presumably be induced to migrate outside the area we label as MAN in adult male zebra finches, which could explain why lesions of MAN become ineffective.

Of course, the resolution of these ideas depends on the outcome of future experiments. However, it is worth noting that either of the two mechanisms we have proposed seems unprecedented, since cell migration and neuronal death are normally features of embryonic or early postnatal development and have never, to our knowledge, been studied in relation to the development of a learned behavior.

In any case, we feel that determination of the relation between song development and the loss of neurons from MAN will greatly increase our understanding of the neural mechanisms of vocal learning.

C. The Central Motor Program

Deafening exerts little or no effect on the song of adult birds that have learned to produce a fixed vocal pattern that does not change during adulthood. This result led Konishi (1965) to hypothesize that vocal learning entailed either the development of a central motor program to control song production independently of peripheral feedback or a switch from reliance on auditory feedback to some other, nonauditory source of feedback. We tested the hypothesis that song production in adult zebra finches is dependent on feedback from the syrinx by selectively lesioning afferent fibers from the syrinx (leaving motor innervation intact). The stereotypical vocal patterns of adult males were examined before and after both deafening and either unilateral or bilateral section of syringeal afferent fibers (Bottjer and Arnold, 1984a). The majority of these birds showed no change in their song patterns; the remaining birds showed changes in the form of some individual notes, but not in the overall pattern and tempo of song (see Figure 7). These results provide direct support for the idea that song performance in adult birds is not immediately dependent on peripheral feedback. Of course, we examined only the contribution of auditory and syringeal feedback and therefore cannot rule out the possibility that feedback from other parts of the periphery is important. However, if peripheral feedback in general contributes significantly to song production, we strongly suspect that input based on hearing and/or the exercise of the vocal organ would be of paramount importance. A second caveat stems from our inability to assess the long-term consequences of eliminating syringeal feedback (because of the problem of nerve regrowth). It seems quite plausible that a central motor program controlling vocal production could be dependent on peripheral feedback for its long-term maintenance.

Very little evidence is available concerning the neural localization of a central motor program for song. McCasland and Konishi (1981) were able to record multiunit activity in HVc of awake, singing birds. They found greatly increased neuronal activity in HVc that was time-locked with song elements and that persisted in deafened birds. These findings led McCasland and Konishi to suggest that HVc either generates or relays learned temporal cues for song. Subsequent multiunit recordings made in nucleus interface (NIf), which projects to HVc, revealed that a similar pattern of song-related activity was present in NIf that appeared to lead that in HVc by several milliseconds (McCasland, 1983). Sectioning the pathway from NIf to HVc disrupted song, whereas sectioning the input to NIF from Uva, a thalamic nucleus, did not disrupt song. It is not yet known whether NIf receives a patterned input from some other source that correlates with song activity. If not, it may be that NIf is the site where efferent commands for song are generated.

SARAH W. BOTTJER
AND ARTHUR P.
ARNOLD

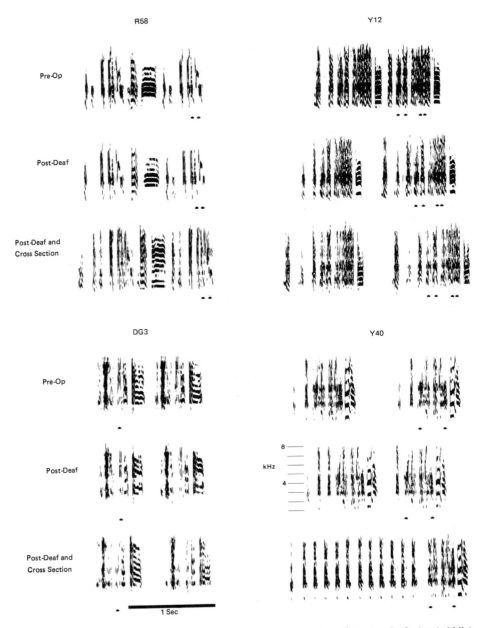

Fig. 7. Sonagrams for four adult male zebra finches preoperatively (top), following deafening (middle), and again following bilateral section of afferent fibers originating in the syrinx. Notes marked by arrows showed minor changes following bilateral section of hypoglossal afferents. In no case was the overall pattern or tempo of song disrupted.

A. DEVELOPMENT OF THE CENTRAL MOTOR PROGRAM

One major question yet to be answered is whether nonauditory sources of feedback contribute to the development of learned song behavior (see Section III.B.). Resolution of this question will enhance our understanding of the neural mechanisms of vocal learning and help to focus attention on the brain pathways carrying critical information. Let us consider two hypothetical situations. In the first, we assume that all nonauditory sources of feedback make no contribution to song development. In this case we know that a motor command somehow originates in the brain and produces a vocalization; this vocal sound feeds back along auditory pathways, presumably relaying at some point to the locus of the auditory template. The vocal sound then must be judged somehow for degree of similarity to the stored memory of song. A highly dissimilar sound would lead to modification of subsequent motor commands (or have no effect), whereas a highly similar sound might "reinforce" the motor command that gave rise to it. One potential problem with this arrangement might be that the motor activity giving rise to the vocalization would be completely dissipated by the time the comparison between auditory feedback and the template is made. However, the pattern of motor activity could be maintained for some time via efference copy or some other mechanism, or auditory feedback could be sufficiently fast to reach the locus of efferent commands before all traces of song-related activity have subsided. Of course, there would be no need for efferent activity to persist if the auditory template were somehow able to transmit directly within the brain a "message" that could alter subsequent efferent activity. However, in this case, one wonders why auditory feedback should be necessary at all. Why couldn't the auditory template simply program the efferent commands that would correspond to its stored representation of sound? Therefore it seems most likely to us that if auditory feedback has the sole responsibility for directing development of crystallized song, then neural activity arising from auditory feedback and from efferent commands for song must be directly compared at some point. Because there are known synaptic connections between (and perhaps within) auditory units and efferent song nuclei (see Section III.A), there are numerous potential sites where such a comparison could transpire.

In the second hypothetical situation, we assume that proprioceptive feedback from the syrinx and/or other organs involved in articulation is necessary for song to develop normally and that such feedback is necessary during part or all of the time that auditory feedback is required. An efferent command giving rise to a vocalization in this case would generate both auditory and proprioceptive feedback. Auditory feedback is relayed to the template for comparison, while proprioceptive input might feed back to the locus where efferent commands are generated, presumably to that specific set of efferent neurons that gave rise to the vocalization. It seems possible that a close match between sound and template could "activate" the template and thereby result in excitation being transmitted to this (efferent) locus at approximately the same time that proprioceptive information is arriving. The enhanced facilitation produced by contiguous input from both proprioceptive feedback and the activated template could strengthen the pattern of motor commands giving rise to that proprioceptive feedback. A mismatch between sound and template would not result in enhanced facilitation (since the template would not be activated), such that motor commands giving rise to inappropriate vocalizations would not be reinforced.

This whole scenario is little more than speculation, but we feel it represents a plausible model that is consistent with current data and that could help focus our experimental investigations. It is also worth noting that the basic idea described here is not without precedent. Hawkins *et al.* (1983) have recently developed a classical conditioning paradigm that produces a learned behavioral change in the nudibranch mollusc *Aplysia* in response to paired presentations of a conditional stimulus (CS) and an unconditional stimulus (US). Hawkins *et al.* found that a US (tail shock) produced facilitation of a monosynaptic EPSP from a sensory neuron onto a motor neuron only if the shock was preceded by spike activity in the sensory neuron engendered by a CS (intracellular stimulation of the sensory neuron). This result suggests that temporally contiguous excitation arising from a CS and a US at a common locus (the sensory neuron) is responsible for a learned behavioral change. Although we are dealing at a completely different level of analysis, we feel it is possible that contiguous excitation arising from auditory and proprioceptive feedback may converge at a common locus (or loci) in the song control system and that this enhanced excitation may induce certain morphological and/or biochemical changes that are critical for song learning.

B. SENSITIVE PERIODS FOR SONG LEARNING

There is apparently a circumscribed period for many species of songbirds during which vocal learning can occur and after which new learning is either difficult or impossible. A central question concerning mechanisms of vocal learning centers on what factors curtail the ability to learn new vocal patterns. What defines the onset and termination of the enhanced ability for vocal learning? Are there separate sensitive periods for perceptual learning and motor learning (Konishi and Nottebohm, 1969; Nottebohm, 1969)? Does crystallization of the song pattern define the end of the sensitive period for song learning (Nottebohm, 1969; 1971)? What is the role of hormones in determining the sensitive period(s) for song learning?

Nottebohm (1969) captured a single male chaffinch during its first winter and castrated it shortly thereafter. The bird had not been heard producing any subsong up until its second spring, when it was implanted with a pellet of testosterone propionate and tutored with two different normal chaffinch songs. This bird developed two song themes, both of which incorporated many elements of one of the tutor songs. Tutoring the same bird with yet another song the following year produced no changes in its crystallized song pattern. This experiment indicates that chaffinches can develop a normal motor pattern of song as well as incorporate new acoustic elements into song long after the sensitive period is normally over. The results also suggest that loss of the ability to learn to imitate new songs may be mediated by high testosterone levels and/or the development of a fixed motor program for a particular song.

Kroodsma and Pickert (1980) have recently provided evidence that the sensitive period for song learning in marsh wrens *(Cistothorus palustris)* is influenced by two environmental factors: daylength (which may correspond with androgen levels) and amount of adult song heard during the hatching year. They found that birds that were exposed to short days and few or no adult songs during the hatching year imitated new tutor tapes the following spring, whereas birds exposed to long days did not imitate any new songs in the spring (irrespective of the amount of adult song heard previously). These results suggest that high testosterone levels, vocal production, and/or acoustic experience contribute to the loss of ability to learn new song patterns.

Both of these experiments clearly show that the end of the sensitive period is not determined by strict maturational factors. Rather, these results are compatible with the idea that various experiential influences (e.g., exposure to hormones, acoustic stimulation, vocal experience) may be at least partly responsible for promoting neural and/or biochemical changes in the song control system. To the extent that such changes are irreversible, further learning might be impossible (i.e., the sensitive period would be curtailed).

It is easy to imagine a role for androgens in motor learning, since syringeal muscles and motor neurons both accumulate testosterone, as do cells in MAN, HVc, and RA (Arnold and Saltiel, 1979; Lieberburg and Nottebohm, 1979). However, Kroodsma and Pickert's (1980) experiment raises the interesting possibility that androgens are also necessary for perceptual learning. Kroodsma (1978) reported that the most effective time for tutoring marsh wrens (i.e., the sensitive period for perceptual learning) was between 15 and 60 days (these birds were maintained on long days). It is remarkable therefore that the marsh wrens in Kroodsma and Pickert's experiment, which were acoustically isolated and maintained on short days, were able to learn to imitate tutor tapes the following spring, when they were 8–10 months old. One possible interpretation of these results is that first-year males exposed to long days have higher testosterone levels than those exposed to short days; these increased androgen levels may act on the brain so as to enable it to encode auditory experiences. If no auditory (song) experience is forthcoming, testosterone still exerts changes in the brain that are irreversible, such that subsequent auditory experience cannot be encoded.

Thus testosterone may act as the "enabling" component during song learning. However, this idea is not supported for all species, since Arnold (1975) demonstrated that castration before perceptual and motor learning did not prevent young male zebra finches from producing a good copy of their fathers' songs.

Nottebohm (1971) suggested that the crystallization of a motor pattern for song may determine the end of the sensitive period for song learning. However, it is still not clear what factors contribute to the end of motor plasticity. Is proprioceptive feedback resulting from vocal experience encoded in the brain, leading to permanent changes such as we have hypothesized in the case of perceptual learning? Are hormonal factors perhaps also important for such motor learning? As mentioned, the crystallization of a song pattern does not always correlate with the immediate curtailment of the end of a sensitive period in the sense that deafening 80- to 85-day-old zebra finches that are already producing a "crystallized" song pattern results in complete disruption of that song (Price, 1979; Bottjer, unpublished data).

Particularly needed in order to begin to understand what is critical about the critical period are descriptive neural and hormonal data during various phases of vocal development. Experiments that measure the overall volume of song control nuclei (and cellular factors such as dendritic morphology and soma size and number) as well as circulating levels of steroid hormones during different periods of song learning would provide invaluable data for generating hypotheses and experimental tests.

C. SONG LEARNING AND BRAIN VOLUME CHANGES

It should be emphasized that not all songbirds lose the ability to acquire new song elements. For example, Nottebohm and Nottebohm (1978) showed that canaries produce more song syllables in their second spring than in their first and that there was considerable turnover in the syllable types used during the first and second years. The number of syllables that were unchanged between the first and second breeding seasons ranged from 11 to 47% for individual birds; the remainder were either modified or new. Nottebohm (1981) found that the volumes of HVc and RA also change with the seasons (see Section II.D), being large in the spring (when a new song repertoire has presumably been learned) and small in the fall (when song elements have presumably been forgotten). Because testosterone induces song as well as enlargement of HVc and RA in adult female canaries (Nottebohm, 1980a) and growth of dendrites in RA of adult females (DeVoogd and Nottebohm, 1981b), Nottebohm hypothesized that overall seasonal volume changes in HVc and RA of adult males reflect cyclical growth and retraction of dendritic segments and, consequently, increases and decreases in number of synapses. He further suggested that the increased volume seen in the spring may somehow enable or facilitate the ability to learn new motor (or auditory–motor) coordinations for song and that the decreased volume seen in the fall correlates with the loss (forgetting) of song elements.

It seemed likely that the seasonal volume changes in canary HVc and RA would correlate more with circulating levels of androgens than with learning ability *per se*. We therefore replicated Nottebohm's (1981) experiment in conjunction with Dr. Myron C. Baker, using "spring" and "fall" white-crowned sparrows (Baker *et al.*, 1984). White-crowned sparrows breed seasonally, like canaries, but apparently do not preserve the ability to learn new song elements. We found, to our surprise, that there were no differences in the size of HVc or RA between birds maintained on either long ("spring") or short ("fall") days, even though the former group had large gonads and were in full song, whereas the latter group had very small gonads and were quiescent. This outcome suggests that seasonal volume changes in HVc and RA are not invariably associated with changing hormonal levels, but seem to be correlated with learned changes in song behavior. (Whether the canary's ability to produce new song elements depends on increased levels of testosterone is not yet known.)

Thus it seems that the remarkable assortment of brain–behavior correlations discovered by Nottebohm and his colleagues (see Section II.D) may all testify to the fact that greater amounts of song learning require greater amounts of neural substrate. It will be particularly interesting to determine the exact factors that may contribute to increased volume in song control nuclei and perhaps to different phases of song learning (see Bottjer and Arnold, 1984b). In addition to dendritic growth, volume changes might reflect increases in the size of neuronal cell bodies, the number of axons projecting into HVc and RA, and/or the actual number of neurons and glial cells. This last possibility may not be unlikely, since Goldman and Nottebohm (1983) showed that neuronal and glial proliferation occurs in HVc (but not RA) of adult canaries. Paton and Nottebohm (1984) demonstrated that intracellular potentials could be recorded from recently generated neurons in HVc, showing that neurons formed in adulthood can be recruited into functional circuits. Goldman and Nottebohm (1983) also reported that testosterone markedly increases proliferation of glial and endothelial cells in HVc of adult female canaries, suggesting that nonneuronal elements may contribute to volume changes and/or song learning.

V. Relations of Avian and Human Vocal Learning

A. Similarities between Vocal Learning in Birds and Humans

Vocal learning in birds exhibits some rather striking parallels to vocal learning in humans. In both cases, species-typical sounds have evolved that serve a communicatory function between individual members of the species. Furthermore, in both cases these sounds seem to be acquired from (or influenced by) experience during the course of development. Some of the specific similarities in vocal ontogeny between birds and humans are as follows (cf. Marler, 1979):

1. Human infants, like nestling songbirds, respond selectively to certain attributes or features of sounds very early in life; these features correspond to perceptual distinctions made by adult animals. Eimas and his colleagues (Eimas et al., 1971; Eimas, 1974) demonstrated that infants as young as 1 month can discriminate between certain speech sounds (such as /b/ versus /p/) in a manner resembling that exhibited by adults (for reviews of infant speech perception see Jusczyk, 1981; Morse, 1979; Pisoni, 1978, 1979). Thus both birds and humans apparently

possess neural mechanisms adapted to respond to the sounds used for vocal communication (cf. Frishkopf *et al.*, 1968). The notion of an auditory "template" or "selective filter" has been invoked in both cases as a prerequisite for the development of normal vocal behavior (Marler, 1970; Studdert-Kennedy, 1976).

2. Birds require specific acoustic stimulation in order to maintain this early perceptual sensitivity and perhaps to develop perceptual capacities not present at birth (Peters *et al.*, 1980). Although no clear-cut data exist for human subjects, it seems unlikely that an otherwise normal human infant subjected to acoustic isolation would exhibit normal phonological development. It is true that although infants can discriminate some phonemic sounds as early as 1 month (this ability itself may be due in part to auditory experience *in utero*), other phonemic distinctions are acquired at a later stage of development (Morse, 1979).

3. Both birds and humans exhibit a preliminary phase of vocalizing ("subsong" or "babbling," respectively), characterized by a progressive differentiation of sound production, resulting in successively closer approximations to normal adult vocal behavior. Studies of speech production in human infants have revealed a sequence of development during which a greater and greater number of phonological features are evident in infants' utterances (e.g., Menyuk, 1971). Vocal development in both cases thus appears to entail a phase of motor learning, a gradual process based in part on earlier perceptual abilities. It is not known whether this phase also involves progressive differentiation and enhancement of perceptual abilities.

4. The effects of deafness at various stages of development appear to be essentially identical in birds and humans. As described earlier, Nottebohm (1968) found that chaffinches showed progressively less disruption of the normal song pattern, the more vocal experience they had had. Lenneberg (1967) reported that sudden deafness before the age of 4 results in severe disruption of speech production in humans. Moreover, children who undergo a sudden loss of hearing prior to the onset of speech are apparently indistinguishable from the congenitally deaf, whereas children who become deaf after the onset of speech "can be trained much more easily in all language arts" (Lenneberg, 1967, p. 155). These data suggest that learning to match self-produced articulatory gestures with their resultant auditory feedback is a crucial component of vocal learning in both birds and humans.

5. Both birds and humans suffer greater deficits in vocal behavior as a result of left rather than right hemisphere damage as adults (see, e.g., Nottebohm, 1979). However, localization of vocal control to the left side of the brain does not seem to be a necessary component of vocal learning, since birds and humans that are unable to use the left hemisphere during early development nevertheless acquire normal vocal behavior using the right side of the brain.

6. Finally, there is evidence that both birds and humans must acquire vocal ability during a restricted period of ontogeny—the so-called critical or sensitive period. Whereas it is not known for certain that birds or humans cannot acquire the ability to communicate vocally after a certain point of maturation, it does seem

clear that, at the very least, the conditions necessary for vocal learning change rather dramatically during the course of development.

B. Relationships between Vocal Perception and Production

1. THE MOTOR THEORY OF SPEECH PERCEPTION. One of the central problems in the study of human speech perception has been the inability to isolate invariant acoustic cues corresponding to the perception of particular speech sounds, or phonemes (see Pisoni, 1978, p. 188). It was partly this problem that prompted the development of the "motor" theory of speech perception (see, e.g., Liberman *et al.*, 1967), which states that speech sounds are perceived by reference to their articulation patterns. Such a theory could explain the problem of acoustic invariance, because the articulatory gesture required to produce a particular phoneme such as /d/ is quite similar even though the acoustic (i.e., spectrographic) representation of that sound may vary greatly as a function of context.

Liberman *et al.* (1967) originally suggested that speech sounds were perceived by reference to motor commands to the speech muscles. However, MacNeilage (1970) pointed out that motor commands to vocal muscles would also vary as a function of context and proposed instead that humans learn to associate particular "target" configurations of the vocal tract with specific speech sounds. MacNeilage and Ladefoged (1976), in their comprehensive review of speech production, argue that the target configuration notion is valid, although the exact mechanisms whereby the nervous system accomplishes a match between the shape of the vocal tract and particular speech sounds are not known (cf. Abbs and Gracco, 1983).

A strong form of the motor theory of speech perception would state that during development it is through the articulation of a given phoneme in different contexts that a child learns to perceive varying acoustic cues as one and the same phoneme, since the motor patterns the child must make to produce that sound are the same (see Pickett, 1980). However, the fact that human infants exhibit adultlike discrimination of speech sounds (see above) poses a problem for this view, since infants are able to discriminate between different phonemic sounds well before they have had any significant articulatory experience. Furthermore, evidence for "feature detectors" sensitive to distinct acoustic attributes has been obtained (for review see Pisoni, 1978), suggesting that there are some invariant acoustic cues underlying perception of speech sounds (even though they have not yet been identified).

The idea that the perception and production of speech sounds are inextricably related is, of course, quite compatible with the data and theoretical framework generated by the study of avian vocal learning (cf. Marler, 1977; 1979; Marler and Peters, 1981b). If auditory feedback is used to gradually produce articulatory gestures that match the species-specific sounds encoded in the template, then it seems quite possible that common neural mechanisms develop that could mediate both production and perception. Margoliash's (1983) finding that some auditory units

SARAH W. BOTTJER
AND ARTHUR P.
ARNOLD

in HVc are maximally sensitive to self-produced vocalizations is quite consistent with this notion. Both birds and humans may recognize specific sounds used for vocal communication by virtue of the fact that they perceive a specific sound by reference to the articulatory pattern that they have learned to associate with that sound. This idea was expressed quite succinctly by Thorpe in 1961 (p. 79):

> If we suppose that the animal is endowed by nature with a number of vocal motor mechanisms which enable it to utter a variety of sounds, then whenever one of these is set in action the animal hears its own voice uttering a corresponding sound. In consequence, the sense impression becomes associated with that motor mechanism. Now suppose the animal hears the same sounds uttered by another. The sound will have the same effect, namely it will already have been associated with the vocal motor mechanism.

One of the major stumbling blocks for the motor theory of speech perception, it seems to us, has been its inability to be subjected to direct test in humans. This is one of several areas where the avian song system could serve as a model for human vocal learning. As a direct test of the motor theory of speech perception, one could interfere with the development of vocal production in birds and measure their subsequent ability to make fine discriminations between species-specific vocal sounds. If such sounds come to be perceived by reference to articulation patterns, then such birds should show a deficit in perceptual ability. We look forward to these and other experiments in the next few years with the hope that our knowledge of the mechanisms underlying vocal learning in both birds and humans will be greatly advanced.

Acknowledgments

This research was supported by USPHS grants NS 18392 to SWB and KS 19645 to APA and by NSF grant BNS 80-06798 to APA. We are indebted to Karen Deutch and Stacey Zeck for aid in manuscript preparation and to Eliot Brenowitz, Sherri Glaessner, Peter Marler, Elizabeth Miesner, Ernie Nordeen, and Dale Sengelaub for their comments. We thank Peter Marler for kindly providing our Figure 1.

References

Abbs, J. H., and Gracco, V. L. Sensorimotor actions in the control of multimovement speech gestures. *Trends in Neurosciences,* 1983, *6,* 391–395.

Arnold, A. P. The effects of castration on song development in zebra finches. *Journal of Experimental Zoology,* 1975, *191,* 261–278.

Arnold, A. P. Quantitative analysis of sex differences in hormone accumulation in the zebra finch brain: Methodology and theoretical issues. *Journal of Comparative Neurology,* 1980, *189,* 421–436.

Arnold, A. P. Neural control of passerine song. In D. E. Kroodsma and E. H. Miller (Eds.), *Acoustic communication in birds.* Vol. 1. New York: Academic Press, 1982.

Arnold, A. P., and Bottjer, S. W. Vocal learning in zebra finches: The role of hypoglossal afferent fibers. *Society for Neuroscience Abstract,* 1983, *9,* 537.

Arnold, A. P., and Bottjer, S. W. Cerebral lateralization in birds. In S. D. Glick (Ed.), *Cerebral lateralization in subhuman species.* New York: Academic Press, 1985.

Arnold, A. P., and Saltiel, A. Sexual difference in pattern of hormone accumulation in the brain of a songbird. *Science,* 1979, *205,* 702–705.

Arnold, A. P., Nottebohm, F., and Pfaff, D. W. Hormone concentrating cells in vocal control and other areas of the brain of the zebra finch *(Poephila guttata). Journal of Comparative Neurology,* 1976, *165,* 487–512.

Baker, M. C., Bottjer, S. W., and Arnold, A. P. Sexual dimorphism and lack of seasonal changes in vocal control regions of the white-crowned sparrow brain. *Brain Research,* 1984, *295,* 85–89.

Bottjer, S. W. Effects of deafening in the zebra finch prior to song crystallization. Unpublished data.

Bottjer, S. W., and Arnold A. P. Afferent neurons in the hypoglossal nerve of the zebra finch *(Poephila guttata):* Localization with horseradish peroxidase. *Journal of Comparative Neurology,* 1982, *210,* 190–197.

Bottjer, S. W., and Arnold A. P. The role of feedback from the vocal organ: I. Maintenance of stereotypical vocalizations by adult zebra finches. *Journal of Neuroscience,* 1984a, *4,* 2387–2396.

Bottjer, S. W., and Arnold, A. P. Hormones and structural plasticity in the adult brain. *Trends in Neuroscience,* 1984b, *7,* 168–171.

Bottjer, S. W., Miesner, E., and Arnold, A. P. Forebrain lesions disrupt development but not maintenance of song in passerine birds. *Science,* 1984, *224,* 901–903.

Bottjer, S. W., Glaessner, S. L., and Arnold, A. P. Ontogeny of brain nuclei controlling song learning and behavior in zebra finches. *Journal of Neuroscience,* 1985, *5,* 1556–1562.

Brenowitz, E. A., Arnold, A. P., and Levin, R. N. Neural correlates of female song in tropical duetting birds. *Brain Research,* 1985 (in press).

Brown, G. L. An exploratory study of vocalization areas in the brain of the red-winged blackbird *(Agelaius phoenicius). Behaviour,* 1971, *39,* 91–127.

DeVoogd, T. J., and Nottebohm, F. Sex differences in dendritic morphology of a song control nucleus in the canary: A quantitative Golgi study. *Journal of Comparative Neurology,* 1981a, *196,* 309–316.

DeVoogd, T. J., and Nottebohm, F. Gonadal hormones induce dendritic growth in adult avian brain. *Science,* 1981b, *214,* 202–204.

Dittus, W. P., and Lemon, R. E. Auditory feedback in the singing of cardinals. *Ibis,* 1970, *112,* 544–548.

Dooling, R., and Searcy, M. Early perceptual selectivity in the swamp sparrow. *Developmental Psychobiology,* 1980, *13,* 499–506.

Eimas, P. D. Auditory and linguistic processing of cues for place of articulation by infants. *Perception and Psychophysics,* 1974, *16,* 513–521.

Eimas, P. D., Siqueland, E. R., Jusczyk, P., and Vigorito, J. Speech perception in infants. *Science,* 1971, *171,* 303–306.

Frishkopf, L. S., Capranica, R. R., and Goldstein, M. H. Neural coding in the bullfrog's auditory system: A teleological approach. *Proceedings of the Institute of Electrical and Electronics Engineers,* 1968, *56,* 969–980.

Goldman, S. A., and Nottebohm, F. Neuronal production, migration, and differentiation in a vocal control nucleus of the adult female canary brain. *Proceeedings of the National Academy of Science,* 1983, *80,* 2390–2394.

Greenewalt, C. H. *Bird song: Acoustics, and physiology.* Washington, D.C.: Smithsonian Institution Press, 1968.

Gurney, M. E. Hormonal control of cell form and number in the zebra finch song system. *Journal of Neuroscience,* 1981, *1,* 658–673.

Gurney, M. E. Behavioral correlates of sexual differentiation in the zebra finch song system. *Brain Research,* 1982, *231,* 153–172.

Gurney, M. E., and Konishi, M. Hormone-induced sexual differentiation of brain and behavior in zebra finches. *Science,* 1980, *208,* 1380–1382.

Hawkins, R. D., Abrams, T. W., Carew, T. J., and Kandel, E. R. A cellular mechanism of classical conditioning in *Aplysia:* Activity-dependent amplification of presynaptic facilitation. *Science,* 1983, *219,* 400–404.

Immelmann, K. Song development in the zebra finch and other Estrildid finches. In R. A. Hinde (Ed.), *Bird vocalizations.* New York: Cambridge University Press, 1969.

Jusczyk, P. Infant speech perception: A critical appraisal. In P. D. Eimas and J. L. Miller (Eds.), *Perspectives on the study of speech.* Hillsdale: Erlbaum, 1981.

Katz, L. C., and Gurney, M. E. Auditory responses in the zebra finch's motor system for song. *Brain Research,* 1981, *211,* 192–197.

Kelley, D. B., and Nottebohm, F. Projections of a telencephalic auditory nucleus—Field L—in the canary. *Journal of Comparative Neurology,* 1979, *183,* 455–469.

Kirsch, M., Coles, R. B., and Leppelsack, H. J. Unit recordings from a new auditory area in the frontal neostriatum of the awake starling *(Sturnus vulgaris). Experimental Brain Research,* 1980, *38,* 375–380.

Konishi, M. The role of auditory feedback in the control of vocalizations in the white-crowned sparrow. *Zeitschrift für Tierpsychologie,* 1965, *22,* 770–783.

Konishi, M., and Gurney, M. E. Sexual differentiation of brain and behavior. *Trends in Neurosciences,* 1982, *5,* 20–23.

Konishi, M., and Nottebohm, F. Experimental studies in the ontogeny of avian vocalizations. In R. A. Hinde (Ed.), *Bird vocalizations.* New York: Cambridge University Press, 1969.

Kroodsma, D. E. Aspects of learning in the ontogeny of bird songs: Where, from whom, when, how many, which, and how accurately? In G. Burghardt and M. Bekoff (Eds.), *Ontogeny of behavior.* New York: Garland, 1978.

Kroodsma, D. E. Learning and the ontogeny of sound signals in birds. In D. E. Kroodsma and E. H. Miller (Eds.), *Acoustic communication in birds. Vol. 2.* New York: Academic Press, 1982.

Kroodsma, D. E., and Pickert, R. Environmentally dependent sensitive periods for avian vocal learning. *Nature,* 1980, *288,* 477–479.

Lenneberg, E. *Biological foundations of language.* New York: Wiley, 1967.

Leppelsack, H. J., and Vogt, M. Responses of auditory neurons in the forebrain of a songbird to stimulation with species-specific sounds. *Journal of Comparative Physiology,* 1976, *107,* 263–274.

Liberman, A. M., Cooper, F. S., Shankweiler, D., and Studdert-Kennedy, M. Perception of the speech code. *Psychological Review,* 1967, *74,* 431–461.

Lieberburg, I., and Nottebohm, F. High affinity androgen binding proteins in syringeal tissues of songbirds. *General and Comparative Endocrinology,* 1979, *37,* 287–293.

MacNeilage, P. Motor control of serial ordering of speech. *Psychological Review,* 1970, *77,* 182–196.

MacNeilage, P., and Ladefoged, P. The production of speech and language. In E. C. Carterette and M. P. Friedman (Eds.), *Handbook of perception. Vol. VII:* Speech and language. New York: Academic Press, 1976.

Margoliash, D. Acoustic parameters underlying the responses of song-specific neurons in the white-crowned sparrow. *Journal of Neuroscience,* 1983, *3,* 1039–1057.

Marler, P. A comparative approach to vocal learning: Song development in white-crowned sparrows. *Journal of Comparative and Physiological Psychology,* 1970, *71* (whole no. 2), 1–25.

Marler, P. Sensory templates in species-specific behavior. In J. C. Fentress (Ed.), *Simpler networks and behavior.* Sunderland, Mass.: Sinauer, 1976.

Marler, P. Development and learning of recognition systems. In T. H. Bullock (Ed.), *Recognition of complex acoustic signals.* Berlin: Dahlem Konferenzen, 1977.

Marler, P. Development of auditory perception in relation to vocal behavior. In M. von Cranach, K. Foppa, W. Lepenies, and D. Ploog (Eds.), *Human ethology: Claims and limits of a new discipline.* New York: Cambridge University Press, 1979.

Marler, P., and Mundinger, P. Vocal learning in birds. In H. Moltz (ed.), *Ontogeny of vertebrate behavior.* New York: Academic Press, 1971.

Marler, P., and Peters, S. Selective vocal learning in a sparrow. *Science,* 1977, *198,* 519–521.

Marler, P., and Peters, S. Sparrows learn adult song and more from memory. *Science,* 1981a, *213,* 780–782.

Marler, P., and Peters, S. Birdsong and speech: Evidence for special processing. In P. D. Eimas and J. L. Miller (Eds.), *Perspectives on the study of speech.* Hillsdale, N.J.: Erlbaum, 1981b.

Marler, P., and Peters, S. Long-term storage of learned birdsongs prior to production. *Animal Behavior,* 1982a, *30,* 479–482.

Marler, P., and Peters, S. Developmental overproduction and selective attrition: New processes in the epigenesis of birdsong. *Developmental Psychobiology,* 1982b, *15,* 369–378.

Marler, P., and Peters, S. Subsong and plastic song: Their role in the vocal learning process. In D. E. Kroodsma and E. H. Miller (Eds.), *Acoustic communication in birds. Vol. 2.* New York: Academic Press, 1982c.

Marler, P., and Sherman, V. Song structure without auditory feedback: Emendations of the auditory template hypothesis. *Journal of Neuroscience*, 1983, *3*, 517–531.

McCasland, J. S. Neuronal control of bird song production. Ph.D. thesis, California Institute of Technology, 1983.

McCasland, J. S., and Konishi, M. Interaction between auditory and motor activities in an avian song control nucleus. *Proceedings of the National Academy of Science*, 1981, *78*, 7815–7819.

Menyuk, P. *The acquisition and development of language*. Englewood Cliffs, N.J.: Prentice-Hall, 1971.

Morse, P. The infancy of infant speech perception: The first decade of research. *Brain Behavior and Evolution*, 1979, *16*, 351–373.

Nottebohm, F. Auditory experience and song development in the chaffinch *(Fringilla coelebs)*. *Ibis*, 1968, *11*, 549–568.

Nottebohm, F. The "critical period" for song learning. *Ibis*, 1969, *111*, 386–387.

Nottebohm, F. Ontogeny of bird song. *Science*, 1970, *167*, 950–956.

Nottebohm, F. Neural lateralization of vocal control in a passerine bird. I. Song. *Journal of Experimental Zoology*, 1971, *177*, 229–262.

Nottebohm, F. Vocal behavior in birds. In J. R. King and D. S. Farner (Eds.), *Avian biology*, Vol. 5. New York: Academic Press, 1975.

Nottebohm, F. Origins and mechanisms in the establishment of cerebral dominance. In M. S. Gazzaniga (Ed.), *Handbook of behavioral neurobiology*. New York: Plenum, 1979.

Nottebohm, F. Testosterone triggers growth of brain vocal control nuclei in adult female canaries. *Brain Research*, 1980a, *189*, 429–437.

Nottebohm, F. Brain pathways for vocal learning in birds: A review of the first ten years. *Progress in psychobiology and physiological psychology*. Vol. 9. New York: Academic Press, 1980b.

Nottebohm, F. A brain for all seasons: Cyclical anatomical changes in song control nuclei of the canary brain. *Science*, 1981, *214*, 1368–1370.

Nottebohm, F. and Arnold, A. P. Sexual dimorphism in vocal control areas of the songbird brain. *Science*, 1976, *194*, 211–213.

Nottebohm, F., and Nottebohm, M. Left hypoglossal dominance in the control of canary and white-crowned sparrow song. *Journal of Comparative Physiology*, 1976, *108*, 171–192.

Nottebohm, F., and Nottebohm, M. Relationship between song repertoire and age in the canary *(Serinus canarius)*. *Zeitschrift für Tierpsychologie*, 1978, *46*, 298–305.

Nottebohm, F., Stokes, T. M., and Leonard, C. M. Central control of song in the canary *(Serinus canarius)*. *Journal of Comparative Neurology*, 1976, *165*, 457–468.

Nottebohm, F., Manning, E., and Nottebohm, M. Reversal of hypoglossal dominance in canaries following unilateral syringeal denervation. *Journal of Comparative Physiology*, 1979, *134*, 227–240.

Nottebohm, F., Kasparian, S., and Pandazis, C. Brain space for a learned task. *Brain Research*, 1981, *213*, 99–109.

Nottebohm, F., Kelley, D. B., and Paton, J. A. Connections of vocal control nuclei in the canary telencephalon. *Journal of Comparative Neurology*, 1982, *207*, 344–357.

Paton, J. A. and Nottebohm, F. Neurons generated in the adult brain are recruited into functional circuits. *Science*, 1984, *225*, 1046–1048.

Peters, S., Searcy, W. A., and Marler, P. Species song discrimination in choice experiments with territorial male swamp and song sparrows. *Animal Behavior*, 1980, *28*, 393–404.

Pickett, J. M. *The sounds of speech communication: A primer of acoustic phonetics and speech perception*. Baltimore: University Park Press, 1980.

Pisoni, D. B. Speech perception. In W. K. Estes (Ed.), *Handbook of learning and cognitive processes*. Hillsdale, N.J.: Erlbaum, 1978.

Pisoni, D. B. On the perception of speech sounds as biologically significant signals. *Brain Behavior and Evolution*, 1979, *16*, 330–350.

Price, P. H. Developmental determinants of structure in zebra finch song. *Journal of Comparative and Physiological Psychology*, 1979, *93*, 250–277.

Studdert-Kennedy, M. Speech perception. In N. J. Lass (Ed.), *Contemporary issues in experimental phonetics*. New York: Academic Press, 1976.

Thorpe, W. H. The learning of song patterns by birds, with especial reference to the song of the chaffinch *(Fringilla coelebs)*. *Ibis*, 1958, *100*, 535–570.

Thorpe, W. H. *Birdsong: The biology of vocal communication and expression in birds*. London: Cambridge University Press, 1961.

Early Development of Olfactory Function

Patricia E. Pedersen, Charles A. Greer, and Gordon M. Shepherd

I. Introduction

Over the past decade, increasing experimental evidence has attested to the importance of olfaction in the early development of rodents. It has become clear that odor cues play a critical role in behaviors essential for survival; these include nipple location and attachment and home and maternal orientation, among others. The predominant role of this sense in developing rodents, its relative maturity at birth in contrast to other sensory systems, and its behavioral plasticity are all properties that invite vigorous neuroanatomical and neurophysiological investigation. This is especially important because olfaction is a sensory system of which our understanding of its basic processes lags behind that of other sensory systems.

The purpose of this chapter is to describe in a systematic way our current understanding of the development of olfactory structure and function, focusing primarily on the main and accessory olfactory systems. We shall attempt to characterize the main stages of normal development and the remarkable degree of plasticity that is expressed even at the earliest stages. We shall also discuss the modified glomerular complex, which appears to comprise a separate subsystem of the olfactory pathway and which has been implicated in processing the odor cues essential for nipple location and attachment in rat pups.

Patricia E. Pedersen, Charles A. Greer, and Gordon M. Shepherd   Sections of Neuroanatomy and Neurosurgery, Yale University School of Medicine, New Haven, Connecticut 06510.

164

PATRICIA E.
PEDERSEN,
CHARLES A. GREER,
AND GORDON M.
SHEPHERD

A main theme of this review is that an ontogenetic pattern appears to be emerging that supports the hypothesis that the olfactory system develops sequentially through at least three subsystems. The accessory olfactory pathway may function prenatally to detect odors of the amniotic milieu; immediately after birth and throughout the nursing period the modified glomerular complex may process the suckling odor; and the main olfactory bulb may develop gradually, through a series of maturational stages, throughout the nursing period. The extent to which this developmental progression is reflected anatomically, functionally, and behaviorally will be addressed in this chapter. Our discussion will be aimed at elucidating some fundamental issues that are of general interest: (1) How does a developing neural substrate mediate functionally critical responses? (2) As each subsystem develops, how does it come to mediate an odor-guided behavior?

II. Main Olfactory Epithelium

The main olfactory epithelium contains receptor cells that project to the main olfactory bulb. This sensory epithelium is comprised of three basic cell types: basal cells, receptor cells, and supporting cells (Graziadei, 1971) (see Figure 1 A and B). Underlying this is the lamina propria, containing blood vessels, connective tissue, and Bowman's glands, the latter providing mucous secretions to bathe the exposed surface of the sensory epithelium. Basal cells, distinguished by their clear, globose nucleus, reside in the deeper portion of the epithelium. These cells divide and differentiate into the bipolar receptor neurons that receive and relay odor information (Graziadei and Monti Graziadei, 1978). Cellular differentiation of these two cell types, basal and receptor, is evident as early as E10 in the mouse (Cuschieri and Bannister, 1975b). The middle third of the epithelium is occupied by the pear-shaped somas of the receptor cells. Extending apically from the soma, a single unbranched dendrite rises toward the surface. At the surface they terminate in dendritic knobs with extensions of cilia. These characteristic terminal swellings and cilia are clearly recognizable on olfactory receptor cells by E12 of gestation (Cuschieri and Bannister, 1975b). They are generally considered to possess receptor sites for odor molecules.

Between E12 and E17 occur major steps in maturation of the olfactory epithelium (Cuschieri and Bannister, 1975a,b). This includes cytological changes in some receptor cells such as an increase in membraneous organelles, Golgi complexes, mitochondria, an increase in the number of cilia, and lysosomes. Similarly, at E17, the final maturation of supporting and basal cells and Bowman glands occurs. Cuschieri and Bannister (1975b) speculate that after this time, E17 (in mice), some receptors are capable of sensory activity.

A fine, unmyelinated axon leaves the bipolar neuron at the deeper aspect, coursing through the epithelium and the lamina propria and eventually passing, along with other receptor cell axons, through the cribiform plate. Caudal to the cribiform plate the olfactory axons enter the bulb and establish first-order synapses

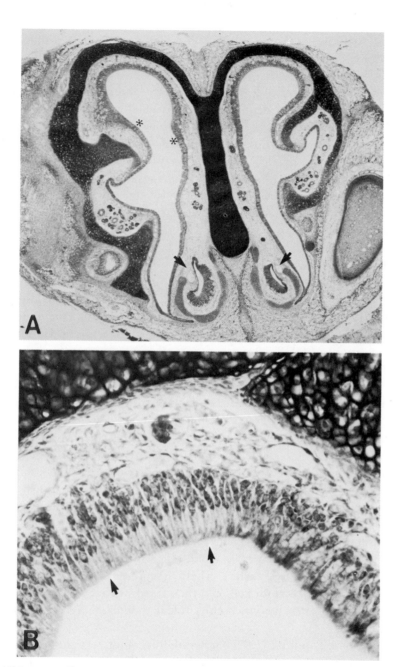

Fig. 1. (A) Low-magnification light micrograph of a transverse 10-μm section through the nasal cavity of a newborn rat. On the left side, asterisks (*) denote the distribution of the olfactory sensory epithelium that contains the receptor cells. It is a continuous band extending through the dorsal aspect of the cavity as illustrated. The vomeronasal organ and associated cartilaginous surround are indicated with arrows. Although the development of the turbinates extending into the cavity become more extensive during development, the basic anatomical organization illustrated here is identical to that found in adults. (B) High-magnification light micrograph of the main olfactory epithelium shown in Figure 1A. Note the laminar distribution of nuclei as described in the text. Arrows denote the free surface of the epithelium along which the cilia of the receptor cells extend.

PATRICIA E.
PEDERSEN,
CHARLES A. GREER,
AND GORDON M.
SHEPHERD

with apical dendrites of mitral, tufted, and periglomerular cells, within characteristic structures called glomeruli. The adult range of axon diameters (0.05–0.3 μm) is achieved by E18 (Cuscheiri and Bannister, 1975b). At the time receptor cells are evident (E10) axon processes leave the epithelium and lamina propria and, 1 day later, enter the bulb. Coincident with axonal contact with the bulb is the appearance of the dendritic processes, described earlier, emanating from receptor cell bodies (Cuschieri and Bannister, 1975b). This finding is of particular interest because it suggests that dendritic formation may be influenced by retrograde influences occurring as a result of successful contact with the bulb.

A protein unique to olfactory receptor cells, olfactory marker protein (OMP), is found only in olfactory receptor cell perikara, dendrites, axons, and axon terminals in the bulb (Margolis, 1972). Although its functional significance is unknown, the appearance of OMP has been suggested to be temporally related to the establishment of synapses from receptor axons to bulbar neurons (Farbman and Margolis, 1980; Monti Graziadei *et al.*, 1980). One day after the onset of receptor–axon synapses in the mouse olfactory bulb, OMP appears in mouse receptor cells. Unfortunately, the comparison has not been made for rats, although it is known OMP staining occurs much later, on E18 in rat olfactory receptor cells. The timely appearance of OMP, after synaptogenesis, suggests that it may be a useful marker for functional connectivity between the bulb and epithelium.

The complex of turbinates in the rat nasal cavity that is covered by olfactory epithelium appears to be quite variable in the density of its cell types. Receptor cells can comprise a single layer in some portions or be several cells thick. Cuschieri and Bannister (1975a) found that early in development, receptor nuerons appear to be most mature in the caudal portions of the nasal cavity. This suggests that different portions of the receptor sheet in the nasal cavity may establish their bulbar connections at different times. This sequence of development may promote orderly topographical arrangements between the nasal mucosa and bulb, although empirical investigations are not available for addressing this particular issue.

Regardless of receptor cell density, the upper third of the epithelium is composed primarily of supporting cells. These cells with rounded clear nuclei terminate in microvilli at the free surface. Their function is unknown, although their proximity to olfactory receptor cells and their dendrites and the presence of vesicles suggest that they may play a secretory role in modulating receptor cell activity (Getchell, 1982).

Cuschieri and Bannister (1975a) noted that receptor cells clustered in groups of two to six early in embryonic development, although this became less evident with age. This juxtaposition even extended to the level of dendrites and terminal knobs. They have suggested, as has Leonard (1981), that these close appositions could promote communication between receptors and thus contribute to receptor specialization for the reception of similar chemical information. This could also contribute to the topographical organization of the projections of the olfactory receptor sheet onto the bulb.

A notable feature of the olfactory epithelium is that, unlike any other area of the nervous system, there is continuous neurogenesis throughout life (Graziadei and Monti Graziadei, 1978). The olfactory receptor cell population is continuously replaced; consequently, new connections are established with the bulb. Receptor cell replacement occurs under normal conditions as well as after nerve or bulb lesions (Doucette *et al.*, 1983; Graziadei *et al.*, 1980; Graziadei and Monti Graziadei, 1978, 1980; Harding *et al.*, 1977; Monti Graziadei and Graziadei, 1979). Biochemical (Graziadei and Metcalf, 1971; Harding *et al.*, 1977) and morphological (Monti Graziadei and Graziadei, 1979) studies have provided strong evidence that olfactory receptor cells are replaced in mice over a 30- to 70-day cycle. This raises some interesting questions, such as: How does receptor cell loss affect odor perception? How does receptor cell loss and replacement affect the memory of prior odor experiences? Do new olfactory axons travel back to the same glomeruli? What is the mechanism that induces this constant degeneration and regeneration? Also intriguing are the findings of Graziadei and his colleagues (Graziadei and Kaplan, 1980; Graziadei *et al.*, 1978; Graziadei *et al.*, 1979; Graziadei and Samanen, 1980; Stout and Graziadei, 1980) that olfactory axons will grow, in the absence of normal olfactory bulbs, into atypical target sites such as transplanted occipital cortex, frontal cortex, olfactory bulb mitral cell layer, and olfactory bulb granule cell layer. Furthermore, the axons appear to induce the formation of glomerular-like structures. Heterologous synapses have been observed in these experiments (Graziadei *et al.*, 1979), and there is evidence that they may be functional (Wright and Harding, 1982).

The fact that olfactory axons can establish contacts with any layer of the bulb, even redirecting mitral cell dendrites to the newly formed glomeruli, supports the hypothesis that olfactory axons induce morphological changes in their target sites. This should not be surprising, since we know that the growth of olfactory axons into basal telencephalon induces olfactory bulb formation in the embryo (Burr, 1916). But it leaves open the question of how orderly growth is achieved during development.

III. Vomeronasal Organ

The vomeronasal organ (VNO) contains receptor cells whose axons gather into large bundles, pass along the nasal septum through the cribiform plate, and enter the glomerular layer of the accessory olfactory bulb (see Figure 2). On E11 in the mouse, a thickening in the epithelium on the medial wall of the invaginated olfactory placode is the site of the future VNO (Cuschiere and Bannister, 1975a). The VNO, like the main epithelium, is composed of epithelial tissue, glands, and blood vessels. It is encased in a partially closed cartilaginous cavity located bilaterally in the inferior portion of the nasal cavity. The organ varies in size and extent as well as in its histological structure along its rostral–caudal axis. This is because of the variable density in the thickness of its epithelia and because the organ is twisted along its axis, changing in relation to the surrounding blood vessels.

PATRICIA E.
PEDERSEN,
CHARLES A. GREER,
AND GORDON M.
SHEPHERD

In cross section the VNO appears as a crescent-shaped lumen. The lumen communicates directly or indirectly with the mouth, depending on the species. In the rat the VNO connects with the nasal fossa through a foramen at the rostral end of the organ (Vaccarezza *et al.*, 1981). Fluids enter and leave this lumen as a result of sympathetic and nonsympathetic activation of the blood vessels, which produces a pumping action (Meredith and O'Connell, 1978; Meredith *et al.*, 1980).

The VNO can be divided into three portions, with each segment composed of different types of epithelium (Vaccarezza *et al.*, 1981). At the most rostral end, the VNO is composed primarily of pseudostratified epithelium; the middle segment

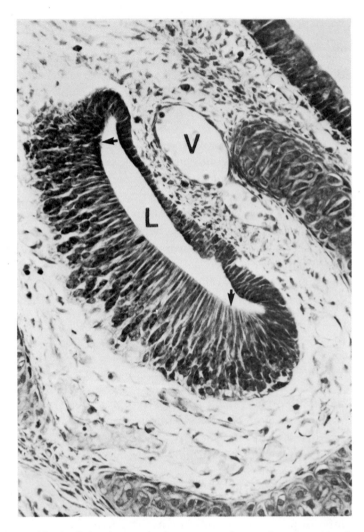

Fig. 2. High-magnification light micrograph of a vomeronasal organ from the section shown in Figure 1A. The lumen (L) of the vomeronasal organ is crescent shaped. A blood vessel (V) located lateral to the lumen in this section appears to be important for the pumping action that allows fluids to access the lumen. Arrows delimit the extent of the sensory epithelium in this section.

contains sensory epithelium on its medial wall and respiratory epithelium on its lateral wall; and the caudal portion is lined with simple columnar epithelium. The sensory epithelium somewhat resembles that of the main olfactory epithelium; the receptor neurons, like the main olfactory neurons, have an axonal process and a dendritic process. The distinctive feature of vomeronasal receptors is that they possess microvilli instead of cilia, supporting the notion that cilia *per se* are not essential for odor detection. A layer of elongated supporting cells, distinguished by their dark oval nuclei and microvilli processes, lies more superficial to the receptor cells between their dendrites. Unlike the main olfactory mucosa, however, the VNO lacks a clearly defined basal layer, or beneath that, Bowman's glands. Vomeronasal glands are found, however, and like the Bowman's glands of the main epithelium, secrete mucus into the overlying lamina. For some time the scarcity of basal cells are taken as evidence that the VNO epithelium does not have regenerative properties. However, a series of ^{3}H thymidine studies (Barber and Raisman, 1975a) and experimental studies following axotomy and accessory olfactory bulb lesions (Barber and Raisman, 1975b) have provided clear evidence that receptor cells are indeed replaced through differentiation of the precursor pool. The reservoir of precursor cells appears to reside not in the basal layer but at the caudal end of the organ near the boundary of the respiratory and sensory epithelium. Whether the remarkable events that take place with regrowth of main olfactory nerves into atypical target sites also occur with vomeronasal axon regrowth is not known.

It appears that the vomeronasal epithelium, like the main mucosa, possesses some mature receptors at birth. In fact, Kratzing's study (1971) on the development of the VNO of suckling rats noted very few structural changes in rats 0–22 days of age. One noteworthy feature was that a single cilium extended from the supporting cells in rats less than 7 days of age. Olfactory marker protein (OMP) staining does not occur until postnatal day 4 in rat vomeronasal receptor cells (Farbman and Margolis, 1980). The amount of antisera required to demonstrate its presence is 20 times that required to identify OMP in main olfactory receptor cells. The significance of this finding and its relation, if any, to function is unknown.

Axons from the VNO receptors penetrate the basal lamina to form the vomeronasal nerve, pass through the cribiform plate, and course along the medial side of the main olfactory bulb until they terminate on dendrites in the accessory olfactory bulb.

IV. MAIN OLFACTORY BULB

A striking feature of the main olfactory bulb is its laminar organization (see Figure 3). At each respective lamina with the bulb, odor information is integrated, processed, and eventually sent to other areas of the brain. The outer nerve layer is composed of bundles of olfactory axons. In the glomerular layer the axon terminals make synaptic contacts with dendrites of second-order neurons: the principal

PATRICIA E.
PEDERSEN,
CHARLES A. GREER,
AND GORDON M.
SHEPHERD

relay neurons of the bulb (mitral and tufted cells) and smaller intrinsic neurons (periglomerular cells) (Pinching and Powell, 1971; Shepherd, 1972; White, 1973). These synaptically rich areas of neuropil form the glomeruli. Considerable convergence occurs within each glomerulus, as can be seen in the ratio of receptor cells (50 million) to glomeruli (2000) in the rabbit (Shepherd, 1979).

The third layer of the bulb, the external plexiform layer, is composed of tufted cell bodies, mitral, tufted, and granule cell dendrites, recurrent axon collaterals, and centrifugal axons. Granule cell dendrites are key modulators of mitral and tufted cell activity, by means of a wealth of reciprocal dendrodendritic synaptic interactions (Rall *et al.,* 1966; Shepherd, 1972). The granule cell dendrites, in turn, are modulated by mitral and tufted cell dendrites in addition to other intrinsic neurons and centrifugal fibers from the forebrain and brainstem (Price and Powell, 1970; Pinching and Powell, 1972).

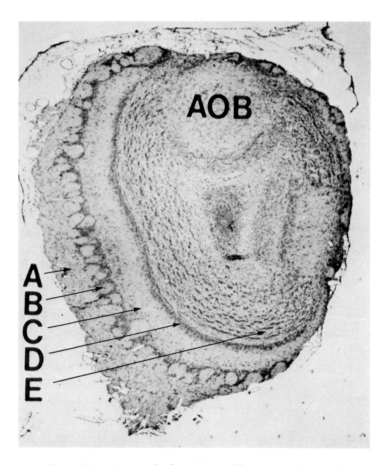

Fig. 3. Low-magnification light micrograph of a transverse 10-μm section from a 21-day-old postnatal rat. The section is from the caudal aspect of the olfactory bulb to include the accessory olfactory bulb. The laminar organization of the main olfactory bulb is evident: (A) olfactory nerve layer, (B) glomerular layer, (C) external plexiform layer, (D) mitral cell layer, and (E) internal granule cell layer.

Beneath the external plexiform layer lies the mitral body layer. The large somas of the mitral cells reside in a single layer from which they send a single prominent dendrite radially to the glomeruli, where serial and dendrodendritic synaptic interactions take place with periglomerular cells (Pinching and Powell, 1971). The secondary dendrites of mitral cells, in some instances extending as far as 1000 μm, reside in the deep half of the external plexiform layer, where they take part in the dendrodendritic synaptic interactions with granule cells (Rall *et al.*, 1966; Scheibel and Scheibel, 1975). Although their distribution may differ, the dendritic distribution and synaptic involvement of tufted cells appears comparable to that of mitral cells (Pinching and Powell, 1971; Rall *et al.*, 1966). Axons of both mitral and tufted cells exit the bulb as the lateral olfactory tract, destined for olfactory cortical areas (Haberly and Price, 1977).

The internal granule layer, deep to the mitral body layer, is composed of granule cell bodies and their dendrites, as well as centrifugal axons. Granule cells are recipients of extensive control of centrifugal fibers, as noted previously.

The olfactory bulb cells have been classified generally as mitral, tufted, granule, and periglomerular, based on the cell shape and size, dendritic structure, and laminar distribution. It now appears, on the basis of studies of morphology, synaptic organization, and neurochemistry, that many subpopulations of cell types exist (Macrides *et al.*, 1981; Macrides and Schneider, 1982; Halaz and Shepherd, 1983). Their striking neurochemical and morphological differences suggest that they may possess different functional properties.

A series of thymidine studies documenting neurogenesis in the olfactory bulbs of mice and rats has demonstrated that much of the olfactory bulb's neuronal development occurs prenatally. Beginning early in embryogenesis, a succession of cell types proliferates from the germinal zone of both the anterior horn of the lateral ventricle and the olfactory ventricle in a manner reflecting the general principle of larger neurons preceding smaller ones (Altman and Das, 1966; Hinds, 1968a). The most prominent cell type is the mitral cell, a secondary relay neuron that receives synapses directly on its terminal dendritic tuft from the olfactory axons and relays the information to the olfactory cortex. In the mouse the peak day of mitral cell production occurs on E12 (Hinds, 1968a,b). Bayer (1983), using a technique that permits estimation of the proportion of neurons originating within a period of 24–48 hr found that in rats over 80% of mitral cells originated on E15–E16, completing their produciton by E18. While mice appear to have an accelerated bulbar development, it is important to remember that their gestation period is at least 3 days shorter than that of rats. By E13, in mice, the full complement of mitral cells is packed into a histologically distinctive layer, even though the olfactory bulb is considerably smaller than that in the adult (Hinds, 1968a,b). Correlated with the emergence of a radial orientation of mitral cell dendrites toward the presumptive glomeruli is the appearance of the first olfactory nerve synapse on E14 (Hinds and Hinds, 1976a,b). This is 2 days after olfactory bundles have penetrated deeply into the bulb and encountered the mitral cell dendrites.

Tufted cells, another principal relay neuron that resides in both the glomer-

PATRICIA E.
PEDERSEN,
CHARLES A. GREER,
AND GORDON M.
SHEPHERD

ular and external plexiform layers, arise between E14 and E17 in the mouse (Hinds, 1968a,b) and like mitral cells, are not seen to proliferate postnatally. In the rat, tufted cells, their diminishing cell body sizes correlated with their distance from the mitral body layer, were also observed to differ in time of thymidine labeling (Bayer, 1983). Internal tufted cells (the largest) arose earlier (E18) by 1 day than smaller tufted cells, which reside in the outer external plexiform layers. Still later, between E20 and E22, the tufted cells of the glomerular layer arise.

These developmental differences support the idea that tufted cells may be comprised of distinct subpopulations. For example, external tufted cells appear as at least three distinct types (Macrides and Schneider, 1982). One type has multiple tufts within the same of adjacent glomeruli, lacks a secondary dendrite, and predominately receives periglomerular cell input. This external tufted cell stands in marked contrast to internal tufted cells, with their larger somas, arborization into a single glomerulus, and synaptic connections predominately with granule cells. In addition, its secondary dendrite extends into the deep third of the external plexiform layer. Thus these latter cells appear more spatially restricted to particular glomeruli and may receive more inhibitory input via their secondary dendrites. Consequently, it may be important to consider what little we know of these cell types developmentally. Indeed, these subpopulations may relay separate olfactory information to different parts of the brain and thus comprise functionally defined parallel pathways.

The last cells to originate in both rats and mouse are the interneurons, the granule cells. Granule cells start to proliferate on E18, the day before birth in the mouse (Hinds, 1968a,b). Periglomerular short-axon cells and granule cells exhibit some neurogenesis prenatally in the rat, but the peak time of their production is during the first two postnatal weeks (Bayer, 183; Roselli-Austin and Altman, 1979). Indeed, in both rat and mouse, this trend continues well into the postnatal period. Granule cells continue to multiply in ependymal and subependymal layers of the ventricles up until at least postnatal days 14–20. At this time the number of granule cells that migrate to neocortex and basal ganglia is moderate (Altman, 1969). There is even evidence of neurogenesis continuing up to day 180 in these cell populations (Bayer, 1983; Kaplan and Hinds, 1977).

In both newborn rats and mice, investigators agree that differentiation of the main olfactory bulb into its distinctive laminae is recognizable by birth, although at that time it is clearly immature (see Figures 4, 5). The protracted development of the main olfatory bulb is most notably characterized by the postnatal proliferation of granule cells. The invasion of these cells no doubt contributes to the rapid growth in the volume of the bulb and width of its respective layers during the first three postnatal weeks. Granule cells appear to migrate uniformly within a given layer in the bulb; however, there are differences between layers (Hinds, 1967; Bayer, 1983). Grannule cells appear last in the internal granule cell layer. Although the granule cells are the only cells emanating from the ventricle postnatally, mitral and tufted cells continue to mature postnatally. For example, for mitral cells, the size of soma increases, dendritic growth and axonal elongation occur, and the

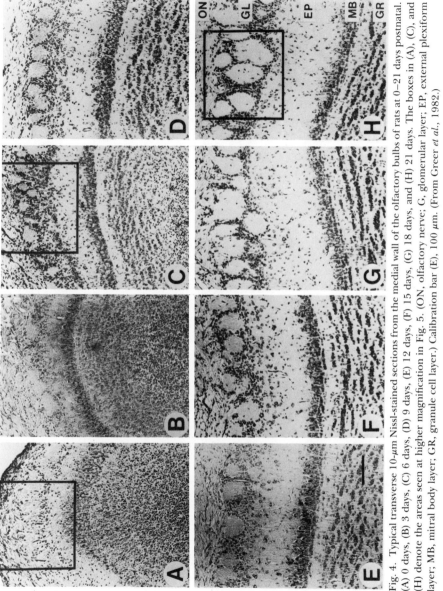

Fig. 4. Typical transverse 10-μm Nissl-stained sections from the medial wall of the olfactory bulbs of rats at 0–21 days postnatal. (A) 0 days, (B) 3 days, (C) 6 days, (D) 9 days, (E) 12 days, (F) 15 days, (G) 18 days, and (H) 21 days. The boxes in (A), (C), and (H) denote the areas seen at higher magnification in Fig. 5. (ON, olfactory nerve; G, glomerular layer; EP, external plexiform layer; MB, mitral body layer; GR, granule cell layer.) Calibration bar (E), 100 μm. (From Greer *et al.*, 1982.)

PATRICIA E.
PEDERSEN,
CHARLES A. GREER,
AND GORDON M.
SHEPHERD

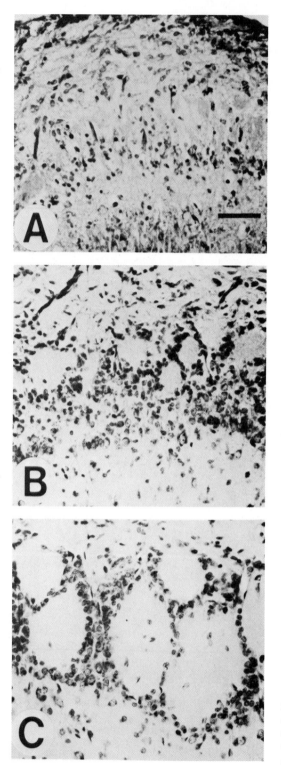

Fig. 5. Photomicrographs illustrating glomerular development in the areas demarcated by boxes in Figure 4. (A) 0 days, (B) 6 days, and (C) 21 days. Calibration bar (A), 50 μm. (From Greer *et al.*, 1982.)

amount of Golgi complex and rough endoplasmic reticulum increases significantly (Hinds and McNelly, 1977; Singh and Nathaniel, 1977).

By postnatal days 12–14, the main olfactory bulb appears adultlike, the overall surface area has increased, the mitral cell layer (most likely due to overall bulbar growth (Smith, 1935)) has thinned out to a monolayer, the external plexiform layer has expanded from 45 to 350 μm in width, and the glomeruli are more defined by the proliferating periglomerular cells. This morphological maturity is substantiated by functional assessments with both behavioral and physiological studies.

V. Accessory Olfactory Bulb

While the accessory olfactory bulb does possess a laminar organization, it does not have the clearly defined layers characteristic of the main olfactory bulb (Allison, 1953). (cf. Figure 3). The glomeruli are poorly defined, the external plexiform layer is rather thin in comparison, and the mitral layer appears indistinct and several cells thick. Despite the "immature" appearance of the accessory olfactory bulb, the peak development of its cell types and their organization into laminae occur before birth and prior to the development of the main olfactory bulb.

In mice, mitral cells arise 1 day earlier (E10) than their homologues in the main olfactory bulb (Hinds, 1968a). In rats a similar sequence of development occurs. Approximately 60% of the mitral cell population originates on E13, significantly earlier than those mitral cells in the main olfactory bulb (Bayer, 1983).

Tufted cells have not been identified in the accessory olfactory bulb. Periglomerular cells peak in cell production between E17 and E18 (Hinds, 1968a). The thick, crescent-shaped layer of granule cells that lay beneath the mitral cell layer completes its development before birth in the mouse. In the rat most of neurogenesis occurs prenatally between E17 and E19; however, there is evidence of 10–13% of granule cells originating over the first 2 postnatal weeks and even beyond postnatal day 20 (Bayer, 1983). Thus by birth the structural features of the accessory olfactory bulb indicate that it is capable of sensory activity. Now we have evidence that this pathway may process odor before birth.

VI. Central Olfactory Projections

The extensive hierarchy of projections from the main and accessory olfactory bulbs influencing many regions in the brain reflects the importance of the sense of smell to rodents as well as the integrative function of the brain. Although we know little of how odor information is processed centrally, spatiotemporal patterns of neuronal activity in olfactory cortical areas have been suggested to play an important role.

The axons of the mitral and tufted cells collect to form the lateral olfactory tract. The lateral olfactory tract spreads over the entire ventral forebrain and ter-

PATRICIA E.
PEDERSEN,
CHARLES A. GREER,
AND GORDON M.
SHEPHERD

minates in the superficial laminae of the molecular layer (Ia) of the olfactory cortex (Luskin and Price, 1983a,b; Powell, *et al.,* 1965; Price, 1973; White, 1965). Specifically, in adult rats, olfactory axons in the lateral olfactory tract terminate in (1) anterior olfactory nucleus, a very rostral olfactory area that makes up the olfactory peduncle and becomes a poorly laminated structure posteriorly (Haberly and Price, 1978); (2) tenia tecta (Haberly and Price, 1978); (3) piriform cortex, a region for which some evidence of topographic distribution from the bulb exists (Scott *et al.,* 1980); (4) olfactory tubercle, a basal region that appears to receive extensive input primarily from tufted cells and from ventral and posterior parts of the olfactory bulb (Price and Powell, 1971; Scott *et al.,* 1980; White, 1965); (5) entorhinal cortex; and (6) amygdala and periamygdaloid cortex. Clearly, these olfactory projection sites are implicated in many behaviors, cognitive, homeostatic, and emotional.

In the rat there is no topography in centrifugal connections (Luskin and Price, 1983a). The centrifugal fibers to the bulb can mediate and/or modulate olfactory perception with respect to both the internal state of the animal and external chemical stimuli. The centrifugal fibers project to the main olfactory bulb from regions that receive bulbar input and from regions that do not. The most prominent site of the latter and a major source of centrifugal fibers is the horizontal limb of the diagonal band (de Olmos *et al.,* 1978; Luskin and Price, 1983a; Price and Powell, 1970b). The fibers from the ipsilateral horizontal limb of the diagonal band pass in close association with those of the medial forebrain bundle and lateral olfactory tract and eventually terminate predominately on granule, periglomerular, and short-axon cells. The inputs from midbrain and hypothalamus to the horizontal limb of the diagonal band suggest that these fibers mediate limbic influences on main olfactory function.

The anterior olfactory nucleus (except pars externa) ipsilaterally projects to the glomerular and deep layers of the bulb (Haberly and Price, 1978; Luskin and Price, 1983a). Olfactory tubercle, cortical and posterolateral amygdala also project to the main olfactory bulb, although these fiber pathways do not appear to be present in the hamster (Macrides *et al.,* 1981). Other prominent sources of centrifugal fibers include piriform cortex, nucleus of the lateral olfactory tract, locus coeruleus, and dorsal raphe nuclei (Price and Powell, 1970; de Olmos *et al.,* 1978).

The piriform cortex sends fibers to two thalamic nuclei, mediodorsal and submedial, which, in turn, project to distinct neocortical areas (Price and Slotnick, 1983). The mediodorsal nuclei appear to be involved in complex odor-guided behavior because lesion of this nucleus results in severe deficits in odor reversal learning and higher-level functioning involving odor cues (Slotnick and Kaneko, 1981). Amygdala lesions had no effect on animals in this experiment, suggesting that thalamic, not limbic, olfactory projections are important in complex learning of olfactory cues.

In contrast to this, the accessory olfactory bulb has no obvious connections with the thalamus and therefore does not access neocortical structures (Scalia and Winans, 1975). Moreover, the accessory olfactory pathway maintains its anatomical distinction from the main olfactory pathway by projecting to nonoverlapping areas of the amygdala. The accessory olfactory bulb terminates in the nucleus of the accessory olfactory tract, medial and posteromedical cortical nuclei of the amygdala, and bed nucleus of the stria terminalis (Scalia and Winans, 1975; Kevetter and Winans, 1981a,b).

In the accessory olfactory bulb, cells of medial and posteromedial nucleus of the amygdala, nucleus of the accessory olfactory tract, and bed nucleus of the stria terminalis project via the stria terminalis to internal granule cells in the accessory olfactory bulb (Barber and Field, 1975; de Olmos *et al.*, 1978). These centrifugal fibers are clearly distinctive from those of the main olfactory bulb because labeled leucine studies show label abruptly stopping at the boundary of main bulb structures.

The few investigations that have been directed toward documenting the development of central olfactory connections have demonstrated that some connections are present before birth, at least in rats, although development extends well into the postnatal period. Schwob and Price (1978), using tritiated leucine injections, noted labeled fibers throughout the piriform cortex (except its lateral edge), lateral entorhinal cortex, and lateral olfactory tubercle by E20 in rats. Similarly, Singh (1977) described the presence of olfactory axons in piriform cortex by birth. Some of the synaptic connections appear mature at this time (Westrum, 1975). The Timm stain, specific for endogenous zinc characteristically associated with axon terminals, has revealed that in rats, by E19, piriform cortex exhibits intense staining in posterior portions, particularly neuropil of layers II and III and deep part of I (Friedman and Price, 1984). The anterior piriform cortex is included by E21. Thus structural features of the main olfactory pathway are evident by birth. During the first postnatal week, development of cortical connections continues and reaches adultlike density and distribution by the third postnatal week.

There is some evidence of topographical gradients in the development of the cells of origin of the lateral olfactory tract. Grafe and Leonard (1982) found maturational gradients in young hamsters (3 and 7 days of age) by observing a distinct localization of horseradish peroxidase–labeled cells in the medial quadrant of the bulb projecting to the lateral olfactory tract. These age differences were not apparent at day 12.

The central connections of the accessory olfactory bulb develop slightly earlier than those of the main olfactory bulb. [^{3}H]leucine labels the medial nucleus and portions of the posterior cortical amygdala nucleus as well as rostral portions of stria terminalis 2–3 days before birth (Schwob and Price, 1978). The Timm stain is positive in most of the posterior cortical amygdala and some periamygdaloid structures by birth (Friedman and Price, 1984). Bayer (1983) demonstrated a strong anterior–posterior gradient in neural development in the amygdala, with the anterior portion of the medial nucleus developing earlier than the anterior cortical portion. Thus accessory olfactory central connections parallel the developmental sequence of the main and accessory olfactory bulbs. Indeed, accessory olfactory bulb central projections appear adultlike by postnatal day 3. Amygdaloid

PATRICIA E.
PEDERSEN,
CHARLES A. GREER,
AND GORDON M.
SHEPHERD

nuclei receiving accessory olfactory fibers thus appear in advance of nuclei receiving fibers from the main olfactory bulb; central connections to these nuclei appear to parallel this. This sequence of development may help establish correct connections by target cortical neurons being ready to receive olfactory projections when they arrive (Bayer, 1983).

Leonard (1975) studied the development of bulbofugal connections in hamsters to see if any temporal or topographical gradient existed that might be correlated with specific olfactory behaviors. She used long-lasting degeneration agyrophilia (LLDA) as a measure of synaptic formation. She found a much later sequence of development of synaptic connections. Piriform cortex exhibited LLDA 5–9 days, olfactory tubercle and medial amygdala 5–9 days, and entorhinal cortex and cortical amygdala 9–13 days. It is surprising in light of the accelerated development of the accessory olfactory bulb that the cortical amygdala is the last place to show LLDA.

VII. Modified Glomerular Complex

The main and accessory olfactory pathways maintain their anatomical distinction both peripherally and centrally. Another region in the olfactory bulb has been identified recently that appears to be histologically distinct from either the main or accessory olfactory bulbs and was first documented by Teicher *et al.* (1980) and further described by Greer *et al.* (1982). This region, shown in transverse section to be comprised of a complex of 4–6 glomeruli, is called the modified glomerular complex (MGC) (see Figure 6). The MGC starts approximately one third posterior to the beginning of the accessory olfactory bulb and extends in a rostral–caudal direction approximately 300–600 μm, depending on the age of the pup (Slotnick and Pedersen, unpublished observations). It is located in the medial dorsal aspect of the olfactory bulb, adjacent to the dorsal extension of the mitral body layer. Its glomeruli are distinctive by the presence of periglomerular cells that serve to separate its lateral edge from the accessory olfactory bulb, and its medial edge from the main olfactory bulb (Greer *et al.*, 1982). Dorsal to this complex is found the vomeronasal nerve coursing to the accessory olfactory bulb. Whether the VNO sends axons into the MGC has not been determined.

Developmentally, the MGC seems to be a significant region both physiologically and anatomically. In contrast to the rest of the bulb, the MGC appears more mature early in development. On the day of birth, the MGC can be identified easily in histological sections (Friedman and Price, 1984; Greer *et al.*, 1982). At that time, glomeruli in both the main and accessory olfactory bulb are indistinct. The accessory olfactory bulb glomeruli remain poorly defined even into adulthood (Allison, 1953). The main olfactory bulb glomeruli do not become defined until approximately postnatal day 3 with the appearance of partial periglomerular cell bodies. The distinctive demarcation of individual glomeruli by a ring of periglomerular cells occurs approximately 6 days postnatally (Greer *et al.*, 1982). These main olfac-

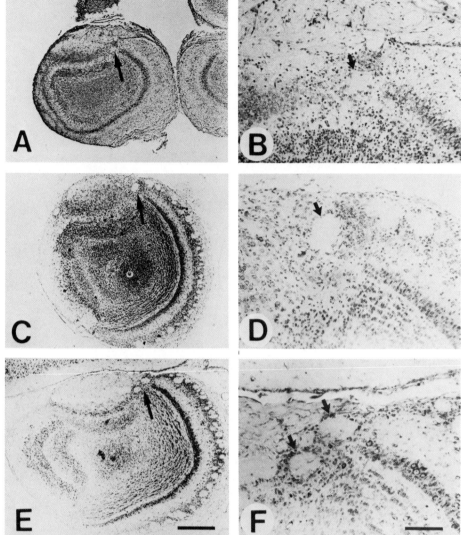

Fig. 6. Photomicrographs illustrating the typical morphology of the MGC. (A) 0 days postnatal, (B) higher magnification of the MGC seen in (A), (C) 6 days postnatal, (D) higher magnification of the MGC seen in (C), (E) 12 days postnatal, and (F) higher magnification of the MGC seen in (E). Calibration bars; (E) 350 μm, (F) 85 μm. (From Greer *et al.*, 1982.)

tory bulb glomeruli become more clear with the increase in glomeruli diameter and number over the next few weeks of the postnatal period.

The MGC was initially recognized because of the dense 2DG uptake that occurred within its glomeruli in suckling pups. The characteristics of 2DG uptake in Teicher *et al.*'s study (1980), as well as the subsequent study by Greer *et al.* (1982), parallel the mature appearance of this region. Unfortunately, because it

180

PATRICIA E.
PEDERSEN,
CHARLES A. GREER,
AND GORDON M.
SHEPHERD

has only been noticed recently, analysis of its cellular components, including their differentiation, migration, and synaptogenesis, has not been done. We are currently injecting HRP into the MGC region to determine if indeed this is part of an anatomically distinct pathway, receiving input from a specific population of receptors and projecting to a specific olfactory cortical region, or, alternatively, if it receives input from the main and/or vomeronasal receptors and projects to one and/or the other cortical region. Undoubtedly, these findings will help us understand the significance of this region.

VIII. THE ONTOGENY OF OLFACTORY FUNCTION: OVERVIEW

Table 1 is a summary of the ontogenetic sequence of odor-guided behaviors. It is impressive that so many behaviors essential for assuring that rodent pups receive adequate nutrition, warmth, and protection are very dependent on odor cues. This should not be so surprising in view of the limited sensory capabilities of young pups dependent on the nest situation. Pups, born blind and deaf, do not see until day 14 or hear until day 9. Thus olfactory and thermotactile stimuli are the major cues carrying environmental information. The marked structural maturity of the olfactory system so early in development, as described earlier, also attests to its importance so early. However, the predominant role of olfaction in behavior continues well into development even when pups are capable of visually guided behavior.

The young rodent and the ubiquitous role of olfaction in its development provides us with a species and a system in which to establish correlations between developing anatomy and the development of function. This is particularly useful since so little is known about neural mechanisms underlying odor-guided behaviors. Indeed, little is understood of the anatomy and/or physiology of odor discrimination in adults. One obvious aspect of Table 1 is that there appears to be a

TABLE 1. ONTOGENY OF ODOR-GUIDED BEHAVIORS IN RATS

E20-E22	Odor aversion learning (Smotherman, 1982)
P1	Suckling (Teicher and Blass, 1977; Pedersen and Blass, 1982)
	Appetitive learning (Johanson and Hall, 1982)
P2	Odor aversion learning (Rudy and Cheatle, 1978)
	Orientation (Bolles and Woods, 1964; Shapiro and Salas, 1970)
P3-4	Preference for home nest shavings (Cornwell-Jones and Sobrian, 1977)
P6	Home odor affects activity and ingestion (Johanson and Hall, 1981)
P10-12	Discrimination of mother versus nonlactative female/male (Nyakas and Endroczi, 1970)
	Discrimination of own versus stranger's nest (Gregory and Pfaff, 1971)
P14	Huddling under olfactory control (Alberts and Brunjes, (1978)
	Caecotrophe attraction (Leon and Moltz, 1971)
P1-14	Improvement in nasal chemosensitivity (Alberts and May, 1980)

temporal sequence in the development of odor-guided behaviors. Of course this is in part due to the development of motor abilties that allow behaviors to be expressed such as locomotion to a goal some distance away. However, even though it is difficult to separate the relative contributions of sensory and motor development, some odor-related behaviors that pups are fully capable of performing occur at different ages. For example, caecotrophe attraction, important for pups maintaining proximity to the nest, does not occur until day 14 (Leon and Moltz, 1971, 1972), well after pups exhibit an odor preference. Huddling, an important behavior for its thermal and social benefits, is not under olfactory control until day 15 (Alberts and Brunjes, 1978). What is the significance of this apparent sequence? It may reflect the maturation of the olfactory system. Indeed, the ontogenetic events within the olfactory system may provide some clue as to the organization of these behaviors. For example, the temporal order of development of bulbar projections to olfactory cortical regions might be related to the onset of behaviors influenced by odor cues.

We do know that odor discrimination abilities do improve postnatally. Clearly, early in development, indeed within hours of birth, rat pups are capable of making a critical odor discrimination (Teicher and Blass, 1977). On days 3–4, pups exhibit a preference for home shavings over clean shavings (Cornwell-Jones and Sobrian, 1977), and by days 10–12 they can discriminate their own shavings from those of a stranger's nest (Gregory and Pfaff, 1971). Moreover, responsiveness to odors improves dramatically during the first 2 postnatal weeks (Alberts and May, 1980). This ontogenetic chronology of odor abilities is paralleled by postnatal maturation of components of the olfactory system.

One obvious approach to determining the underlying neural substrates of olfactory function has been to lesion components of the olfactory system. The limitations of this are obvious (Alberts, 1974; Hofer, 1976). It not only results in the production of nonsensory deficits but also typically involves gross areas, thus making it difficult to distinguish the contribution of the respective subsystems. Although electrophysiological analyses have made some contribution to our understanding of the development of receptor odor selectivity in neonates, technical difficulties have prevented precise experimental analyses, much less recording in the awake, behaving animal. The 2DG method has provided us with an ideal technique for relating the ontogeny of structure and function in the olfactory bulb. It is a noninvasive method that does not rely on pathological changes and can be used in the behaving animal. Indeed, with the application of 2DG to both adults and neonates, significant information has been obtained regarding the organization of function in the olfactory system.

The 2DG method is based on the use of ^{14}C-labeled 2DG in tracer amounts (Kennedy *et al.,* 1975). 2DG, an analogue of glucose, is taken up into physiologically active neurons and neuropil by the same active mechanism that takes up glucose. However, 2DG is not metabolized beyond 2-deoxy-D-glucose 6-phosphate. Consequently, it is essentially "trapped" in the region into which it was originally actively transported. At the termination of the experimental procedure, an auto-

PATRICIA E.
PEDERSEN,
CHARLES A. GREER,
AND GORDON M.
SHEPHERD

radiographic analysis reveals the topographical distributions of the (^{14}C) label attached to the 2DG-6-phosphate. The distribution of label within the CNS reveals the relative degree of functional or physiological activity between injecting the 2DG and sacrificing the animal, typically 30–45 min. The technique has been employed successfully to corroborate previous electrophysiological analyses of the functional organization of several sensory systems, including visual, auditory, and somatosensory systems.

The use of 2DG procedures in mammalian adult olfactory systems has demonstrated punctate foci of high 2DG uptake in the olfactory bulb following odor stimulation (Sharp *et al.,* 1975; Sharp *et al.,* 1977; Stewart, *et al.,* 1979). These foci are histologically associated with olfactory bulb glomeruli. There is strong evidence that the 2DG procedures actually monitor the distribution of odor-responsive receptor cell axons among the glomeruli. (Green *et al.,* 1981). Furthermore, topographical distribution of glomerular foci in the olfactory bulb differed when the patterns induced by the odor of amyl acetate were compared with those following camphor, or nonodorized air, or room air, etc. Thus it appears that superimposed upon the anatomical organization of the olfactory bulb is a functional organization that appears to distinguish among odors.

The next three sections will discuss the three subsystems—main olfactory bulb, accessory olfactory bulb, and modified glomerular complex—in terms of their functional role during development. Because much of what we know is from application of the 2DG method,[1] most of the findings will be from those studies, including our own. However, we will also include, where pertinent, other methods that will help us to understand the nerual substrate of odor discrimination, including odors that are behaviorally significant.

IX. PRENATAL ACTIVITY IN THE ACCESSORY OLFACTORY BULB

To understand developing sensory functions, it behooves us to begin with the fetus and the intrauterine environment of each sensory system. There is ample evidence that components of the olfactory system, at least in mice and rats, are structurally mature enough prior to birth to be capable of sensory activity. This includes the modified glomerular complex and components within the peripheral and central divisions of the main and accessory olfactory pathways. Furthermore, adequate

---

[1]The 2-deoxyglucose technique is presently the only procedure available for monitoring the metabolic activity indicative of neuronal function throughout the entire neuroaxis. However, several caveats must be put forth regarding its use and interpretation (Hand, 1981). These include (1) the relative contribution of excitatory and inhibitory synaptic activity to 2DG uptake, (2) the cytoplasmic neuronal component responsible for 2DG uptake and accumulation, (3) the changes in 2DG uptake that may accompany the biochemical development of the CNS, (4) the influence of uncontrolled environmental stimuli on 2DG uptake, (5) the potential misinterpretation of alterations of 2DG uptake by polyneuronal circuits, and (6) the critical importance of recognizing that for accurate interpretation, the 2DG autoradiographs must be precisely correlated with their corresponding histological activity.

stimuli for olfaction are present prenatally (Bradley and Mistretta, 1975). Amniotic fluid is an odorous substance, in part because of the presence of citric, lactic, and fatty acids. In addition, amniotic fluid changes constantly as a consequence of the mother's physiological status and as a result of fetal swallowing and excreting *in utero* (Lev and Orlic, 1972). Thus amniotic fluid may provide a rich source of chemosensory stimulation for the fetus.

Behavioral evidence is accumulating that fetuses not only detect the prenatal milieu but also learn about it. Changing the intrauterine milieu by injecting odorous substances in amniotic sacs subsequently modifies the cue for the first nipple attachment (Pedersen and Blass, 1982). Pups will avoid odors that they experience *in utero* in conjunction with an i.p. injection of lithium chloride as late as postnatal day 16 (Stickrod *et al.*, 1982; Smotherman, 1982a). Even adult odor preferences can be altered by intrauterine manipulations of the amniotic fluid (Smotherman, 1982b). Since these postnatal responses involve odor detection, it suggests that olfactory pathways are one means of detecting the amniotic milieu.

We also recognize the possibility that chemicals in the amniotic fluid may be adequate stimuli for the gustatory system. This, in fact, appears to be the case. Taste buds develop prenatally in both sheep (Bradley and Mistretta, 1973, and this volume) and rats (Farbman, 1963; Mistretta, 1972). Moreover, in sheep, electrophysiological responses to various salts can be recorded from the chorda tympani before term and these responses change over time as do membrane properties of the taste receptor cells (Mistretta and Bradley, 1983). Indeed, these studies provide direct physiological evidence that the fetus detects its prenatal environment.

Encouraged by these anatomical and behavioral studies, we have explored the hypothesis that olfactory pathways might be functionally active *in utero* (Pedersen *et al.*, 1983). We chose to use the 2DG method, already applied so successfully to the study of postnatally developing olfactory pathways (see previous discussion) and recently to *in utero* investigations of functional activity in suprachiasmatic nuclei of fetal rats (Reppert and Schwartz, 1983). Rather than inject fetuses directly with 2DG and risk experimental interference with the mother and/or her litter, we decided to inject the mother intravenously. Based on their common circulation, fetuses have access to the pool of labeled 2-deoxyglucose.

Pregnant dams were given intravenous injections of 300 μCi/kg of 14 C2DG in the morning of their last day of a timed gestation (E22). For 1 hr the dam was permitted to move freely. She was then decapitated and her fetuses removed immediately by caesarean section. The brains of five fetuses from each of four litters (n = 20) were removed and processed for autoradiographic analyses.

All autoradiographs of fetal brains exhibited 2DG uptake; therefore 2DG passed from maternal to fetal circulation. Eighteen of the 20 fetuses exhibited relatively high levels of 2DG uptake in dorsolateral regions of the bulb. This is depicted in Figure 7, which is a representative autoradiograph of the 18 fetuses. In order to determine the exact placement of this uptake in the bulb, an outline of the corresponding histological section (B) was drawn including the inner borders of the mitral cell layer of the main olfactory bulb and mitral–granule layer of

PATRICIA E.
PEDERSEN,
CHARLES A. GREER,
AND GORDON M.
SHEPHERD

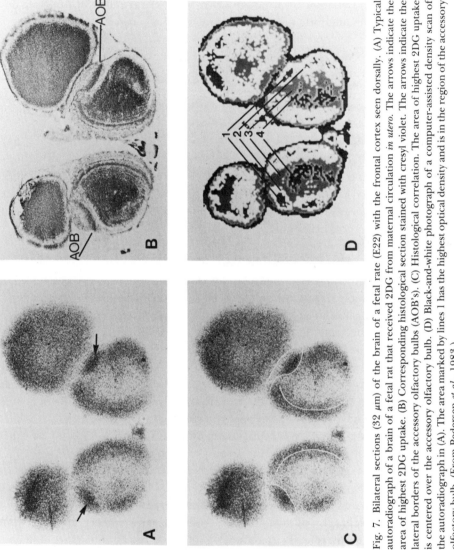

Fig. 7. Bilateral sections (32 μm) of the brain of a fetal rate (E22) with the frontal cortex seen dorsally. (A) Typical autoradiograph of a brain of a fetal rat that received 2DG from maternal circulation *in utero*. The arrows indicate the area of highest 2DG uptake. (B) Corresponding histological section stained with cresyl violet. The arrows indicate the lateral borders of the accessory olfactory bulbs (AOB's). (C) Histological correlation. The area of highest 2DG uptake is centered over the accessory olfactory bulb. (D) Black-and-white photograph of a computer-assisted density scan of the autoradiograph in (A). The area marked by lines 1 has the highest optical density and is in the region of the accessory olfactory bulb. (From Pedersen *et al.*, 1983.)

the accessory olfactory bulb. This drawing was then superimposed on autoradiograph. This histological correlation with the autoradiograph is depicted in (C). Clearly, the high uptake evident in the bulb in (A) is confined within the histological borders of the accessory olfactory bulbs.

Less dense areas of 2DG uptake are present in the superficial laminae of the remainder of the bulb. Although the glomerular layer of the main olfactory bulb is only present medially in these sections, no evidence of differential uptake appears within the main bulb glomerular sheets or in the modified glomerular complex in this or any other sections. We found this to be the case for 16 of the 20 fetuses. However, in four fetuses, we saw evidence of high 2DG uptake within the borders of the main olfactory bulb. Of particular interest is that the increased uptake appeared to be localized to the dorsomedial regions of the olfactory bulb. This is the same region that usually shows uptake when young pups are given 2DG and stimulated with odors. There appeared to be no consistent factors, such as litter (sex was not determined), that could account for just these four pups exhibiting main bulb uptake. Another region in which relatively high levels of uptake were present in all pups was the anterior olfactory nucleus. The extent to which other, more central, areas of the brain exhibit increased uptake is currently being investigated.

To confirm our initial observations and conduct a more objective analysis, each autoradiograph was scanned by a computer image analysis system, as described elsewhere. (Goochee *et al.*, 1980). In brief, this system scans the autoradiograph optically and is programmed to color-code four different levels of optical density (i.e., 2DG uptake), proceeding from the highest to lowest. A black and white photograph of a computer scan of the autoradiograph in (A) is depicted in (D). The areas marked by the numbered arrows indicate the highest regions of 2DG uptake. As is evident from the corresponding histological section, the highest uptake is within the borders of the accessory olfactory bulb. This analysis confirms our subjective observations that the accessory olfactory bulb has the highest 2DG uptake.

2DG uptake is considered to reflect neuronal activity (Greer *et al.*, 1981; Sharp *et al.*, 1975; Schwarz, *et al.*, 1979). Thus our findings support behavioral evidence that neuronal activity, and consequently olfactory function, occurs *in utero*. There are, however, several alternative interpretations of the observed uptake in the accessory olfactory bulb. One hypothesis is that uptake may be present as a result of neurogenesis, an increase in mitotic activity, and/or cell migration of the cells prenatally. These are unlikely possibilities because 2DG has not been demonstrated to be a sensitive indicator of these events in the central nervous system. Furthermore, we would have expected to see uptake in the main olfactory bulb, especially since cellular development and migration are extensive in this region in the latter part of gestation, during the time of this analysis. In addition, other areas of high mitotic activity, such as the ependymal layer of the ventricles, do not exhibit differential levels of 2DG uptake. Second, uptake may represent the neural activity of centrifugal fibers that have established synaptic contact with the mitral–granule

PATRICIA E.
PEDERSEN,
CHARLES A. GREER,
AND GORDON M.
SHEPHERD

cells of the accessory olfactory bulb. This is an intriguing possibility because centrifugals are known to modulate afferent activity and it invites investigation as to why this pathway would be active at this time in gestation and under these circumstances. The most parsimonious explanation appears to be that activation of vomeronasal receptors by odor molecules in the amniotic fluid may stimulate the vomeronasal nerves, thus inducing high uptake at their terminal sites in the accessory olfactory bulb. From results obtained in the main olfactory bulb (see the following discussion) when giving pups postnatal odor stimulation, this seems a very likely possiblity. It is important to note, however, that uptake did appear to be confined mostly over the mitral–granule layer of the accessory olfactory bulb instead of over the glomerular layer. Monti Graziadei *et al.* (1980) have observed that early in development, main olfactory nerves terminate not just in the glomerular layer of the main olfactory bulb but well into the external plexiform layer. Presumably, with development, these far-reaching afferents recede to assume their final position within the glomeruli. This situation could be present in the accessory olfactory bulb, that is, some vomeronasal nerves may reach into the deeper layers of the accessory olfactory bulb. Without further knowledge of the synaptic organization of the developing bulb, it is difficult to assess this. The preceding possibilities do not detract from the major finding, since uptake is still indicative of "function" in this pathway. They do relate to the limitations of the 2DG method in interpretation of data.

It is paradoxical that postnatal suckling odors are processed by the MGC and main olfactory bulb within hours of birth, but prenatal odors, which presumably influence suckling, are processed over the accessory olfactory pathway. What is the significance of this? Could information relayed over one anatomically distinct chemosensory pathway be related to another? There is some evidence that this may happen. Both Meredith (1983b) and Wysocki and colleagues (Wysocki *et al.*, 1982) have found that rats and guinea pigs need intact vomeronasal organs for experiencing odors related to reproductive activities but, once the animals are sexually experienced, not for the actual expression of reproduction-related behaviors. For example, guinea pigs sexually experienced will subsequently emit ultrasounds to urine. If their vomeronasal organ is lesioned (or removed) after their sexual experience, they will continue to emit ultrasounds to urine. If, however, their vomeronasal organ is removed prior to their sexual experience, regardless of the amount of sexual experience the guinea pigs subsequently have, they never ultrasound to urine.

These findings suggest (as discussed by Meredith, 1983, pp. 241–244) that initially, behaviorally significant odors are processed by the vomeronasal pathway as well as by the main olfactory pathway. Subsequently, by virtue of associations made between odor stimuli processed by both these pathways, the main olfactory pathway is sufficient to process relevant stimuli. The animal would then not need to rely on vomeronasal cues to initiate the appropriate behavior. This process may also be occurring in the suckling situation; that is, the accessory olfactory pathway may be important for processing amniotic fluid odors but not for the actual expression of suckling to amniotic odors. The communication between these two pathways could occur at several levels, for example, as a result of intraamygdaloid connections (Kevetter and Winans, 1981a,b) that would allow influence on one or the other pathway. Another possibility is that the MGC, strategically located between the main and accessory olfactory bulbs, receives input from both main and vomeronasal receptors. We have not yet seen evidence of high levels of 2DG uptake in the MGC region *in utero* but are currently testing the preceding hypothesis using HRP tract-tracing methods.

It is curious that uptake occurred in the accessory olfactory pathway and that no differential uptake was noted in the MGC and/or main olfactory bulb. Surely amniotic fluid, by virtue of its presence throughout the nasal cavity, was capable of stimulating those chemoreceptors and thus stimulus access was equivalent for all receptors. In fact, the requirement of a pumping action to suck odorants into the vomeronasal chamber (Meredith and O'Connell, 1978; Meredith *et al.*, 1980) make the latter a more inaccessible area. One possibility for this finding may be that nonvolatile stimuli activate vomeronasal receptors but not those of the main epithelium and that these stimuli may be present in amniotic fluid. There is strong evidence in the adult to suggest that nonvolatile stimuli stimulate vomeronasal receptors (Wysocki *et al.*, 1980). Furthermore, a number of naturally occurring biologically active substances appear to be nonvolatile. This may also help explain why this laboratory and others have not noted any evidence of differential 2DG densities within the region of the accessory olfactory bulb *ex utero*. This is most striking when comparisions are made between the bulbs of the *in utero* fetuses and pups of the same gestational age, delivered by caesarean section, given 2DG, and placed in a citral odor environment as well as a suckling situation. High 2DG uptake in these rat pups remains confined to portions of the glomerular layer of the main olfactory bulb and/or MGC (Pedersen *et al.*, 1982).

The finding of 2DG uptake in the accessory olfactory bulb appears unique to fetuses being *in utero*. Ideally, one would like to deprive the fetuses of the odorous aspects of amniotic fluid and then see lack of differential uptake in the accessory olfactory bulb. Toward this end we are currently attempting to deprive fetuses of amniotic fluid input by unilateral nares closure. In this way, high 2DG uptake should occur only on one side.

X. EARLY POSTNATAL ACTIVITY IN THE MGC

We have chosen for a number of reasons to focus on one olfactory guided behavior that is characteristically mammalian—suckling. It is a well-defined, discrete behavior that is present from birth and predominant in the behavioral repertoire of developing rat pups. Moreover, suckling is a behavior in which the characteristics of its odor control are well documented. Rat pups rely on odor cues both to locate and to attach to their mothers' nipples. This odor dependency starts at birth with the first nipple attachment (Teicher and Blass, 1977) and extends

188

PATRICIA E.
PEDERSEN,
CHARLES A. GREER,
AND GORDON M.
SHEPHERD

throughout the entire nursing period (Blass *et al.*, 1977; Bruno *et al.*, 1980; Hofer *et al.*, 1976; Teicher and Blass, 1976).

The types of cue(s) to which pups will attach are limited. If the mother's nipples are lavaged with an organic solvent so as not to interfere with nipple morphology, nipple attachement is severely disrupted. Attachment can be restored, however, provided the appropriate substance is painted in the nipple. Amniotic fluid, parturient mother saliva, or nipple wash extract is effective for newborns (Teicher and Blass, 1977); pup saliva or nipple wash extract is similarly effective for pups 2 or more days old (Teicher and Blass, 1976). Suckling is predominately under this external sensory control in pups younger than 15 days of age, since deprived and nondeprived pups of these ages attach with similar latencies to saliva-coated nipples. After 15 days, deprived pups attach less frequently and with longer latencies than nondeprived pups (Hall *et al.*, 1975, 1977). Presumably, internal control mechanisms related to "hunger" are significant factors. However, if the appropriate cue is not present, pups will not suckle.

The discrete nature of this behavior as well as its well-documented odor dependence encouraged Teicher and colleagues (Teicher *et al.*, 1980) in our laboratory to explore its neural basis. The 2DG method was especially well suited for this purpose. As described earlier, in adult animals it reveals patterns of activity in the olfactory bulb after stimulation with a chemical odor. Moreover, these patterns can be produced by odors at very weak concentrations, near threshold for human perception (Stewart *et al.*, 1979). This indicated that it should permit monitoring of activity elicited by natural odors in behavioral contexts, in which the odor concentration would be expected to be low. An additional condition was that pups will readily attach to nipples provided the appropriate odor cue is present.

Ten-day-old rat pups given 2DG and permitted to locate and attach to nipples of an anesthetized dam reliably and consistently exhibit 2DG uptake in the MGC (Teicher *et al.*, 1980). This finding suggested that the MGC was involved in processing the odor cues associated with suckling. To extend this initial finding, Greer and colleagues (Greer *et al.*, 1982) studied pups 0, 3, 6, 9, 12, 15, and 18 days of age under conditions in which they received 2DG and suckled their respective dams. All pups of all ages, similar to Teicher's original finding, exhibited dense foci within the borders of the MGC.

These findings not only revealed a previously unrecognized region in the olfactory bulb but suggested for the first time a specific site in the bulb that was of defined functional importance. One of its distinctive characteristics was that it deviated from the general developmental characteristics of odor-induced 2DG in rat pups. 2DG foci appeared distinct and well defined on the day of birth. This most likely was related to its histological maturity and precocious development (see the preceding discussion of MGC anatomy). 2DG uptake did occur in other regions of the olfactory bulb in young rat pups—specifically, along the caudal medial aspect of the bulb. In contrast to 2DG uptake in MGC, their uptake did not appear focal and was rather poorly defined. This correlates closely with the immature histology of the rest of the olfactory bulb. The diffuse appearance of 2DG foci in the main olfactory bulb of neonates may reflect the lack of modulatory local circuits mediated via periglomerular cells (Greer *et al.*, 1982). The initial appearance of well-defined 2DG uptake in response to odor stimulation occurs in conjunction with histological maturation and demarcation of glomeruli by surrounding periglomerular cells. This latter finding applies not only for pups in the suckling situation but also for pups in conditions in which they are exposed to amyl acetate (Greer *et al.*, 1982).

2DG uptake in areas other than the MGC may reflect the complex odor conditions present in the experimental situations. Olfactory receptors are stimulated not only by odors on and or near the nipple but also by those odors on the mother, in room air, and even those associated with the experimenter. The odor cues present during suckling, however, do appear to be processed by the MGC. Pups under other odor conditions in which they are exposed to amyl acetate (Greer *et al.*, 1982), citral (Pedersen *et al.*, 1982), pure air, and/or room air (Greer *et al.*, 1982; Pedersen *et al.*, 1982) do not show uptake in the MGC region.

Despite the strict dependency of pups on the presence of a specific cue on the nipple in order for them to locate and to attach to the nipple, they can learn to suckle alternate cues. The special circumstances that induce pups to suckle a cue other than the one that normally coats the nipple provide some clues as to how the natural odor gains control over suckling. For the first attachment, pups will suckle a citral-scented, washed nipple, if they were exposed to citral in the last days of gestation and were stroked in its presence for 1 hr postparturition (Pedersen and Blass, 1982). Moreover, these "citral" pups do not even attach to normally scented nipples. Experience with citral exclusively prenatally, or postnatally, is not sufficient to subsequently induce attachment to citral-scented nipples. Thus the prenatal exposure to amniotic fluid and its presence in the nest postnatally during active maternal stimulation (licking, retrieving, etc.) most likely account for amniotic fluid's effectiveness as a suckling cue.

For pups 2–3 days of age, the conditions for learning to suckle an artificial odor are less stringent. Stimulation, either by mechanical or pharmacological means in the presence of the odor, is sufficient for that odor to subsequently induce attachment (Pedersen *et al.*, 1982). Either stimulation or odor exposure conditions alone are insufficient to induce attachment to the odor. These findings suggest that saliva, placed on the nipple by the pup itself, becomes potent by arousal components occurring around attachment time in presence of the saliva odor. That this itself is a learning process occurring during the first 2 days in the nest is indicated by the fact that stimulation in the presence of citral immediately after birth is not sufficient for citral to subsequently induce attachment.

These studies are intriguing because they demonstrate that one of the most critical behaviors necessary for survival has learning components in which some form of stimulation plays a necessary role. What is the significance of this learning for what we know thus far about odor processing in young pups? Our knowledge of the bulb under normal conditions may allow us to find the neural basis for odors becoming behaviorally significant.

PATRICIA E.
PEDERSEN,
CHARLES A. GREER,
AND GORDON M.
SHEPHERD

Maturation of the main olfactory bulb is particularly evident in the first 2 weeks postnatally. The formation, differentiation, and migration of olfactory bulb neurons, as well as synaptogenesis that establishes and expands the areas to which these neurons project, account for much of the events that transform the newborn bulb into one remarkably adultlike at day 14. These anatomical changes are clearly reflected in the physiology of the bulb.

Physiological changes in the developing main olfactory bulb are apparent in electrophysiological analyses. Math and Devrainville (1980) studied unit activity in the developing rat and reported spontaneous activity within 12 hr of birth. Their analysis further suggested that local circuits between mitral and granule cells did not become functionally apparent until approximately 3 days postnatal, when inhibition was first observed.

Their single unit observations are in accord with electroencephalographic analyses of Salas *et al.*, (1969) and Iwahara *et al.* (1973), in which the large granule cell component of field potentials was not prominently observed in the olfactory bulb until 3–5 days postnatal. By this rough measure, processing of afferent odor information by local circuits in the olfactory bulb does not appear prior to 3 days postnatal. This is in accord with considerable information, already reviewed, on the later development of granule cell interneurons and their synaptic connections with mitral cells in the mouse (Hinds and Hinds, 1976a,b), and rat (Mair *et al.*, 1982). As a consequence, the specificity of the information relayed by the second-order neurons, the mitral and tufted cells, may be determined primarily by the relative degree of maturity in the receptor cells whose axons connect to a given subset of second-order neurons (cf. Mair and Gesteland, 1982).

There is some evidence of response selectivity in olfactory receptor cells to different odors rather early in development. Gesteland *et al.* (1982) analyzed extracellular single-unit activity of olfactory receptor cell responses to odors in rats between the ages E16 and P14. Selectivity to odors was not observed at the earliest stages of development. Between E16 and E18, 100% of the neurons tested responded to all odorous stimuli. At E21–E22, only 6% of the cells responded to all stimuli, and by P1 no receptor cell responded to all odors. These studies suggested to the authors that by birth olfactory receptor cells may develop selective response spectra. The extent to which the *in utero* or immediate postnatal environment determines or influences receptor cell tuning is of major interest but is presently unknown.

This is a first attempt at a difficult problem of recording from embryonic olfactory cells. At this stage the interpretation is limited, because the recorded cells are unidentified, it is unknown if their axons reach the bulb, the more immature cells may be more susceptible to injury, and the portion of the receptor cell popualtion recorded from may be small. Furthermore, the manner in which these findings relate to the odor-specific domains (Stewart *et al.*, 1979) found in the olfactory bulb of pups with the 2DG methods remains to be explored.

2DG Studies

191

EARLY
DEVELOPMENT OF
OLFACTORY
FUNCTION

Pups of various ages from birth to weaning have been injected with 2DG and their main olfactory bulbs subsequently examined for evidence of odor-induced activity. Several major findings have resulted from these studies that suggest trends in the functional development of the main olfactory bulb.

First, it is apparent that early in development, prior to 6 days postnatal, there is little evidence of patterns of activity related to specific odors. In adults, individual glomeruli respond to specific odor stimulation while immediately adjacent glomeruli are inactive (Stewart *et al.*, 1979). Thus patterns of glomerular activity related to specific odors (e.g., camphor) can be distinguished. This glomerular specificity is poor in neonates. In fact, bulbar patterns in young rat pups after odor exposure to either nest odor or ethyl acetoacetate (EAA) are virtually indistinguishable (Astic and Saucier, 1982). 2DG uptake appears diffuse in neonatal bulb under these and other odor conditions (Astic and Saucier, 1982; Greer *et al.*, 1982; Pedersen *et al.*, 1982). This has been suggested to reflect the immaturity of the bulb, specifically, the lack of modulatory local circuits via periglomerular cells (Astic and Saucier, 1984; Greer *et al.*, 1982). This is supported further by findings on guinea pigs, a precocial species both anatomically and behaviorally in which 2DG patterns at birth appear more complex and numerous than in rat pups (Astic and Saucier, 1984). Indeed, the poorly defined 2DG foci in rats may prevent us from discriminating glomerular specificity. Spatial patterns do become evident after day 6, however, at a time when histological demarcation of glomeruli by periglomerular cells is perceptiible. By day 9, spatial patterns of different odor stimuli are overlapping but distinct (Astic and Saucier, 1982). Therefore adultlike 2DG patterns are apparent when the bulb has achieved the greater part of its growth.

The second major finding that 2DG analyses have revealed is that the relative amount of functional activity induced by odor exposure increases dramatically during ontogeny (Astic and Saucier, 1982; Greer *et al.*, 1982). During 0–3 days postnatal, only a few areas of increased 2DG uptake were observed. By day 15 the physiologically active areas had increased to near adult levels. The increase in the number of glomeruli with age most likely accounts for this finding.

Thirdly, uptake appears to be confined to specific areas of the bulb in early development and appears in other regions later. For example, a prominent area that virtually always exhibits uptake in newborns (even in the suckling situation) is the caudal medial portion of the bulb. Between 0 and 3 days, amyl-acetate-induced uptake was limited to the caudomedial aspect of the olfactory bulb (Greer *et al.*, 1982). By 12–15 days, however, the rostral lateral quadrant also showed active foci, as seen in the adult (Greer *et al.*, 1982). There is evidence that this region may be more anatomically advanced than the rest of the bulb. Grafe and Leonard (1982) found that cells that send efferents to olfactory cortical regions in the first week of life tend to be concentrated in the medial quadrant of the bulb of hamsters. In addition, receptors that project to this portion of the main olfactory bulb may have made their synaptic contacts early, and/or nasal airflow in neonates may allow eas-

PATRICIA E.
PEDERSEN,
CHARLES A. GREER,
AND GORDON M.
SHEPHERD

ier access to these receptors projecting to caudal medial bulb. Regardless of the basis of this finding, it is clear that this region is important for olfactory processing in the neonate, and this importance continues into the adult (see Stewart *et al.,* 1979). Heterogenous development of the main olfactory bulb could result in some neurons becoming functionally mature before others. Such maturational gradients might suggest the existence of further subsystems within the main olfactory bulb. This possibility has not been addressed.

XII. DISCUSSION

A main conclusion arising from the work reviewed here is that the developing olfactory system is composed of several subsystems specialized for detecting and processing different chemical stimuli. We have focused on three subsystems, two of which—AOB and MOB—are well known to be anatomically and functionally distinct. The third, MGC, appears thus far on morphological and functional grounds to be a separate subsystem. These three subsystems comprise anatomically segregated frameworks for processing odor information in the developing neonate.

A striking feature of these three neural pathways is that they are also temporally segregated in their development; that is, they develop according to different time schedules. This conclusion is primarily based on the anatomical and 2DG studies and is consistent with the available physiological data. Figure 8 illustrates this concept in terms of the 2DG results. Briefly, the earliest 2DG foci are seen in the AOB on E22, just before birth. Within an hour of delivery, the neonatal pup exhibits a clear focus in the MGC. In the MOB the earliest distinct foci appear around postnatal day 6. This sequence correlates with the maturation of these regions as revealed by the anatomical studies reported previously. Although no physiological studies have been carreid out on the AOB or MGC, the later postnatal development of activity in the MOB parallels the later anatomical development of the MOB relative to the other subsystems.

This principle of sequential function of subsystems in the early development of the olfactory system now requires testing with further work. We emphasize that the schema in Figure 8 is based solely on the 2DG results. It delineates the earliest time of onset of function, as tested thus far, and does not preclude an earlier onset in the AOB. We have used the appearance of foci as a criterion of activity; this is only a qualitative measure and does not rule out the likelihood of some degree of function represented by less dense 2DG patterns.

The developmental time schedule for the different subsystems is likely crucial for mediation of neonatal odor-guided behavior. During development the neural substrate for odor processing is comprised of both mature and immature elements, both between and within subsystems. Thus the most mature component of the olfactory system at a given stage of early development—the AOB prenatally, for example—is most likely to be the one that is engaged in actively processing odor

RAT

193

EARLY
DEVELOPMENT OF
OLFACTORY
FUNCTION

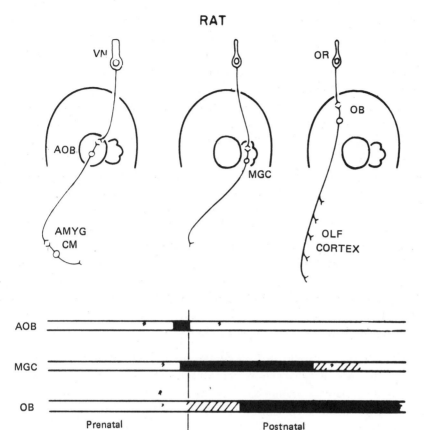

Fig. 8. Schematic illustrating the sequence of functional activity that may occur within subcomponents of the olfactory system during development. The filled areas of the bands, derived from 2DG data, suggest the time period during which these systems may begin to function.

information at that stage. Development progressing in this manner allows the addition of new capacities and abilities onto the existing ones. For example, the emergence of the MGC and the classes of odors it processes, added to those already processed by the AOB, increases the diversity of odors to which the animal can respond. This progression also increases the possibility of plasticity in each subsystem as it interacts with the environment and influences the subsequent subsystem. Furthermore, on an anatomical level, a temporal sequence of development might, in coordination with the development of central cortical regions, assure the appropriate connectivity with those areas concerned with the motor behavior at a stage in development (cf. Bayer, 1983). A possible example, as noted previously, is that those parts of the amygdala that receive input from the AOB appear to mature early, in advance of those parts of the amygdala receiving input from the MOB (Bayer, 1983). In line with this, therefore, the implication is that those central areas

PATRICIA E.
PEDERSEN,
CHARLES A. GREER,
AND GORDON M.
SHEPHERD

involved in the motor response of suckling presumably mature early enough to receive odor input; indeed, their early maturation may attract developing axons.

As each subsystem develops, how does it come to mediate a particular odor-guided behavior? Of particualr significance is that the behavioral literature to date suggests that responsiveness to odors is a function of learning. There are numerous demonstrations of this. For example, pups by day 15 prefer to huddle with a con-specific bearing the same odor cues that were experienced earlier on the mother's ventrum in the nest situation (Brunjes and Alberts, 1979). As described earlier, odors become potent elicitors of nipple attachment by their prior exposure to pups under certain arousal conditions (Pedersen and Blass, 1982; Pedersen *et al.*, 1982). Pups orient to and approach shavings scented with the odor that they had previously experienced in conjunction with intraoral infusions of milk (Brake, 1981; Johanson and Hall, 1982) or for extended periods of time in the nest situation (Rodriguez Enchadia *et al.*, 1982). In fact, there have been no reports of odor-guided behaviors that are not subject to experience (see Rosenblatt, 1983).

What is the mechanism by which experience brings about this association between a particular odor and a particular behavior during early development? The following list delineates several factors that may play a significant role in contributing to the formation of this association:

1. Exposure to an odor with potential stimulating properties.
2. Peripheral receptors competent to respond or be induced to respond.
3. Central pathway competent to transmit or be induced to transmit.
4. Central arousal system for selective validation of the appropriate central connections.
5. Coordination between subsystems.

With respect to the first two factors, we must assume that odors can have informative value. Odors provide information by, for example, signaling the presence of an object (food, nipple), indicating an animal of a particular status (estrus female), conferring familiarity (nest odors), or even guiding animals to a particular animal or place (mother). However, the presence of a large number of odor compounds that interact with the signaling odor(s) and other nonodorous sensory stimuli make the identification of the specific odors involved with a behavior a difficult endeavor. Furthermore, the mechanism underlying odor discrimination is unknown. On the one hand, analogous with the insect situation, there may be a single active chemical constituent for the signal and a specific receptor that detects the molecule (Boeckh *et al.*, 1965). Alternatively, odor "quality" may be a complex integration of a number of odor constituents that require stimulation of a number of receptors with overlapping spectra (see Lettvin and Gesteland, 1965).

The receptors are presumably competent to respond to these various odor conditions. Normally reared animals are capable of physiologically responding at the bulbar level to odors regardless of their history of odor exposure. Thus neonates exhibit 2DG patterns to amyl acetate (Greer *et al.*, 1982) or EAA (Astic and Saucier, 1982) on their first exposure. However, some exposure to odors appears to be a necessary condition for subsequent odor learning. Thus peripheral receptors may be induced to respond by virtue of changes that occur as a result of this exposure. For example, activation of a subset of receptors by odor(s) may accelerate their maturation. In fact, the consequent functional activity that would ensue could affect synaptic connections in the glomeruli, one mechanism being biochemical stabilization of synapses (Changeux and Danchin, 1976).

With regard to the central pathway, we will discuss the possible mechanism of odor–behavior association using as an example the subsystems mediating the first nipple attachment in suckling. As the 2DG (Greer *et al.*, 1982; Teicher *et al.*, 1981) and horseradish peroxidase (Jastreboff *et al.*, 1984) studies indicate, information about the normal odor cue is transmitted by a pathway from a set of receptors through the MGC to the olfactory cortex, as well as through a parallel pathway through the main olfactory bulb. This has already been illustrated in Figure 8 and is shown in somewhat greater detail in Figure 9. In order for the normal odor cue to be associated with initial attachment behavior, there must be a set of receptors stimulated by the cue, and there must be functional synapses in the pathway to transmit the information. However, this is not sufficient to explain how the information about the odor cue comes to elicit the very first nipple attachment.

This is a difficult problem, and the studies we have reviewed do not provide evidence for answering it. There are probably several factors involved. A factor that we suggest is likely to be critical is arousal. As already discussed in the section on the MGC, a critical timing of arousal with the odor input is necessary for linking that odor input with the appropriate behavior.

Arousal is a complex state and to define and discuss it adequately would go beyond the scope of this review. For our purposes we note that arousal can be defined operationally (see section on MGC). It has been linked to the neural substrate provided by the locus coeruleus and its noradrenergic fibers that project widely throughout the brain (Campbell *et al.*, 1969; Campbell and Mabry, 1973). For example, it is believed that the locus coeruleus appears early in development (E12) (Lauder and Bloom, 1974) and that innervation of the forebrain occurs prenatally (Levitt and Moore, 1979). This projection system thus is present, and presumably functional, by the time the first nipple attachment occurs after birth. We postulate that the linking of the sensory pathway for the odor cue and the motor response of nipple attachment requires the conjunctive action of this projection system, acting at one or more levels in the pathway from sensory reception to motor output. Some possibilities for these sites of action are indicated in the diagram of Figure 9 (see Chapter 3).

Further work will be necessary to document in detail the development of the noradrenergic projection fibers in relation to each of these levels and to assess their functional abilities at early ages. Further work will also be needed to assess other systems that may contribute to the arousal state as it affects the olfactory pathway. These may include other transmitter systems, such as the 5HT fibers from the dorsal raphe (Lauder and Bloom, 1974), other central centrifugal fiber systems from the forebrain, and circulating hormonal substances that could act in the

PATRICIA E.
PEDERSEN,
CHARLES A. GREER,
AND GORDON M.
SHEPHERD

olfactory bulbs or olfactory cortical neurons and synapses. The rich variety of neurotransmitter substances and receptors in the olfactory bulb suggest that the bulb may be a sensitive target for these humoral factors (Halasz and Shepherd, 1983).

These factors that control development of the normal suckling response also provide a framework for understanding the events in which an artificial odor cue is able to elicit the intitial nipple attachment. As discussed in the section on the MGC, arousal combined with pre- and postnatal exposure to the artificial odor is essential for inducing attachment to the artificial odor. It may therefore be pos-

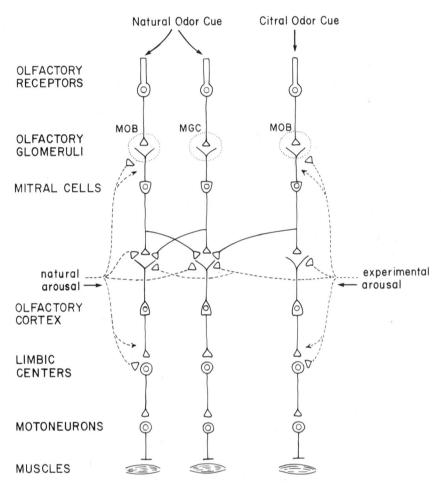

Fig. 9. Theoretical schematic depicting potential circuits via which natural and artificial odor cues may elicit the motor responses subserving nipple attachment. The natural odor cue appears to activate the MGC preferentially as well as other regions of the main olfactory bulb. The natural odor cue is believed to be subsequently processed through the higher-order olfactory centers, including olfactory cortex and limbic centers, and finally to influence motor responses. As illustrated by the broken lines in the schematic, arousal can influence the processing of the odor cue at several levels. The artificial cue appears to be processed independent of the MGC and, consequently, must gain access to the circuits subserving suckling at higher centers such as olfactory cortex. As in the case of the natural odor cue, experimental arousal may influence the processing of the artificial odor cue at several levels.

tulated, as indicated in the diagram of Figure 9, that the action of the arousal system fibers provides the means by which the part of the olfactory pathway carrying the information about the artificial odor cue becomes linked to the pathway controlling the motor of nipple attachment. A possible analogy for this is provided by the experiments of Kasamatsu and Pettigrew (1979), who demonstrated that visual cortical plasticity depends on the presence of an intact noradrenergic system.

The experimental work with citral thus provides forceful evidence for the active role of arousal systems in validating the central connections mediating olfactory learning in the neonate. This raises a number of questions to be tested by future work. One is the use of 2DG to label the activity under artificial odor conditions. Preliminary reports suggest that the plasticity does not occur at the bulbar level (Pedersen *et al.*, 1982), suggesting that it may take place in the olfactory cortex, or more centrally. Furthermore, this schema also needs to be tested in other subsystems as well, to account for the possibility of their integrated function.

Acknowledgments

We are grateful to the National Institutes of Neurological Disorders, Communicative Diseases, and Strokes for support by grants NS-06978 to P.E.P., NS-19430 to C.A.G., NS-16993 to G.M.S. and by Basil O'Connor Starter Research Grant #5-420 from March of Dimes Birth Defects Foundation to C.A.G.

REFERENCES

Alberts, J. R. Producing and interpreting experimental olfactory deficits. *Physiology and Behavior,* 1974, *12,* 657–670.

Alberts, J. R. Olfactory contribution to behavioral development in rodents. In R. L. Doty (Ed.), *Mammalian olfaction, reproductive processes and behavior.* New York: Academic Press, 1976.

Alberts, J. R. Ontogeny of olfaction: Reciprocal roles of sensation and behavior in the development of perception. In R. N. Aslin, J. R. Alberts, and M. R. Petersen (Eds.), *Development of perception: Psychobiological perspectives.* Vol. I. New York: Academic Press, 1981.

Alberts, J. R., and Brunjes, P. C. Ontogeny of thermal and olfactory determinants of huddling in the rat. *Journal of Comparative and Physiological Psychology,* 1978, *92,* 897–906.

Alberts, J. R., and May, B. Ontogeny of olfaction: Development of the rat's sensitivity to urine and amyl acetate. *Physiology and Behavior,* 1980, *24,* 965–970.

Allison, A. C. The morphology of the olfactory system in the vertebrates. *Biological Reviews,* 1953, *28,* 193–244.

Altman, J. Autoradiographic and histological studies of postnatal neurogenesis. IV. Cell proliferation and migration in the anterior forebrain, with special reference to persisting neurogenesis in the olfactory bulb. *Journal of Comparative Neurology,* 1969, *137,* 433–458.

Altman, J., and Das, G. D. Autoradiographic and histological studies of postnatal neurogenesis. I. A longitudinal investigation of the kinetics, migration, and transformation of cells incorporating triated thymidine in neonate rats, with special reference to postnatal neurogenesis in some brain regions. *Journal of Comparative Neurology,* 1966, *126,* 337–390.

Astic, L., and Saucier, D. Ontogenesis of the functional activity of rat olfactory bulb: Autoradiographic study with the 2-deoxyglucose method. *Developmental Brain Research,* 1982, *2,* 1243–1256.

Astic, L., and Saucier, D. Ontogenesis of the functional activity of guinea pig olfactory bulb: Autoradiographic study with the 2-deoxyglucose method. *Developmental Brain Research,* 1983, *16,* 257–263.

198

PATRICIA E.
PEDERSEN,
CHARLES A. GREER,
AND GORDON M.
SHEPHERD

Barber, P. C., and Field, P. M. Autoradiographic demonstration of afferent connections of the accessory olfactory bulb in the mouse. *Brain Research,* 1975, *85,* 201–203.

Barber, P. C., and Raisman, G. Cell division in the vomeronasal organ of the adult mouse. *Brain Research,* 1978a. *141,* 57–66.

Barber, P. C., and Raisman, G. Replacement of receptor neurons after section of the vomeronasal nerves in the adult mouse. *Brain Research,* 1978b, *147,* 297–313.

Bayer, S. A. Quantitative ^{3}H-thymidine radiographic analyses of neurogenesis in the rat amygdala. *Journal of Comparative Neurology,* 1980, *194,* 845–875.

Bayer, S. A. ^{3}H-thymidine-radiographic studies of neurogenesis in the rat olfactory bulb. *Experimental Brain Research,* 1983, *30,* 329–340.

Blass, E. M., Hall, W. G., and Teicher, M. H. The ontogeny of suckling and ingestive behaviors. In J. M. Sprague and A. N. Epstein (Eds.), *Progress in psychobiology and physiological psychology.* Vol. 8. New York: Academic Press, 1979.

Blass, E. M., and Teicher, M. H. Suckling. *Science,* 1980, *210,* 15–22.

Blass, E. M., Teicher, M. H., Cramer, C. P., Bruno, J. P., and Hall, W. G. Olfactory, thermal and tactile controls of suckling in preauditory and previsual rats. *Journal of Comparative and Physiological Psychology,* 1977, *91,* 1248–1260.

Boeckh, J., Kaissling, K. E., and Schneider, D. Insect olfactory receptors. In *Cold Spring Harbor Symposia on Quantitative Biology,.* Vol. XXX. New York: Cold Spring Harbor, 1965.

Bradley, R. M., and Mistretta, C. M. The gustatory sense in foetal sheep during the last third of gestation. *Journal of Physiology,* 1973, *231,* 271–282.

Bradley, R. M., and Mistretta, C. M. Fetal sensory receptors. *Physiological Reviews,* 1975, *55,* 352–375.

Brake, S. C. Suckling infant rats learn a preference for a novel olfactory stimulus paired with milk delivery. *Science,* 1981, *211,* 506–508.

Brunjes, P. C., and Alberts, J. R. Olfactory stimulation induces filial huddling preferences in rat pups. *Journal of Comparative and Physiological Psychology,* 1979, *93,* 548–555.

Bruno, J. P., Teicher, M. H., and Blass, E. M. Sensory determinants of suckling behavior in weanling rats. *Journal of Comparative and Physiological Psychology,* 1980, *94,* 115–127.

Burr, H. S. The effects of the removal of the nasal pits in amblystoma embryos. *Journal of Experimental Zoology,* 1916, *20,* 27–57.

Campbell, B., and Mabry, P. The role of catecholamines in behavioral arousal during ontogenesis. *Psychopharmacologia,* 1973, *31,* 253–264.

Campbell, B., Lytle, L., and Fibiger, H. Ontogeny of adrenergic arousal and cholinergic inhibitory mechanisms in the rat. *Science,* 1969, *166,* 635–637.

Changeux, J. P., and Danchin, A. Selective stabilization of developing synapses as a mechanism for the specification of neuronal networks. *Nature,* 1976, *264,* 705–712.

Cheal, M. L. Social olfaction: A review of the ontogeny of olfactory influences on vertebrate behavior. *Behavioral Biology,* 1975, *15,* 1–25.

Cornwell-Jones, C., and Sobrian, S. K. Development of odor guided behavior in Wistar and Sprague-Dawley rat pups. *Physiology and Behavior,* 1977, *19,* 685–688.

Cuschieri, A., and Bannister, L. H. The development of the olfactory mucosa in the mouse: Light microscopy. *Journal of Anatomy,* 1975a, *119,* 277–286.

Cuschieri, A., and Bannister, L. H. The development of the olfactory mucosa in the mouse: Electron microscopy. *Journal of Anatomy,* 1975b, *119,* 471–498.

de Olmos, J., Hardy, H., and Heimer, L. The afferent connections of the main and accessory olfactory bulb formations in the rat: An experimental HRP study. *Journal of Comparative Neurology,* 1978, *181,* 213–244.

Doucette, J. R., Kiernan, J. A., and Flumerfelt, B. A. The re-innervation of olfactory glomeruli following transection of primary olfactory axons in the central or peripheral nervous system. *Journal of Anatomy,* 1983, *137,* 1–19.

Farbman, A. I. Electron microscope study of the developing taste bud in rat fungiform papillae. *Developmental Biology,* 1965, *11,* 110–135.

Farbman, A. I., and Margolis, F. L. Olfactory marker protein during ontogeny: Immunohistochemical localization. *Developmental Biology,* 1980, *74,* 205–215.

Friedman, B., and Price, J. L. Fiber systems in the olfactory bulb and cortex: A study in adult and developing rats, using the Timm method with light and electron microscope. *Journal of Comparative Neurology,* 1984, *223,* 88–109.

Gesteland, R. C., Yancey, R. A., and Farbman, A. I. Development of olfactory receptor neuron selectivity in the rat fetus. *Neuroscience*, 1982, *7*, 3127–3136.

Getchell, T. Flourescent studies of the primary olfactory pathway in amphibia. *Society of Neuroscience*, 1982, *8*, 10.

Goochee, C., Rasband, W., and Sokoloff, L. Computerized densitometry of [^{14}C]deoxyglucose autoradiographs. *Annals of Neurology*, 1980, *7*, 359–370.

Grafe, M. R., and Leonard C. M. Developmental changes in the topographical distribution of cells contributing to the lateral olfactory tract. *Developmental Brain Research*, 1982, *3*, 387–400.

Graziadei, P. P. C. The olfactory mucosa of vertebrates. In M. Jacobson (Ed.), *Handbook of Sensory Physiology*. Vol. IV. New York: Springer-Verlag, 1977.

Graziadei, P. P. C., and Kaplan, M. S. Regrowth of olfactory sensory axons into transplanted neural tissue. I. Development of connections with the occipital cortex. *Brain Research*, 1980, *201*, 39–44.

Graziadei, P. P. C., Kaplan, M. S., Monti Graziadei, G. A., and Bernstein, J. J. Neurogenesis of sensory neurons in the primate olfactory system after section of the fila olfactoria. *Brain Research*, 1980, *186*, 289–300.

Graziadei, P. P. C., Levine, R. R. and Monti Graziadei, G. A. Regeneration of olfactory axons and synapse formation in the forebrain after bulbectomy in neonatal mice. *Proceedings of the National Academy of Sciences (U.S.A.)*, 1978, *75*, 5230–5234.

Graziadei, P. P. C., Levine, R. R., and Monti Graziadei, G. A. Plasticity of connections of the olfactory sensory neuron: Regeneration into the forebrain following bulbectomy in the neonatal mouse. *Neuroscience*, 1979, *4*, 713–727.

Graziadei, P. P. C., and Monti Graziadei, G. A. Continuous nerve cell renewal in the olfactory system. In M. Jacobsen (Ed.), *Handbook of Sensory Physiology*. Vol. IX. New York: Springer-Verlag, 1978.

Graziadei, P. P. C., and Monti Graziadei, G. A. Neurogenesis and neuron regeneration in the olfactory system of mammals. III. Deafferentation and reinnervation of the olfactory bulb following section of the fila olfactoria in the rat. *Journal of Neurocytology*, 1980, *9*, 145–162.

Graziadei, P. P. C., and Samanen, D. W. Ectopic glomerular structures in the olfactory bulb of neonatal and adult mice. *Brain Research*, 1980, *187*, 467–472.

Greer, C. A., Stewart, W. B., Kauer, J. S., and Shepherd, G. M. Topographical and laminar localization of 2-deoxyglucose uptake in rat olfactory bulb induced by electrical stimulation of olfactory nerves. *Brain Research*, 1981, *217*, 279–293.

Greer, C. A., Stewart, W. B., Teicher, M. H. and Shepherd, G. M. Functional localization of the olfactory bulb and a unique glomerular complex in the neonatal rat. *Journal of Neuroscience*, 1982, *2*, 1744–1759.

Gregory, E. H., and Pfaff, D. W. Development of olfactory guided behavior in infant rats. *Physiology and Behavior*, 1971, *6*, 573–576.

Haberly, L. B., and Price, J. L. The axonal projection patterns of the mitral and tufted cells of the olfactory bulb in the rat. *Brain Research*, 1977, *129*, 152–157.

Haberly, L. B., and Price, J. L. Associational and commissural fiber systems of the olfactory cortex of the rat. *Journal of Comparative Neurology*, 1978, *181*, 781–808.

Halasz, N., and Shepherd, G. M. Neurochemistry of the vertebrate olfactory bulb. *Neuroscience*, 1983, *10*, 579–619.

Hall, W. G., Cramer, C. P., and Blass, E. M. Developmental changes in suckling of rat pups. *Nature* (London), 1975, *258*, 319–320.

Hall, W. G., Cramer, C. P., and Blass, E. M. The ontogeny of suckling in rats: Transitions toward adult ingestion. *Journal of Comparative and Physiological Psychology*, 1977, *91*, 1141–1155.

Hand, P. J. The 2-deoxyglucose method. In L. Heimer and M. J. Robards (Eds.), *Neuroanatomical tract-tracing methods*, New York: Plenum Press, 1981.

Harding, J., Graziadei, P. P. C., Monti Graziadei, G. A., and Margolis, F. L. Denervation in the primary olfactory pathway of mice. IV. Biochemical and morphological evidence for neuronal replacement following nerve section. *Brain Research*, 1977, *132*, 11–28.

Hinds, J. W. Autoradiographic study of histogenesis in the mouse olfactory bulb. I. Time of origin of neurons and neuroglia. *Journal of Comparative Neurology*, 1967, *134*, 287–304.

Hinds, J. W. Autoradiographic study of histogenesis in the mouse olfactory bulb. I. Time of origin of neurons and neuroglia. *Journal of Comparative Neurology*, 1968a, *134*, 287–304.

Hinds, J. W. Autoradiographic study of histogenesis in the mouse olfactory bulb. II. Cell proliferation and migration. *Journal of Comparative Neurology*, 1968b, *132*, 305–322.

200

PATRICIA E.
PEDERSEN,
CHARLES A. GREER,
AND GORDON M.
SHEPHERD

Hinds, J. W., and Hinds, P. L. Synaptogenesis in the mouse olfactory bulb. I. Quantitative studies. *Journal of Comparative Neurology,* 1976a, *169,* 15–40.

Hinds, J. W., and Hinds, P. L. Synapse formation in the mouse olfactory bulb. II. Morphogenesis. *Journal of Comparative Neurology,* 1976b, *169,* 41–62.

Hinds, J. W., and McNelly, N. A. Aging of the rat olfactory bulb: Growth and atrophy of constituent layers and changes in size and number of mitral cells. *Journal of Comparative Neurology,* 1977, *171,* 345–368.

Hofer, M. A. Olfactory denervation: Its biological and behavioral effects in infant rats. *Journal of Comparative and Physiological Psychology,* 1976, *90,* 829–838.

Hofer, M. A., Shair, H., and Singh, P. Evidence that maternal ventral skin substances promote suckling in infant rats. *Physiology and Behavior,* 1976, *17,* 131–136.

Jastreboff, P. J., Pedersen, P. E., Greer, C. A., Stewart, W. B., Kauer, J. S., Benson, T. B., and Shepherd, G. M. Specific olfactory receptor populations projecting to identified glomeruli in the rat olfactory bulb. *Proceedings of the National Academy of Sciences (U.S.A.,)* 1984, *81,* 5250–5254.

Johanson, I. B., and Hall, W. G. The ontogeny of feeding in rats: V. Influence of texture, home odor, and sibling presence on ingestive behavior. *Journal of Comparative and Physiological Psychology,* 1981, *95,* 837–847.

Johanson, I. B., and Hall, W. G. Appetitive conditioning in neonatal rats. Conditioned orientation to a novel odor. *Developmental Psychobiology,* 1982, *15,* 379–397.

Iwahara, S., Oishi, H., Sano, K., Yang, K., and Takahashi, T. Electrical activity of the olfactory bulb in the postnatal rat. *Japanese Journal of Physiology,* 1973, *23,* 361–370.

Kaplan, M. S., and Hinds, J. W. Neurogenesis in the adult rat: Electron microscopic analysis of light radioautoradiographs. *Science,* 1977, *197,* 1092–1094.

Kasamatsu, T., and Pettigrew, J. D.. Preservation of binocularity after monocular deprivation in the striate cortex of kittens treated with 6-hydroxydopamine. *Journal of Comparative Neurology,* 1979, *185,* 139–162.

Kennedy, C. M., DesRosiers, H., Jehle, J. W., Reivich, M., Sharp, F., and Sokoloff, L. Mapping of functional neural pathways by autoradiographic survey of local metabolic rat with [^{14}C] deoxyglucose. *Science,* 1975, *187,* 850–853.

Kevetter, G. A., and Winans, S. S. Connections of the corticomedial amygdala in the golden hamster. I. Efferents of the "vomeronasal amygdala." *Journal of Comparative Neurology,* 1981a, *197,* 81–98.

Kevetter, G. A., and Winans, S. S. Connections of the corticomedial amygdala in the golden hamster. II. Efferents of the "olfactory amygdala". *Journal of Comparative Neurology,* 1981b, *197,* 99–111.

Kratzing, J. The fine structure of the sensory epithelium of the vomeronasal organ in suckling rats. *Australian Journal of Biological Sciences,* 1976, *24,* 787–796.

Lauder, J. M., and Bloom, F. E. Ontogeny of monamine neurons in the locus coeruleus, raphe nuclei and substantia nigra of the rat. I. Cell differentiation. *Journal of Comparative Neurology,* 1974, *155,* 469–482.

Leon, M., and Moltz, H. Maternal pheromone: Discrimination by pre-weaning albino rats. *Physiology and Behavior,* 1971, *7,* 265–267.

Leon, M., and Moltz, H. The development of the pheromonal bond in the albino rat. *Physiology and Behavior,* 1972, *8,* 683–686.

Leonard, C. M. Developmental changes in olfactory bulb projections revealed by degeneration agyrophilia. *Journal of Comparative Neurology,* 1975, *162,* 467–486.

Leonard, C. M. Neurological mechanisms for early olfactory recognition. In R. N. Aslin, J. R. Alberts, and M. R. Peterson (Eds.) *Development of perception: Psychobiological perspectives.* Vol. I. New York: Academic Press, 1981.

Lettvin, J. Y., and Gesteland, R. C. Speculations on smell. In *Cold Spring Harbor Symposia on Quantitative Biology.* Vol. XXX. New York: Cold Spring Harbor, 1965.

Lev, R., and Orlic, D. Protein absorption by the intestine of the fetal rat *in utero. Science,* 1972, *177,* 522–524.

Levitt, P., and Moore, R. Y. Development of the noradrenergic innervation of neocortex. *Brain Research,* 1979, *162,* 243–259.

Loizou, L. A. The postnatal ontogeny of monoamine-containing neurons in the central nervous system of the albino rat. *Brain Research,* 1972, *40,* 395–418.

Luskin, M. B., and Price, J. L. The topographic organization of associational fibers of the olfactory system in the rat, including centrifugal fibers to the olfactory bulb. *Journal of Comparative Neurology,* 1983a, *216,* 264–291.

Luskin, M. B., and Price, J. L. The laminar distribution of intracortical fibers originating in the olfactory cortex of the rat. *Journal of Comparative Neurology*, 1983b, *216*, 292–302.

Macrides, F., Davis, B. J., Young, W. M., Nadi, N. S., and Margolis, F. L. Cholinergic and catecholaminergic afferents to the olfactory bulb in the hamster: A neuroanatomical, biochemical and histochemical investigation. *Journal of Comparative Neurology*, 1981, *203*, 495–514.

Macrides, F., and Schneider, S. P. Laminar organization of mitral and tufted cells in the main olfactory bulb of the adult hamster. *Journal of Comparative Neurology*, 1982, *208*, 419–430.

Mair, R. G., and Gesteland, R. C. Response properties of mitral cells in the olfactory bulb of the neonatal rat. *Neuroscience*, 1982, *7*, 3117–3125.

Mair, R. G., Gellman, R. L., and Gesteland, R. C. Postnatal proliferation and maturation of olfactory bulb neurons in the rat. *Neuroscience*, 1982, *7*, 3105–3116.

Margolis, F. L. A brain protein unique to the olfactory bulb. *Proceedings of the National Academy of Sciences (U.S.A.)*, 1972, *69*, 1221–1224.

Math, F., and Devrainville, J. L. Electrophysiological study on the postnatal development of mitral cell activity in the rat olfactory bulb. *Brain Research*, 1980, *190*, 243–247.

Meredith, M. Vomeronasal lesions before sexual experience impair male behavior in hamsters. Paper presented at Association for Chemoreception Sciences, Sarasota, Florida, 1983.

Meredith, M. Sensory physiology of pheromone communication. In J. G. Vandenbergh (Ed.), *Pheromones and reproduction in mammals*. New York: Academic Press, 1983.

Meredith, M., and O'Connell, R. J. Efferent control of stimulus access to the hamster vomeronasal organ. *Journal of Physiology (London)*, 1979, *286*, 301–316.

Meredith, M., Marques, D. M., O'Connell, R. J., and Stern, F. Vomeronasal pump: Significance for male hamster sexual behavior. *Science*, 1980, *207*, 1224–1226.

Mistretta, C. M. Topographical and histological study of the developing rat tongue, palate, and taste buds, In J. F. Bosma (Ed.), *Third symposium on oral sensation and perception: The mouth of the infant*. 1972, Springfield, Ill.: Charles C Thomas.

Mistretta, C. M., and Bradley, R. M. Neural basis of developing taste sensation: Response changes in fetal, postnatal, and adult sheep. *Journal of Comparative Neurology*, 1983, *215*, 199–210.

Monti Graziadei, G. A., and Graziadei, P. P. C. Neurogenesis and neuron regeneration in the olfactory system of mammals. II. Degeneration and reconstitution of the olfactory sensory neurons after axotomy. *Journal of Neurocytology*, 1979, *8*, 197–213.

Monti Graziadei, G. A., Stanley, R. S., and Graziadei, P. P. C. The olfactory marker protein in the olfactory system of the mouse during development. *Neuroscience*, 1980, *5*, 1239–1252.

Nyakas, C., and Endroczi, E. Olfaction guided approaching behavior of infantile rats to the mother in maze box. *Acta Physiologica Academiae Scientarum Hungaricae*, 1970, *38*, 59–65.

Pedersen, P. E., and Blass, E. M. Olfactory control over suckling in albino rats. In R. N. Aslin, J. R. Alberts, and M. R. Peterson (Eds.), *Development of perception: Psychobiological perspectives*. Vol. I. New York: Academic Press, 1981.

Pedersen, P. E., and Blass, E. M. Prenatal and postnatal determinants of the first suckling episode in the albino rat. *Developmental Psychobiology*, 1982, *15*, 349–356.

Pedersen, P. E., Williams, C. L., and Blass, E. M. Activation and odor conditioning of suckling behavior in 3-day-old albino rats. *Journal of Experimental Psychology: Animal Behavior Processes*, 1982, *8*, 329, 341.

Pedersen, P. E., Greer, C. A., Stewart, W. B., and Shepherd, G. M. A 2DG study of behavioral plasticity in odor dependent suckling. Paper presented at Association for Chemoreception Sciences, Sarasota, Florida, April, 1982.

Pedersen, P. E., Stewart, W. B., Greer, C. A., and Shepherd, G. M. Evidence for olfactory function *in utero*. *Science*, 1983, *221*, 478–480.

Pinching, A. J., and Powell, T. P. S. The neuropil of the glomeruli of the olfactory bulb. *Journal of Cell Science*, 1971, *9*, 347–377.

Pinching, A. J., and Powell, T. P. S. The terminations of the centrifugal fibers in the glomerular layer of the olfactory bulb. *Journal of Cell Science*, 1972, *16*, 621–625.

Powell, T. P. S., Cowan, W. M., and Raisman, G. The central olfactory connections, *Journal of Anatomy*, 1965, *99*, 791–813.

Price, J. L., and Powell, T. P. S. An electron microscopic study of the termination of the afferent fibers to the olfactory bulb from the cerebral hemispheres. *Journal of Cell Science*, 1970a, *7*, 157–187.

Price, J. L., and Powell, T. P. S. An experimental study of the origin and the course of the centrifugal fibers to the olfactory bulb in the rat. *Journal of Anatomy*, 1970b, *107*, 215–237.

202

PATRICIA E.
PEDERSEN,
CHARLES A. GREER,
AND GORDON M.
SHEPHERD

Price, J. L., and Powell, T. P. S. Certain observations on the olfactory pathway. *Journal of Anatomy,* 1971, *110,* 105–126.

Price, J. L. An autoradiographic study of complementary laminar patterns of termination of afferent fibers to the olfactory cortex. *Journal of Comparative Neurology,* 1973, *150,* 87–108.

Price, J. L., and Slotnick, B. M. Dual olfactory representation in the rat thalamus: An anatomical and electrophysiological study. *Journal of Comparative Neurology,* 1983, *215,* 63–77.

Rall, W., Shepherd, G. M., Reese, T. S., and Brightman, M. W. Dendodendritic synaptic pathway for inhibition in the olfactory bulb. *Experimental Neurology,* 1966, *14,* 44–56.

Reppert, S. M., and Schwartz, W. J. Maternal coordination of the fetal biological clock *in utero. Science,* 1983, *220,* 969–971.

Rodriguez Enchandia, E. L., Foscolo, M., and Broitman, S. T. Preferential nesting in lemon-scented environment in rats reared on lemon scented bedding from birth to weaning. *Physiology and Behavior,* 1982, *29,* 47–49.

Roselli-Austin, L., and Altman, J. The postnatal development of the main olfactory bulb of the rat. *Journal of Developmental Physiology,* 1979, *1,* 295–313.

Rosenblatt, J. S. Olfaction mediates developmental transition in the altricial newborn of selected species of mammals. *Developmental Psychobiology,* 1983, *16,* 347–375.

Rudy, J. W., and Cheatle, M. D. Odor-aversion learning in neonatal rats. *Science,* 1977, *198,* 845–846.

Salas, M., Guzman-Flores, C., and Schapiro, S. An ontogenetic study of olfactory bulb electrical activity in the rat. *Physiology and Behavior,* 1969, *4,* 699–703.

Scalia, F., and Winans, S. S. The differential projections of the olfactory bulb and accessory olfactory bulb in mammals. *Journal of Comparative Neurology,* 1975, *161,* 31–56.

Schapiro, S., and Salas, M. Behavioral response of infant rats to maternal odor. *Physiology and Behavior,* 1970, *5,* 815–817.

Scheibel, M. E., and Scheibel, A. B. Dendrite bundles, central programs and the olfactory bulb. *Brain Research,* 1975, *95,* 407–421.

Schwarz, W. J., Smith, C. B., Davidsen, L., Savaki, H., Sokoloff, L., Mata, M., Fink, D. J., and Gainer, H. Metabolic mapping of functional activity in the hypothalamo-neuohypophysial system of the rat. *Science,* 1979, *205,* 723–725.

Schwob, J. E., and Price, J. L. The cortical projections of the olfactory bulb: Development in fetal and neonatal rats correlated with quantitative variations in adult rats. *Brain Research,* 1978, *151,* 369–374.

Scott, J. W., McBride, R. L., and Schneider, S. P. The organization of projections from the olfactory bulb to the piriform cortex and olfactory tubercle in the rat. *Journal of Comparative Neurology,* 1980, *194,* 519–534.

Shapr, F. R., Kauer, J. S., and Shepherd, G. M. Local sites of activity-related glucose metabolism in rat olfactory bulb during olfactory stimulation, *Brain Research,* 1975, *98,* 596–600.

Sharp, F. R., Kauer, J. S., and Shepherd, G. M. Laminar analysis of 2-deoxyglucose uptake in olfactory bulb and olfactory cortex of rabbit and rat. *Journal of Neurophysiology,* 1977, *40,* 800–813.

Shepherd, G. M. Synaptic organization of the mammalian olfactory bulb. *Physiological Reveiws,* 1972, *52,* 864–917.

Shepherd, G. M. *The synaptic organizaton of the brain.* New York: Oxford University Press, 1979.

Singh, S. C. The development of olfactory and hippocampal pathways in the brain of the rat. *Anatomical Embryology,* 1977, *151,* 183–199.

Singh, D. N. P., and Nathaniel, E. J. H. Postnatal development of mitral cell perikaryon in the olfactory bulb of the rat. I. A light and ultrastructural study. *Anatomical Record,* 1977, *189,* 413–432.

Skeen, L. C. Odor induced patterns of deoxyglucose consumption in the olfactory bulb of the tree shrew: *Tupaia glis. Brain Research,* 1977, *24,* 147–153.

Slotnick, B. M., and Kaneko, N. Role of mediodorsal thalamic nucleus in olfactory discrimination learning in rats. *Science,* 1981, *214,* 91–92.

Smith, C. G. The change in volume of the olfactory and accessory olfactory bulbs of the albino rat during postnatal life. *Journal of Comparative Neurology,* 1935, *61,* 477–508.

Smotherman, W. P. Odor aversion learning by the rat fetus. *Physiology and Behavior,* 1982a, *29,* 769–771.

Smotherman, W. P. *In utero* chemosensory experience alters taste preferences and corticosterone responsiveness. *Behavioral and Neural Biology,* 1982b, *36,* 61–68.

Stewart, W. B., Kauer, J. S., and Shepherd, G. M. Functional localization of rat olfactory bulb analyzed by the 2-deoxyglucose method. *Journal of Comparative Neurology,* 1979, *185,* 715–734.

Stickrod, G., Kimble, D. P., and Smotherman, W. P. *In utero* taste odor aversion conditioning in the rat. *Physiology and Behavior,* 1982, *28,* 5–7.

Stout, R. P., and Graziadei, P. P. C. Influence of the olfactory placode on the development of the brain in *Xenopus laevis. Neuroscience,* 1980, *5,* 2175–2186.

Teicher, M. H., and Blass, E. M. Suckling in newborn rats: Eliminated by nipple lavage, reinstated by pup saliva. *Science,* 1976, *193,* 422–425.

Teicher, M. H., and Blass, E. M. The role of olfaction and amniotic fluid in the first suckling response of newborn albino rats. *Science,* 1977, *198,* 635–636.

Teicher, M. H., Stewart, W. B., Kauer, J. S., and Shepherd, G. M. Suckling pheromone stimulation of a modified glomerular region in the developing rat olfactory bulb revealed by the 2-deoxyglucose method. *Brain Research,* 1980, *194,* 530–535.

Vaccarezza, O. L., Sepich, L. N., and Tramezzani, J. H. The vomeronasal organ of the rat. *Journal of Anatomy,* 1981, *132,* 167–185.

Westrum, L. E. Electron microscopy of synaptic structures in olfactory cortex of early postnatal rats. *Journal of Neurocytology,* 1975, *4,* 713–732.

White, L. E. Olfactory bulb projections of the rat. *Anatomical Record,* 1965, *152,* 465–480.

White, L. E. Synaptic organization of the mammalian olfactory glomerulus: New findings including an intraspecific variation. *Brain Research,* 1973, *60,* 299–313.

Wright, J. W., and Harding, J. W. Recovery of olfactory function after bilateral bulbectomy. *Science,* 1982, *216,* 322–324.

Wysocki, C. J., Nyby, J., Whitney, G., Beauchamp, G. K., and Katz, Y. The vomeronasal organ: Primary role in mouse chemosensory gender recognition. *Physiology and Behavior,* 1982, *29,* 315–327.

Wysocki, C. J., Wellington, J. L., and Beauchamp, G. K. Access of urinary non-volatiles to the mammalian vomeronasal organ. *Science,* 1980, *207,* 781–783.

Development of the Sense of Taste

CHARLOTTE M. MISTRETTA AND ROBERT M. BRADLEY

I. INTRODUCTION

Since 1980 several studies have appeared on development of neurophysiological taste responses. These investigations preview an area of expanding research that provides a basis for understanding development of behavioral taste responses and essential mechanisms of gustatory sensation. The goals of this chapter are to describe complementary results from recent neurophysiological, anatomical, and behavioral studies and to propose molecular and cellular substrates for changing taste function during development.

The early literature on gustatory development primarily contained morphological observations. From the nineteenth century an intrauterine appearance of taste buds was identified in humans (see reviews by Bradley, 1972; Bradley and Mistretta, 1975). However, it was not until 1967 that the presence of presumptive taste buds in 7- to 9-week fetuses was systematically documented with study of serial, histological sections and that the idea of an early appearance and subsequent disappearance of fetal taste buds was dispelled (Bradley and Stern, 1967).

Early behavioral studies of human newborn facial expressions indicated an ability to discriminate among taste stimuli (Peiper, 1925). Therefore human taste

CHARLOTTE M. MISTRETTA   Center for Human Growth and Development; Center for Nursing Research, School of Nursing; and Department of Oral Biology, School of Dentistry, University of Michigan, Ann Arbor, Michigan 48109.   ROBERT M. BRADLEY   Department of Oral Biology, School of Dentistry; and Department of Physiology, Medical School, University of Michigan, Ann Arbor, Michigan 48109.

buds apparently were functional at birth. In addition, it was suggested that the human fetus responded to flavored amniotic fluid by swallowing large volumes (De Snoo, 1937). Several decades elapsed, though, before human newborn taste function was evaluated quantitatively or extensive animal studies were made (see reviews by Bradley and Mistretta, 1975; Cowart, 1981; Weiffenbach, 1978; Weiffenbach *et al.*, 1980).

Much of the current work on gustatory development involves neurophysiological experiments with rats and sheep. Taste buds appear on the twentieth day of the 21-day gestation in laboratory rat, and therefore development of the taste system is essentially a postnatal event. Taste buds appear in sheep at about 7 weeks of the 5-month gestation; the time course of early prenatal formation of taste buds in this species is similar to that in humans. Therefore in sheep and humans there is major development of the taste system both pre- and postnatally. As a result of varying time courses of receptor formation, the developing taste buds are not exposed to the same set of external stimuli in each species (Mistretta, 1981; Mistretta and Bradley, 1978a). Amniotic fluid, breast milk, and fetal and postnatal oral–tracheal secretions form a background environment for the developing gustatory sense, but for different lengths of time in different species.

Finally, it is important to recall some unique, general aspects of the adult taste system before discussing development. Taste buds are located in various structures in the oropharynx and, depending on location, are innervated by three different cranial nerves (facial, glossopharyngeal, and vagus). Fungiform papillae on the anterior tongue contain from 1 to about 20 taste buds each (depending on species), innervated by the chorda tympani branch of the facial nerve. Several hundred taste buds are located in foliate papillae on the posterior, lateral tongue border; these are innervated by the chorda tympani and by the lingual branch of the glossopharyngeal nerve. Circumvallate papillae, found on the posterior tongue dorsum, contain hundreds of taste buds that are innervated by the lingual branch of the glossopharyngeal nerve. The soft palate also contains over 100 taste buds, innervated by the greater superficial petrosal branch of the facial nerve. The hundreds of taste buds on the epiglottis are innervated by the superior laryngeal nerve branch of the vagus nerve. Most studies of taste development have concerned buds in fungiform papillae on the anterior tongue, although there are some reports on taste buds in the circumvallate papillae. Because taste buds are found in such different locations and are innervated by different cranial nerves, generalizations to all taste buds must be drawn carefully.

This chapter is organized with separate sections that address the current experimental work on gustatory development in rats and sheep. The section on sheep includes some results from human studies, to emphasize mammals with a lengthy gestation in which development of taste has major *pre-* and *post*natal components. A subsequent section contains a discussion of possible molecular and cellular changes underlying development of taste responses. Finally, the concluding section presents a general summary of the chapter.

A. Changes in Electrophysiological Responses from Peripheral Taste Nerves

Fungiform and circumvallate papillae first appear in the rat tongue at 15 days of gestation, and the topographical distribution of papillae in the fetus is similar to that in postnatal rat (Figure 1). Taste buds are observed in these papillae on day 20 of the 21-day gestation (Mistretta, 1972). However, anatomical development and acquisition of total number of taste buds occur postnatally (Farbman, 1965; Mistretta, 1972). It is clear now that the taste buds are functional from at least 6

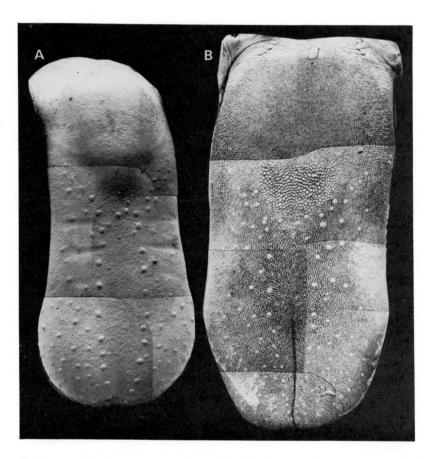

Fig. 1. Comparison of scanning electron micrographs of fetal and newborn rat tongues. (A) Tongue from a fetus at 16 days of gestation (100X). (B) Tongue from a newborn at 8 days postnatal (30X). Fungiform and circumvallate papillae first appear on the rat tongue at 15 days of gestation. The distribution of gustatory papillae in the fetus (A) is similar to that in postnatal rat (B). However, the addition of filiform (nongustatory) papillae alters tongue topography. Taste buds do not appear in fungiform or circumvallate papillae until 20 days of gestation. (From Mistretta, 1972.)

CHARLOTTE M.
MISTRETTA AND
ROBERT M. BRADLEY

days after birth and that morphological maturation is accompanied by extensive neurophysiological development.

In 1980 and 1981 three studies appeared, conducted with different experimental designs in separate laboratories. All three of these investigations led to the same general conclusion: the gustatory system in postnatal rat does not respond initially with adult neurophysiological characteristics. Integrated responses from the whole chorda tympani nerve demonstrated that response magnitudes to several salts, acids, and sucrose altered substantially from about 6 days postnatal to 100 days (Ferrell *et al.*, 1981; Hill and Almli, 1980; Yamada, 1980). Changes were large enough to effect reversals in the order of "best" or most effective stimulus (i.e., that stimulus eliciting the largest response magnitude). For example, Ferrell *et al.* (1981) listed salt stimuli in order of effectiveness.

In 12-day rats: $NH_4Cl > LiCl > NaCl > KCl$.
In adults: $LiCl > NaCl > NH_4Cl > KCl$.

Other stimuli also changed order of effectiveness.

In 12-day rats: HCl > citric acid > NaCl > sucrose.
In adults: NaCl > HCl > sucrose > citric acid.

The most impressive developmental changes were in the marked, increased effectiveness of NaCl and LiCl (Figure 2) and the reversal in relative effectiveness of

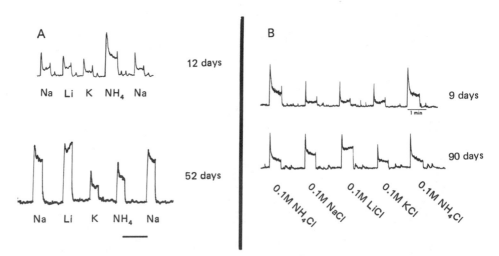

Fig. 2. Integrated records of neural responses from the chorda tympani nerve in young postnatal rats (9 and 12 days) and adults (52 and 90 days of age) from two independent studies: (A) responses in Wistar rats from Ferrell *et al.*, 1981; (B) responses in Sprague Dawley rats from Hill *et al.*, 1982. Chemical stimuli applied to the anterior tongue were 0.1 *M* NaCl, LiCl, KCl, and NH_4Cl, in the order indicated on the figures. The height of pen deflection from the baseline during stimulation periods is a measure of response magnitude. The time marks represent 60 sec. Both sets of data demonstrate that *relative* to NH_4Cl, NaCl and LiCl elicit small responses in young rats and large responses in adults.

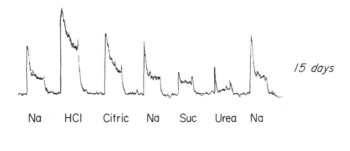

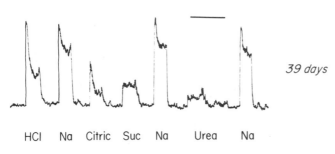

Fig. 3. Integrated records of neural responses from the chorda tympani nerve in rats aged 15 and 39 days after birth. Chemical stimuli applied to the anterior tongue were 0.1 M NaCl, 0.01 N HCl, 0.0025 M citric acid, 0.5 M sucrose, and 1.0 M urea. The concentrations are different among chemicals because they were selected at magnitudes known to elicit large responses in adults, rather than on the basis of equal molarity. The time mark indicates 60 sec. Sucrose and citric acid reverse in order of stimulating effectiveness during development; that is, citric acid elicits larger responses in young rats, smaller responses in older rats, compared to sucrose. (From Ferrell *et al.*, 1981.)

sucrose and citric acid (Figure 3). Sucrose became relatively more effective, citric acid less effective.

The remarkable change in responsiveness to NaCl was reported also by Hill and Almli (1980) and by Yamada (1980) and was substantiated in experiments using a wide range of salt concentrations. Slopes of response–concentration functions for NaCl and LiCl almost doubled between 21 days and 84 days, whereas those for NH_4Cl did not change significantly (Yamada, 1980).

It should be noted that Ferrell *et al.* (1981) reported a decrease in sucrose responses during development, *relative to NaCl,* whereas Yamada (1980) reported a small increase in sucrose responses. The results are not contradictory. Ferrell *et al.* expressed responses relative to NaCl. Since the NaCl responses in fact increase tremendously, no change or a slight increase in sucrose responses would be masked by the large NaCl increase. Yamada, on the other hand, expressed responses to each chemical normalized to an internal measure for that chemical; thus the sucrose increase was not masked. The increasing effectivensss of sucrose relative to other chemicals is observed, also, in the integrated records in Figure 3 and in the order of effective stimuli listed above (Ferrell *et al.*, 1981).

Recordings from the whole chorda tympani nerve also provided an indication of the time course of maturation of the taste response. Ratios of chemical responses *relative to* the NaCl response were calculated and plotted as a function of age (Ferrell *et al.*, 1981). As illustrated in Figure 4, for all chemical responses the period between 10 and 40 days is a time of major transitions in neurophysiological responses.

To clarify and extend the integrated response results, Hill and co-workers (1982) studied responses from single chorda tympani nerve fibers in three age groups: 14–20 days, 24–35 days, and adult. Single fibers at all ages responded to all four monochloride salt stimuli (Figure 5). However, it became apparent that average response frequencies to NaCl increased remarkably during development in the rat; at 24–35 days frequencies were nearly double those at 14–20 days (Figure 6). Conversely, NH_4Cl average response frequencies did not change significantly (Figure 6), and even when a wide concentration range of NH_4Cl was used, developmental differences were not observed. KCl response frequencies altered at

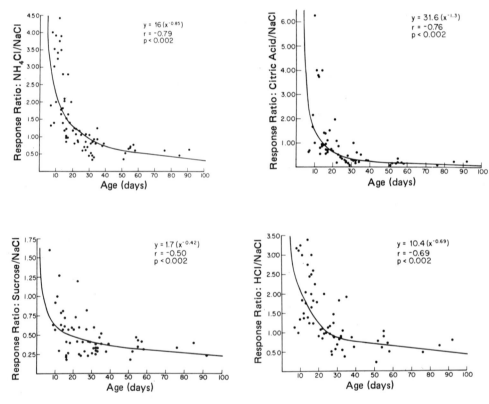

Fig. 4. Ratios of responses to NH_4Cl, sucrose, citric acid, and HCl, all expressed relative to the response to NaCl, as a function of postnatal age in rat. *Relative* to the NaCl response, responses to each chemical decrease during development. The power functions describing developmental changes for each chemical (solid lines on figures) demonstrate that 10–40 days postnatal is a period of major transitions in taste responses. (From Ferrell *et al.*, 1981.)

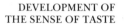

14-20 Days

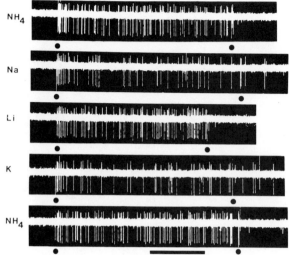

ADULT

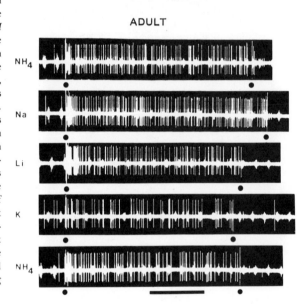

Fig. 5. Responses from a single chorda tympani fiber in a rat aged 14–20 days and an adult. Chemical stimuli, applied to the tongue at the time marked by a dot, were 0.1 M NH$_4$Cl, NaCl, LiCl, and KCl, in the order indicated. Subsequent dots in each record denote water rinses. The time bar represents 5 sec. Generally, chorda tympani fibers at all ages responded to all four salt stimuli. (From Hill *et al.,* 1982). Readers might wonder how the nervous system distinguishes among salt stimuli when each elicits a different response in different single afferent fibers. This question of stimulus coding is the source of lively debate in the field of taste. Some investigators propose that there are specific fiber types for various stimuli and others suggest that it is necessary to examine the response to any one salt in the pattern across all fibers. For a discussion of taste coding see Bartoshuk (1978).

0.5 M concentrations only,[1] not at 0.1 M (Figure 6). Furthermore, the number of single fibers that responded "best" (with highest frequency) to NaCl, rather than NH$_4$Cl or KCl, increased during development (Hill *et al.,* 1982).

Therefore whole nerve responses to NaCl were of low magnitude in the youngest postnatal rats because average, single-fiber response frequencies to NaCl were

[1]For investigators who routinely express concentration in units of percent rather than in molarity, conversions are listed for the four monochloride salts frequently referred to in this chapter. NaCl: 0.1 M = 0.6%; 0.5 M = 2.9%. LiCl: 0.1 M = 0.4%; 0.5 M = 2.1%. NH$_4$Cl: 0.1 M = 0.5%; 0.5 M = 2.7%. KCl: 0.1 M = 0.7%; 0.5 M = 2.1%.

CHARLOTTE M.
MISTRETTA AND
ROBERT M. BRADLEY

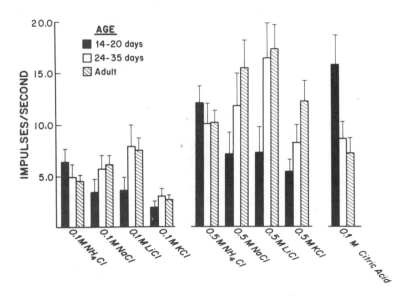

Fig. 6. Mean response frequencies of chorda tympani fibers in three age groups of rats to 0.1 and 0.5 M salts and 0.1 M citric acid. Standard errors are noted above each bar. Average frequencies in response to lingual stimulation with NaCl and LiCl increased significantly during development, whereas response frequencies to NH$_4$Cl did not change. Although the response frequencies during 0.1 M KCl stimulation did not alter, response frequencies to 0.5 M KCl increased significantly as a function of age. Average response frequencies to citric acid decreased developmentally. (From Hill *et al.*, 1982.)

low, and relatively few fibers responded "best" to NaCl. An explanation emerged also for decreasing whole nerve responses to citric acid in comparison to other stimuli, since average single-fiber response frequencies to this acid decreased developmentally (Figure 6).

Additionally, important data were obtained on taste response latencies. Latencies for salt responses from fibers in rats aged 14–20 and 24–35 days were longer than those in adult rats (Hill *et al.*, 1982). Latencies for the two younger groups were not significantly different from each other, however. Decreasing response latencies are characteristic of developing sensory systems, and it is not surprising, therefore, that they were observed in the gustatory sense.

It is impressive that separate studies of neurophysiological taste responses in postnatal rats have led investigators to such similar conclusions. The various experiments included different rat strains, sexes, and age groups; different stimulus concentrations; both whole nerve and single-fiber recordings; analysis of phasic and tonic portions of the response; and various ways of analyzing integrated response data (Ferrell *et al.*, 1981; Hill and Almli, 1980; Hill *et al.*, 1982; Yamada, 1980). Yet all of these experiments demonstrated that responses to NaCl (and LiCl) gradually increased postnatally while those to NH$_4$Cl remained the same, and those to sucrose increased and citric acid decreased. These are clearly robust changes in neurophysiological responses that could have extensive consequences for development of behavioral taste responses.

Development of rat taste buds is essentially an early postnatal phenomenon, so perhaps it is not surprising that major neurophysiological changes take place in the rat gustatory system after birth. It is clear, however, that neurophysiological changes do not simply reflect acquisition of the adult complement of taste bud *numbers*. In Table 1 are summarized data in progress from our laboratory, demonstrating that by 1 day postnatal, 95% of all fungiform papillae contained a taste bud and that by 15 days postnatal, 99% contained a taste bud (Gottfried, *et al.,* 1984). These histological observations substantiated an earlier report that by 14 days postnatal at least 80% of papillae contained a taste bud (Mistretta, 1972). In the earlier study, presence of taste buds was established by counting taste pores in fungiform papillae using scanning electron microscopy. Farbman (1965) has observed also that rat taste buds characteristically contained a pore by 14 days postnatal.

Although numbers of taste buds were established by about 2 weeks postnatal, morphological differences were observed up to 30 days (Gottfried *et al.,* 1984). Taste buds were staged in four groups according to several anatomical characteristics, including orientation of cells, presence of more than one cell type, distinction of taste bud cells from surrounding epithelial cells, and presence of a taste pore. Between 15 and 30 days postnatal, the proportion of immature taste buds continued to decrease as mature buds increased (Table 1). Acquisition of morphologically "mature" taste buds and neurophysiologically "mature" taste responses follow very similar time courses.

Since functional changes in the taste system do relate not only to acquisition

TABLE 1. Acquisition of Taste Buds (TB) in
Fungiform Papillae and Percentage of Taste Buds at
Each of Two Morphological Stages 1 and 4 in
Postnatal Rat[a]

Age (days)	No. fungiform papillae	Fungiform papillae with a TB (%)	Stage 1 TB (%)	Stage 4 TB (%)
1	185	95	75	0
5	212	96	42	8
15	188	99	5	64
30	224	100	1	84
100	206	100	2	86

[a]Taste buds were categorized in one of four stages based on morphological criteria, and data on stages 1 and 4 are presented here. Stage 1 taste buds were most immature and were characterized as a collection of cells without orientation, covered by a number of epithelial cell layers. Stage 4 taste buds were most mature and were characterized as a collection of lightly and darkly staining cells, arranged in an ovoid shape, with a taste pore at the apex. Data are mean values from four to five animals in each age group.

of total taste bud numbers in the developing organism but also to altering taste bud morphology, several categories of receptor change are likely correlates of neurophysiological development. Changes in taste bud cell membranes, in taste bud cell turnover, in synapses between taste cells and afferent fibers, and in the afferent fibers themselves could occur during this period, and all of these phenomena might relate to various aspects of the developing taste response (Hill *et al.*, 1982).

It has been suggested that various taste bud cell types develop at different times in the rat (Farbman, 1965). Whether the types are independent cell lines or stages of one line is not clear (Delay *et al.*, 1984). Different morphological types of synapses have been described in mouse taste buds (Kinnamon *et al.*, 1983). The synaptic types could function differently and might have a different developmental sequence. Future studies on ultrastructural development of the taste bud are crucial for resolving such issues and understanding neurophysiological changes. A more complete discussion of possible cellular correlates of neurophysiological response changes is included in Section IV of this chapter.

C. Changes in Electrophysiological Responses from Central Nervous System Taste Neurons and Behavioral Implications

For the developing mammal to respond to and process information from taste stimuli, not only must the peripheral gustatory system transmit responses reliably, but the neural responses must also cross at least the first central nervous system synapse. As discussed in a previous section, responses from peripheral nerves alter substantially during postnatal development. How are the peripheral response alterations reflected in the central nervous system?

Single-neuron responses have been recorded from second-order cells in the rostral part of the nucleus of the solitary tract (NST), which receives direct projections from the chorda tympani nerve (Hill *et al.*, 1983). Five age groups of rats were studied: 5–7 days postnatal, 14–20 days, 25–35 days, 50–60 days, and adult. Responses were obtained as early as 5–7 days postnatal; however, compared to older rats, neurons in such young animals usually responded to NH_4Cl, KCl, and acids only. Even high concentrations of NaCl were generally ineffective in eliciting responses. Moreover, second-order neurons in these young rats did not respond to sucrose or quinine hydrochloride. By 14–20 days postnatally, responses to NaCl were consistently obtained and sucrose and quinine hydrochloride responses were obtained from a large proportion of neurons. Therefore as development progressed, NST neurons tended to respond to more chemicals and to lower concentrations of chemicals.

Throughout development, average response frequencies for NH_4Cl, citric and hydrochloric acids, and quinine hydrochloride remained constant (Figure 7). However, response frequencies for NaCl, LiCl, KCl, sucrose, and sodium saccharin increased after 35 days postnatal (Figure 7). The increases in response frequencies from NST neurons occurred comparatively later in development than increases in chorda tympani nerve responses (Figure 8). Whereas NST neural responses

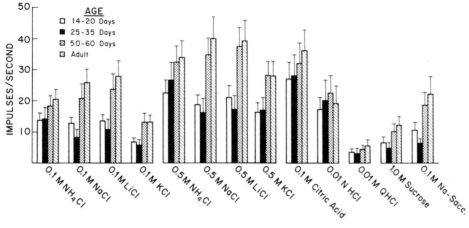

Fig. 7. Mean response frequencies of NST neurons in four age groups to 0.1 and 0.5 M NH$_4$Cl, NaCl, LiCl, and KCl and to 0.1 M citric acid, 0.01 N HCl, 0.01 M quinine hydrochloride, 1.0 M sucrose, and 0.1 M sodium saccharin. Standard errors are shown above each bar. Response frequencies to NH$_4$Cl, citric and hydrochloric acids, and quinine hydrochloride remained constant during development, whereas frequencies increased significantly during development to NaCl, LiCl, KCl, sucrose, and sodium saccharin. (From Hill *et al.*, 1983.)

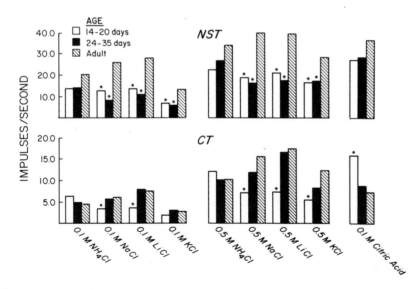

Fig. 8. Mean response frequencies of neurons in the nucleus of the solitary tract (NST, top) and chorda tympani nerve (CT, bottom) from three age groups after stimulation of the tongue with 0.1 M and 0.5 M NH$_4$Cl, NaCl, LiCl, KCl and with 0.1 M citric acid. The asterisks denote mean frequencies that are significantly different from the adult mean. Note that the ordinates are scaled differently for NST and CT responses. Whereas NST neural responses "mature" between 24–35 days and adult ages, chorda tympani responses mature between 14–20 days and 24–35 days. (From Hill *et al.*, 1983.)

"matured" between 24–35 days and adult ages, chorda tympani responses matured between 14–20 days and 24–35 days. It is interesting that the period of increasing response frequencies for NST neurons corresponds with an increase in density of presynaptic boutons in rat NST after 30 days postnatal (Miller *et al.*, 1983).

Changes in central nervous system taste characteristics resembled those from peripheral taste nerves in that mature responses to NaCl and LiCl at moderate concentrations appeared later in development than NH_4Cl and KCl responses; furthermore, response frequencies for NaCl and LiCl increased developmentally. In addition, in studies of NST neurons it was observed that sucrose response frequencies increased. These results lead to predictions that the rat should be initially very responsive to NH_4Cl and KCl and should become progressively more responsive to NaCl and sucrose during postnatal development.

Very recent experiments on behavioral taste responses provide some support for the predictions. For example, when NaCl was infused into the anterior mouth of rat pups from 6 days postnatal to 12 days, pups were indifferent to or preferentially ingested very high concentrations (0.25–0.50 *M*), which would be aversive to adults (Moe and Epstein, 1983). The results demonstrated that the young rat pup can taste NaCl but is not as responsive or "sensitive" to this salt as is the adult.

Keohe and Blass (1984) found that 5-day rat pups suckling on an anesthetized dam ingested little fluid when 0.5 *M* NH_4Cl was infused via an oral cannula, compared to the amount ingested when water was infused. Compared to water, only small amounts of 0.43 *M* NaCl were ingested, but NaCl was ingested in greater amounts than NH_4Cl.

Other new experimental data on very young rats do not fit the prediction as well, however. For example, Moe (1984) has now extended her observations on rat pups to animals aged 3 days. Whereas neurophysiological data might predict that the pups would be almost indifferent to NaCl, the 3-day rats rejected NaCl from an anterior mouth catheter at a concentration that older pups either preferred or found neutral. They also rejected quinine and NH_4Cl solutions at concentrations that were preferred or neutral to older pups. Moe proposes that the trigeminal, rather than the gustatory, sense mediates these responses. However, as discussed in Section IV of this chapter, development of different types of salt receptor components at different developmental stages could account for these data also. NaCl might interact with receptors primarily responsive to NH_4Cl or KCl at this early age; thus NaCl might taste more bitter and sour than pure salty. As NaCl receptors develop, the salty taste of NaCl might emerge.

In addition, Moran *et al.* (1983) have demonstrated that stimuli that elicit a variety of motivational responses (including mouthing, licking, probing, rolling, lordosis) in 3- to 10-day rats are less effective in eliciting all these responses at 15 days. In 15-day pups the behavioral responses are more organized and might not include the total array of motivational responses. Neurophysiological taste responses from 3- to 5-day rat pups are not available and the behavioral results remain confusing without neural correlates. By 5 days of age, 96% of fungiform papillae on the rat tongue contain a taste bud, but only about 8% of these taste buds have morphological characteristics similar to the adult (Table 1; Gottfried *et al.*, 1984). Therefore the taste buds are extremely immature at this age.

In preference studies with *postweaning* rats, other data that are compatible with neurophysiological results have been reported. Midkiff and Bernstein (1983) measured daily intake of 0.154, 0.30, and 0.60 M NaCl versus water in rats from 25 to 65 days of age. They demonstrated that at 28 days postnatal, 0.30 and 0.60 M NaCl are much less aversive than at 56 days (Figure 9). The extremely high concentrations of NaCl may be tolerated by young rats because the neural response is of relatively low frequency, compared to the adult. Based on responses from NST neurons, behavioral responses to NaCl should resemble those in the adult by 50–60 days after birth. Data from Midkiff and Bernstein (1983) substantiate this prediction.

Studies of responses to sucrose from the rabbit also correlate with neurophysiological development. In observations of pups at 2–4 days, 6–8 days, and 14–16 days postnatal, preferences for 0.3 M sucrose versus water were not expressed until 14–16 days (Ganchrow and Matzner, 1979). Moreover, after this early developmental period, sucrose preferences continued to increase. Presumably in rabbit as in rat, development of the peripheral gustatory system is essentially a postnatal event (Ganchrow and Matzner, 1979). Neurophysiological results in rat might explain the rabbit responses, since the preference for 0.30 M sucrose may not be expressed fully until NST response frequencies reach maximum values. (This would occur only in rats older than 35 days, as illustrated in Figure 7.)

It should be noted, however, that in rats there is a decreasing preference for a very high sucrose concentration postnatally (Wurtman and Wurtman, 1979). Postweaning rats had access simultaneously to diets containing 0.70 M sucrose and no sucrose. In males and females, sexual maturation (about 40 days) coincided with

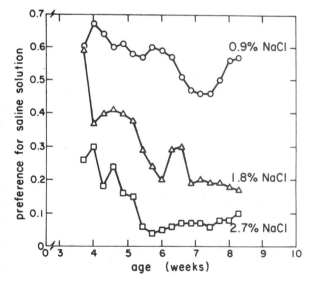

Fig. 9. NaCl preferences in rats from 25 to 58 days of age. Rats in three groups had continuous access to either 0.15 (0.9%), 0.31 (1.8%), or 0.46 M (2.7%) NaCl and water. NaCl and water intakes were measured daily. NaCl preferences were calculated by dividing NaCl intake by total fluid intake. Throughout postnatal development, high concentrations of NaCl (0.31 and 0.46 M) become increasingly aversive. (From Midkiff and Bernstein, 1983.)

decreased consumption of the diet containing 0.70 M sucrose. One could use the same argument from the previous paragraph and suggest that mature behavioral responses to 0.70 M sucrose would not occur until neurophysiological responses mature. The high sucrose concentration eventually may become too sweet, as the neurophysiological response frequency increases. Data from single NST cells indicate that sucrose responsiveness should increase between 25 and 35 days compared to 50–60 days after birth (Figure 7). However, since ingestion of sucrose in solid diets was measured by Wurtman and Wurtman (1979), other metabolic and dietary factors might be influencing results also. For example, around the time of puberty in rats, biological rhythms in response to photoperiodic cues are established, as are related periodicities in meal eating time, meal duration, and other feeding behaviors (Ramaley, 1980). Postingestational effects also might have a role in responses to the high sucrose diet (Mook, 1963).

Of course, development of behavioral taste characteristics should not be related to responses from NST neurons alone, but other central nervous system taste regions must be considered. Hill (1983) has collected data on responses from cells in the pontine taste area in rats aged 14–20 days, 25–36 days, and more than 80 days (adults). He has demonstrated that although NST and pontine neural responses reflect changes in peripheral nerve responses to some extent, there are developmental changes in central neural taste responses that cannot be explained on the basis of the peripheral nerve. For example, mature average response frequencies to NaCl, LiCl, and KCl are attained later in NST than in the chorda tympani, and at all ages there is an overall loss of response specificity to salts in NST neurons. In contrast to chorda tympani and NST neurons, responses to NH_4Cl from pontine taste cells increase during development. For all stimuli, response frequencies mature later in the pontine taste area than in NST. Each nervous system level has unique response characteristics and developmental patterns. All of these must be understood before the basis for behavioral responses is finally clarified.

D. Attempts to Modify the Rat Gustatory System through Early Experience

Since major changes take place in the rat gustatory system after birth (morphologically, neurophysiologically, and behaviorally), this presumably is an optimal time to alter future gustatory function through early experience. However, several attempts have not had long-lasting effects that extend into adulthood (see discussions in Mistretta, 1981; Mistretta and Bradley, 1978a). Two recent experiments directed to this question are illustrative.

Rats aged 16–30 days postnatal were fed a diet artificially high in sucrose (Wurtman and Wurtman, 1979). There was no relationship between experience with diets containing 0, 0.35, or 1.40 M sucrose prior to 30 days of age and the amount of sucrose subsequently eaten in a choice of the same three diets, from 30 to 60 days of age. In other studies (Midkiff and Bernstein, 1983) postweaning exposure to high concentrations of NaCl did not alter adult preference for NaCl in solution or solid food.

From these negative results it cannot be concluded that early salt or sucrose experience has no effect on future preference–aversion behavior, since the experimental designs do not allow exclusion of a specific hypothesis. Choice of stimulus and concentrations used for early exposure, and the method of testing for subsequent effects, can be crucial. Furthermore, development of specific enzyme systems that alter metabolism of various chemicals in the organism must be considered. Enzyme or regulating systems that develop after the early exposure period might alter the general metabolic effect of the stimulus. For example, enzymes involved in carbohydrate and lipid metabolism, including glucokinase (Ballard, 1978) and acetoacetic acid and glycerol kinase (Hahn, 1978), do not reach adult levels until after weaning in rat.

The timing and length of the early experience also are critical. For the taste system in rats it is probably important to include the first four postnatal weeks in the early exposure period, since this is the time of most rapid and substantial change in the peripheral taste system. Taste bud structure and mature neurophysiological response characteristics are being established at this time. Changes in brainstem responses occur later, during postweaning periods. Therefore a postweaning exposure component is probably important also.

In a recent report female rats were maintained on either a high sucrose (50% by weight) or a high glucose (50% by weight) diet during gestation and lactation (Marlin, 1983). The pups were maintained on the same diet postweaning until 42 days and then were fed laboratory chow. Preference tests were conducted weekly from 21 to 84 days of age, between high sucrose and glucose diets. During the initial preference tests (21–42 days) all rats preferred the sucrose diet. However, beginning at age 49 days the rats began to express a preference for the diet (either sucrose or glucose) on which they had been raised and maintained. Marlin's (1983) experiment illustrates the importance of long postnatal exposure periods and lengthy postexposure testing periods to demonstrate an effect of early diet experience on later taste preference.

In our laboratory we attempted to manipulate the neurophysiological response to various monochloride salts via diet alterations (Bradley *et al.*, 1982, 1984). In an initial experiment (Bradley *et al.*, 1982) pregnant rats were fed lab chow without NaCl from 3 days of gestation to term and after birth for 12 days. Nursing dams were then given a diet with 1% NaCl and weaned pups continued on the 1% NaCl diet. Recordings from the chorda tympani nerve at 50–60 days of age in these rats (deprived of NaCl via the maternal diet from 3 days of gestation to 12 days postnatal) demonstrated no differences in responses to monochloride salts and other chemicals, compared to rats that had never experienced NaCl deprivation but had the same diet with 1% NaCl added.

However, if the salt depletion was continued postnatally to 28–48 days and chorda tympani nerve responses were recorded, *with no intervening period on NaCl-replete diet,* neurophysiological responses were altered (Bradley *et al.*, 1984). Compared to control rats, response–concentration functions for NaCl were flattened at higher concentrations. NH$_4$Cl and KCl response–concentration functions were not different than those in controls.

Although a continuous NaCl deprivation to at least 28 days altered chorda tympani responses, deprivation to 12 days postnatal with a subsequent return to NaCl-replete diet did not affect chorda tympani responses. It will be interesting to determine whether very brief (e.g., a few days) exposures to a NaCl-replete diet after deprivation can reverse the neurophysiological alterations in NaCl responses.

There are other examples to illustrate the importance of timing in such experiments. Although *postweaning* exposure to high-NaCl diet did not alter NaCl preference behavior in adult rats (Midkiff and Bernstein, 1983), Contreras and Kosten (1983) demonstrated differences in NaCl preference in adults exposed to 0.50 M NaCl prenatally and for 30 days postnatally. NaCl preference was elevated in these rats; however, the effect apparently related to decreased water intake rather than increased NaCl intake.

As emphasized before (Mistretta, 1981; Mistretta and Bradley, 1978a), the taste system might be especially "plastic" or susceptible to environmental modification. In the context of the general preferences and aversions that are present from birth to ensure selection of appropriate nutrients and avoidance of toxic substances, the organism should be prepared to alter taste behavior according to food availability. Furthermore, through acquired preference for initially unpalatable bitter substances such as those in coffee, humans can expand their taste world (Rozin, 1976). Beauchamp and Moran (1982) have data to suggest that early experience with sucrose effectively maintains the human infant's preferences for sugar; in the absence of experience, the sweet preference is decreased. In other sensory systems examples of neuroanatomical and neurophysiological alterations after environmental manipulations are not uncommon (e.g., Mower *et al.*, 1983; Sanes and Constantine-Paton, 1983). Flexibility might be expected in the taste system through similar avenues of central nervous system synapse modification and possibly through the process of constant renewal or turnover of taste bud cells (Beidler and Smallman, 1965).

III. Development of the Sheep Gustatory System and Applications to Development of Taste Responses in Humans

The sheep has a rather lengthy gestation (about 150 days or 21 weeks), and taste buds appear on the anterior and posterior tongue *in utero*, as they do in the human fetus. In sheep, taste buds are found in fungiform papillae as early as 7–8 weeks of gestation (Bradley and Mistretta, 1972, 1973), and in humans they appear at about 8–9 weeks (Bradley and Stern, 1967; Bradley, 1972). By about 12 weeks of gestation in both species, taste buds possess well-developed pores, which form direct channels between taste bud cell apices and the external, oral environment. Since there is an essentially parallel morphological development of taste buds, comparisons can be made between sheep and human gustatory systems. Therefore in this section we shall discuss neurophysiological and anatomical development of the sense of taste in sheep and then relate these data to development of human, behavioral taste responses.

Several years ago we first recorded from the fetal chorda tympani nerve, innervating taste buds in fungiform papillae on the anterior tongue, and reported that the peripheral gustatory system is functional for at least the last third of gestation in sheep (Bradley and Mistretta, 1973). Responses were obtained to lingual stimulation with NH_4Cl, KCl, NaCl, LiCl, acetic acid, glycerol, sodium saccharin, glycine, and quinine hydrochloride. Recently, we have focused on responses to monochloride salts during development and have recorded from the chorda tympani nerve in five age groups: (1) fetuses at 110 days of gestation; (2) 130 days of gestation; (3) perinatal animals, 1 week before or after birth; (4) lambs; and (5) adult sheep (Mistretta and Bradley, 1983a). Developmental changes in salt responses from the chorda tympani nerve were observed that were quite similar to those in rat. For example, as illustrated in Figure 10, in fetuses beginning the last third of gestation (110 days of gestation), NaCl and LiCl elicited much smaller, integrated response magnitudes than NH_4Cl and KCl (Mistretta and Bradley, 1983a). Throughout the rest of gestation and postnatally, the NaCl and LiCl responses gradually increased in magnitude relative to NH_4Cl and KCl. In adults, NaCl, LiCl, and NH_4Cl all elicited similar response magnitudes, and KCl was less effective as a taste stimulus.

Based on studies of a more limited number of age groups, we originally reported that over a wide range of stimulus intensities, the shapes of salt response–concentration functions were similar (Mistretta and Bradley, 1983a). However, when additional data were obtained from young fetuses (110 and 130 days of gestation), it was obvious that shapes of the response–concentration functions for NH_4Cl were flatter at high concentrations in these early fetuses, compared to other ages (Mistretta and Bradley, 1983b). Thus the response–concentration curves demonstrated that there are changes in NH_4Cl responses in addition to the striking NaCl and LiCl changes.

Studies of developing taste responses from receptors on the back of the tongue might be expected to yield somewhat different results, since these taste buds have a different innervation and respond differently in adults when compared to receptors on the front of the tongue. Therefore we recorded salt responses from multifiber bundles of the glossopharyngeal nerve innervating taste buds in circumvallate papillae (Mistretta and Bradley, 1983b). In the youngest fetuses KCl elicited very large response magnitudes, relative to NH_4Cl; KCl responses then decreased by about 50%, relative to NH_4Cl, in older fetuses (Figure 11). Therefore the order of effective stimulation reversed and NH_4Cl elicited larger responses than KCl. Responses to NaCl and LiCl increased in relative magnitude, but the magnitude of the change was small. For taste buds on the back of the tongue these two salts remained relatively ineffective stimuli over the life span.

Salt response ratios across age groups were compared for glossopharyngeal and chorda tympani nerves (Figure 12), to illustrate developmental differences in responses from taste buds on the anterior and posterior tongue. Anterior tongue responses to NaCl and LiCl, relative to NH$_4$Cl, changed most substantially, and those to KCl changed very little. The reverse was observed for posterior tongue; responses to NH$_4$Cl and KCl changed most substantially. Furthermore, responses to HCl and citric acid altered developmentally on posterior tongue, but not on anterior (Mistretta and Bradley, 1983b).

In summary, several general points should be made, based on neurophysiological recordings from chorda tympani and glossopharyngeal nerves in fetuses, lambs, and adults. Taste responses from both nerves alter substantially during development. However, developmental changes for anterior and posterior taste bud responses are different. NaCl and LiCl changes are most substantial on ante-

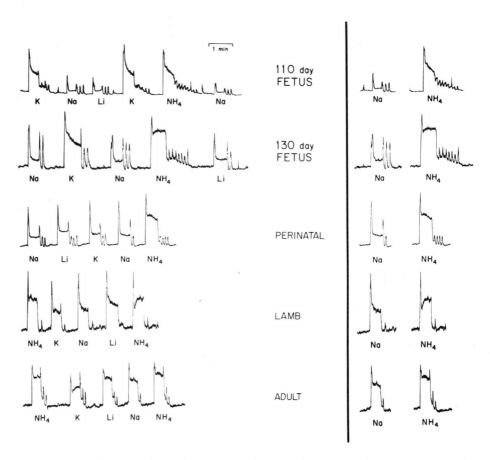

Fig. 10. Integrated records of multifiber responses from the chorda tympani nerve in representative animals from each of five groups of sheep. Chemical stimuli applied to the anterior tongue were 0.5 M NH$_4$Cl, KCl, NaCl, and LiCl. On the right side of the figure the responses to NH$_4$Cl and NaCl have been selected from the continuous recordings to emphasize the increasing NaCl response during development, relative to NH$_4$Cl. For quantitative analysis of these data, the magnitude of each response was expressed as a ratio relative to the NH$_4$Cl response. NH$_4$Cl was used as the standard because it elicits a large-magnitude response throughout development. The response ratios relative to NH$_4$Cl are presented in Figure 12. (From Mistretta and Bradley, 1983b.)

110 day

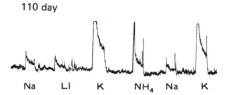

Na   Li   K   NH₄   Na   K

130 day

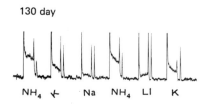

NH₄   ↓   Na   NH₄   Li   K

Perinatal

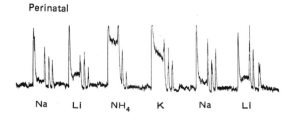

Na   Li   NH₄   K   Na   Li

Fig. 11. Integrated records of multifiber responses from the glossopharyngeal nerve in an animal from each of five age groups of sheep. Chemical stimuli applied to the posterior tongue were 0.5 M NaCl, LiCl, KCl, and NH₄Cl. Although KCl was the most effective salt stimulus in fetuses at 110 days of gestation, NH₄Cl was most effective in older fetuses, lambs, and adults. For this figure the pen recorder gain was set at a high level to illustrate the relatively small-magnitude responses to NaCl and LiCl; therefore in some records the early, transient portions of responses to KCl and NH₄Cl are flattened because they briefly exceeded the recorder gain. Response ratios relative to NH₄Cl for these data are presented in Figure 12. (From Mistretta and Bradley, 1983b.)

Lamb

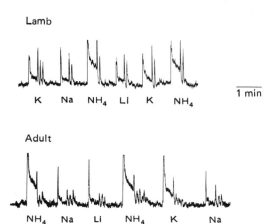

K   Na   NH₄   Li   K   NH₄

$\overline{\text{1 min}}$

Adult

NH₄   Na   Li   NH₄   K   Na

rior tongue, compared to more striking KCl and NH₄Cl response alterations on posterior tongue. Responses to acids change also on posterior tongue. In addition, for anterior and posterior tongue taste buds, neurophysiological responses to some salts continue to change after birth (between 30 and 90 days postnatal and adult ages); developmental alterations are not solely a prenatal phenomenon in this species.

The developmental changes indicate that there are different, specific membrane components interacting with the various salts, since responses to the salts do not change in parallel. Furthermore, different proportions of these membrane components must be present at various developmental stages in taste buds from fungiform compared to circumvallate papillae.

B. ANATOMICAL CORRELATES OF RESPONSE CHANGES IN PERIPHERAL TASTE NERVES

In seeking correlates for neurophysiological response changes in sheep, an important distinction must be made from the rat. Fungiform papillae on the anterior rat tongue usually contain one taste bud each, and an essentially full complement of buds is present by 15 postnatal days. However, morphological changes continue to occur in rat taste buds up to about 30 days postnatal, although new

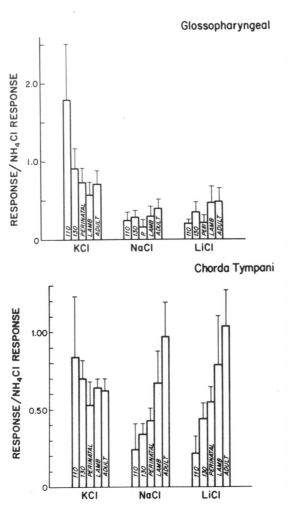

Fig. 12. Top: Mean response ratios and standard deviations for KCl, NaCl, and LiCl, relative to NH_4Cl, recorded from the glossopharyngeal nerve in fetuses (110 and 130 days of gestation) and perinatal animals, lambs and adults. For each salt a response equal in magnitude to that elicited by NH_4Cl yields a ratio of 1.00. KCl response ratios decrease substantially during development, whereas NaCl and LiCl ratios increase slightly. Bottom: Mean response ratios and standard deviations for chorda tympani nerve responses from the same age groups. When recording from the anterior tongue, KCl ratios decreased only slightly during development, whereas NaCl and LiCl ratios increased greatly. Changes in NaCl and LiCl ratios were statistically significant at pre- and postnatal ages. Note that the scales for glossopharyngeal and chorda tympani nerve response ratios are different. (From Mistretta and Bradley, 1983b.)

taste buds are not added to the anterior tongue. These morphological alterations closely parallel maturation of salt taste responses (Gottfried *et al.,* 1984).

In sheep fetuses younger than 100 days of gestation each fungiform papilla contains usually one taste bud, but after this age the number of taste buds per papilla increases; eventually adult ranges of one to eight buds per papilla are attained (Bradley and Mistretta, 1972, 1973). Therefore taste buds are added to the anterior sheep tongue, prenatally and after birth.

Taste buds are added also in circumvallate papillae on the posterior sheep tongue. There are 20–30 circumvallate papillae on each lateral edge of the back of the tongue. About 45–80 taste buds are found per papilla in 110-day fetuses and about 50–200 per papilla in adults (Mistretta and Bradley, 1983b).

Thus the changing neurophysiological responses in sheep fetuses, lambs, and adults are accompanied by an addition of new taste buds that continues into adulthood. This presumably provides a basis for the prolonged, developmental alterations in taste responses in sheep, as will be discussed in Section IV of this chapter.

Not only are new taste buds added throughout the life cycle, but also, from light microscopic and ultrastructural studies, there are data on changing morphological characteristics of sheep taste buds. Structurally immature taste buds are present on the anterior tongue by 50 days of gestation in the fetus (Bradley and Mistretta, 1972, 1973). These buds are collections of cells that stain distinctly from the surrounding epithelium and are separated from the oral cavity by a layer of superficial epithelial cells. By 80 days of gestation, taste bud cells are oriented perpendicularly to the basement membrane, cell apices are arranged around a taste pit, short microvilli and club-shaped processes are present on cell apices, and tight junctions are observed between apices of taste bud cells (Figure 13; Mistretta and Bradley, 1983a). Furthermore, at least two morphological cell types are present (Figure 13) and numerous nerve profiles are distributed throughout the taste bud.

By 100–110 days of gestation, microvilli and club-shaped processes on cell apices have increased in length, presumably augmenting the area of accessible membrane available for initial interaction with chemical stimuli (Mistretta and Bradley, 1983a). During the rest of gestation and postnatally, increasing layers of cornified cells are added on the surface of fungiform papillae. Therefore a taste "pit" region is defined that communicates with the oral environment via a narrow channel or pore. The pit becomes filled with a densely staining, homogeneous material that surrounds the microvilli; it is not yet known when this material first appears, but it is not apparent at 80–110 days of gestation.

In summary, early fetal taste buds in the sheep have apical cellular extensions and microvilli, intercellular tight junctions, and diverse cell types. These structural characteristics are present from at least 80 days of gestation. Therefore the electrophysiological responses that we have described from 110-day fetuses are mediated by taste buds that already possess several structural features of mature taste buds. Thus the initial stimulus–receptor membrane contacts in the taste pore are probably similar to those in adults.

Recent ultrastructural studies of fetal macaque taste buds have contributed

CHARLOTTE M.
MISTRETTA AND
ROBERT M. BRADLEY

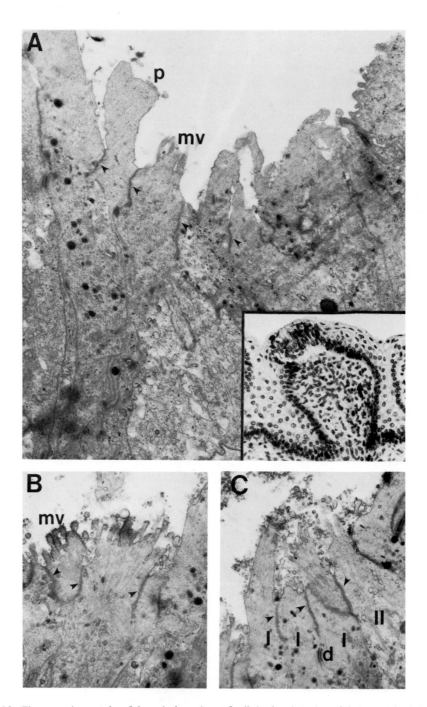

Fig. 13. Electron micrographs of the apical portions of cells in the pit region of three taste buds (A, B, and C) from fetuses aged 80 days of gestation (X20,000, X22,000, X22,000). The inset is a light micrograph of a taste bud and fungiform papilla from an 80-day fetus (X250). The cells terminate in short microvilli (mv) or blunt, club-shaped processes (p). All cell apices are joined by tight junctions (arrowheads). At least two cell types, designated I and II, are already present at this age as illustrated in taste bud (C). Also in (C), desmosomes (d) are labeled. (From Mistretta and Bradley, 1983a.)

new data on development of gustatory synapses in primates (Zahm and Munger, 1983). In early gestation numerous axoaxonic synapses are observed in the taste bud and these are much reduced in number by midgestation. Changes in synaptic types might provide one basis for changing neurophysiological taste responses in the fetus. Furthermore, during the last half of gestation in the macaque taste bud there is a transition in various cell types (Zahm and Munger, 1983). The functional significance of the cell types is not apparent without neurophysiological experiments, but once again a probable basis for changing function is provided in taste bud morphology.

C. Changes in Electrophysiological Responses from Central Nervous System Taste Neurons and Implications for Human Behavioral Responses

Changes in peripheral nerve taste responses from the anterior and posterior tongue presumably will be reflected in central nervous system responses to chemical stimuli. Maturation of central taste pathways could compound the developmental alterations in the periphery to result in substantial limitations of central, neural taste responses. To determine the extent of these limitations, responses recorded from second-order neurons are informative. Recordings were made from fetuses, lambs, and adults in the rostral projection area for chorda tympani afferents of the nucleus of the solitary tract (Mistretta and Bradley, 1978b; Bradley and Mistretta 1980). Chemical stimuli were 0.5 M NH_4Cl, KCl, NaCl, and LiCl; 0.01 N HCl, and 0.1 M citric acid.

Central nervous system responses to chemical stimuli emerged in a systematic progression that reflected development of the peripheral gustatory system (Mistretta and Bradley, 1983a). In fetuses younger than 114 days of gestation, usually only three chemicals were effective stimuli (NH_4Cl, KCl, and citric acid). There were no responses to stimulation of the anterior tongue with NaCl or LiCl in fetuses younger than 114 days (Figure 14). Therefore central neurons in younger fetuses not only responded to fewer chemicals, but also did not acquire the ability to respond to NaCl and LiCl until later in gestation. Gradually, in older fetuses and in lambs, responses to NaCl and LiCl occurred more frequently. As in rat, the data demonstrate that neural response mechanisms for NH_4Cl and KCl are different from those for NaCl and LiCl.

Although the later developmental emergence of responses to NaCl and LiCl from medullary neurons reflects the gradual increase in responses to these salts from taste afferent fibers, it should be noted that at ages when *no* responses to NaCl and LiCl were obtained centrally (about 114 days of gestation), small responses were obtained from the chorda tympani nerve (Mistretta and Bradley, 1983a). Therefore the ability of the central neural taste system to respond to these salts must relate to immaturity of synapses between first-order afferents and second-order cells, or to the second-order cells *per se*.

Since second-order neurons in the gustatory pathway of the fetus are limited in ability to respond to NaCl and LiCl and gradually become responsive to an

CHARLOTTE M.
MISTRETTA AND
ROBERT M. BRADLEY

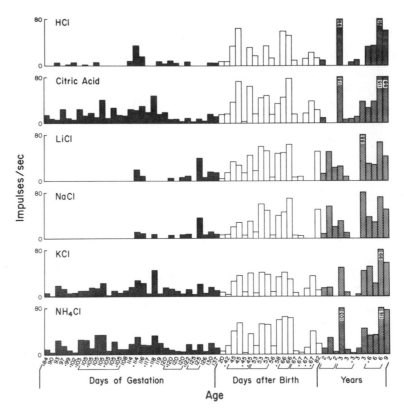

Fig. 14. Frequency histograms for responses from 61 neurons in the sheep nucleus of the solitary tract to stimulation of the anterior tongue with NH$_4$Cl, KCl, NaCl, LiCl, citric acid, and HCl. Ages for fetal units (black bars) are in days of gestation, for lambs (white bars) in days after birth, and for adults (stippled bars) in years. Single units are marked by a small dot next to the age; in other cases, responses from two or three units were recorded at once. Bracketed ages denote units recorded from one animal. The response frequencies for some adult neurons exceeded 80 impulses/sec; the correct frequencies are written within the histogram bar. Note that no responses to NaCl and LiCl were recorded in fetuses younger than 114 days of gestation. However, responses to NH$_4$Cl, KCl, and citric acid were obtained consistently in these fetuses. (From Bradley and Mistretta, 1980.)

increasing number of chemicals throughout development, changes in behavioral taste responses would be predicted based on neurophysiological observations. From the combined data on peripheral and central nervous system taste responses in sheep, we make several predictions about the development of behavioral taste responses in humans, since there is a similar time course for prenatal morphological development of taste buds. (1) The fetal taste system should be able to discriminate among a variety of chemicals, including salts, acids, sugars, and bitter chemicals. Therefore premature infants should be able to detect and possibly selectively respond to some taste stimuli. (2) The full-term newborn also should be able to detect and discriminate among taste stimuli. (3) Since neurophysiological response magnitudes to various chemical stimuli continue to change during postnatal development, behavioral taste responses to chemicals also should change after birth. (4)

In particular, since response magnitudes to NaCl undergo a dramatic developmental increase, newborns should be less responsive to, or less able to discriminate, NaCl than other sour–bitter salts or acids. That is, relative to older infants, the newborn should have more limited information about the NaCl in its environment.

There are many studies on the development of behavioral taste responses in humans to relate to the preceding predictions. Reports from Steiner (1979) support prediction 1, that premature infants (sixth to seventh gestational month) can detect some gustatory stimuli, specifically citric acid. Prediction 2, that full-term newborns should respond to and discriminate among a variety of chemical stimuli, is documented in many studies. Newborns not only distinguish among various chemicals, but based on facial expression, heart rate, tongue movement, and intake measures, apparently exhibit a preference for sugars and aversion to acidic and bitter chemicals (Crook and Lipsitt, 1976; Desor *et al.*, 1973; Nowlis and Kessen, 1976; Steiner, 1979; Weiffenbach and Thach, 1973).

In prediction 3 we suggest that behavioral taste responses should change after birth. Since neurophysiological taste responses continue to alter between lamb and adult stages, behavioral changes would presumably take place in children compared to infants, and possibly in adults compared to children. Cowart (1981) has provided a comprehensive review of the literature on development of taste perception in humans and has described reports of differences between taste preferences of children and adults. Evidence generally indicates that the response to salt alters from hedonically neutral or aversive in infants and young children to preferred in late childhood. In addition, some data suggest that children have a greater preference for sweet than adults.

The fourth prediction proposes decreased responsivity to NaCl in very young infants, compared to older infants or children. The literature on behavioral taste responses to salt in babies is complicated by application of a variety of measures in different laboratories, and therefore data are not totally consistent. Discussions by Cowart (1981) and Crook (1978) are detailed and useful in understanding the breadth of investigations. Human newborns can reportedly detect NaCl over a concentration range from 0.015 to 0.60 M when highly sensitive measures such as suckling pattern and tongue movements are used (Crook, 1978; Nowlis, 1973). However, although NaCl will shorten the length of sucking bursts compared to water, there is no difference in degree of shortening over a concentration range of 0.10–0.60 M (Crook, 1978). In addition, from intake measurements it was concluded that newborns did not discriminate between 0.20 M NaCl and water (Maller and Desor, 1973). In an unfortunate poisoning incident the relative insensitivity of infants to NaCl was again illustrated. Salt was used instead of sugar in the feeding formula for infants in a hospital nursery and the infants consumed the salty formula to an extent that made them extremely ill (Finberg *et al.*, 1963). Finally, Beauchamp has presented data demonstrating that infants at 3 months of age will ingest quantities of 0.10–0.50 M NaCl equal to water in choice tests (Beauchamp and Maller, 1977); at 1.5 years of age these concentrations are rejected.

Therefore in several respects the behavioral predictions that derive from neurophysiological experiments in fetal, newborn, and adult sheep are substantiated by existing studies on human postnatal taste responses. However, measuring taste responses to concentration ranges of various chemicals is difficult in human newborns and infants. More extensive animal experiments could provide the control and flexibility that will be essential for understanding the emergence of behavioral taste responses.

IV. Proposed Membrane Changes Underlying Development of Taste Responses

Over 20 years ago Beidler (1961) predicted changes in taste responses during development, when he discussed the process of taste bud cell turnover or renewal that occurs in adult life (Beidler and Smallman, 1965). Beidler proposed that newly formed taste cells might respond differently to certain stimuli than older taste cells.

Our neurophysiological studies of the developing taste system suggest that Beidler's (1961) original proposition relating taste cell age to response characteristics may be correct. For example, young taste cells might be more responsive to NH_4+, relative to $Na+$ (Mistretta and Bradley, 1983a). Conversely, older cells might be more responsive to $Na+$ (Figure 15). Thus a newly formed taste bud presumably would have a greater proportion of young, highly NH_4+-responsive cells. An "established" taste bud, that has not formed recently, would have old, $Na+$-responsive cells as well as young cells that have just turned over. As an animal ages, fewer new taste buds are formed, and thus the taste system would become proportionately more responsive to NaCl.

In sheep, taste buds are added throughout fetal and postnatal development, into adulthood. This could account for the continuing, large NH_4Cl response in this species. In rats, new taste buds are not added after the first postnatal week. As

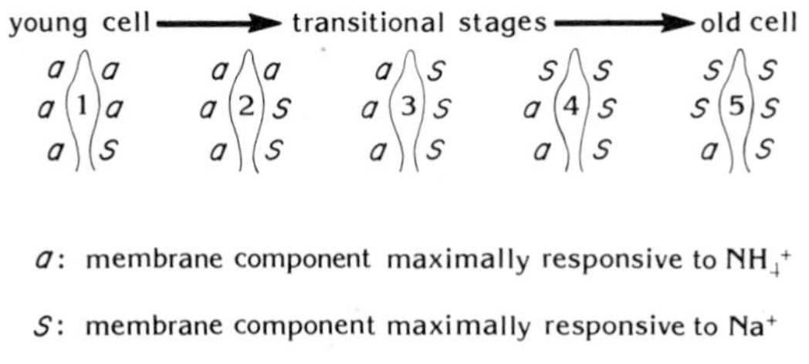

a: membrane component maximally responsive to $NH_4{}^+$

s: membrane component maximally responsive to Na^+

Fig. 15. Diagram to illustrate proposed maturation of taste bud cells and cell membrane components. Membrane components labeled a are responsive primarily to NH_4Cl, and s is responsive to NaCl. It is suggested that younger cells have a greater proportion of membrane component a and that older cells have a greater proportion of s. Thus younger cells would be more responsive to NH_4Cl, older cells to NaCl. In newly formed taste buds, younger cells presumably would be present in greater proportion, and therefore responsiveness to NH_4Cl would be high. In established taste buds, more older cells would be present and therefore responsiveness to NaCl also would be high.

the established set of taste buds matures, the large NaCl response emerges and is maintained.

The basis of NH_4+ or $Na+$ sensitivity in taste bud cells of course must relate to receptors in the cells. Our developmental data demonstrate that different monovalent cations interact with different receptor membrane components in taste cells, since response changes to the various salts do not occur in parallel (Hill *et al.*, 1982; Mistretta and Bradley, 1983a,b). Thus we predict that taste bud cells contain different proportions of various cation-responsive membrane components at different developmental stages. The membrane components might be available negative charges in particular molecular settings that render the charges more responsive to $Na+$ than to NH_4+ or $K+$, or the components might be different receptor molecules present at various developmental stages or changing ion transport channels involved in transduction (Beidler and Gross, 1971; Brand and Bayley, 1980; Heck *et al.*, 1984; Kashiwayanagai, *et al.*, 1983).

Furthermore, each of the cations we have studied might react with more than one receptor "component" or "site" in the taste cell membrane (Mistretta and Bradley, 1983a,b). In adult rats, NH_4+ and $K+$ apparently interact with two different, independent sets of sites, with very different binding constants (Beidler, 1961). NH_4+ and $K+$ might initially interact with one set of sites in fetal sheep and later with both the original set and an added, $Na+$ site. Extending this reasoning, the early fetal sheep or postnatal rat responses to NaCl might be via interactions between NaCl and the NH_4+ or $K+$ receptor components. Therefore NaCl might initially taste more sour or bitter to the developing organism. When the $Na+$ receptor component *per se* develops, NaCl would taste salty.

Studies of changes in specific, fetal receptors or receptor sites are not plentiful. However, receptors responding to B-adrenergic stimuli in rabbit heart have been shown to change developmentally (Hatjis and McLaughlin, 1982). There was a progressive increase in the density of receptor sites with increasing gestation, from 21 to 31 days (term), but the affinity and specificity of B-adrenoreceptors did not change.

Ion channels specific for calcium and potassium develop at different times in Drosophila flight muscles (Salkoff and Wyman, 1983). A fast, transient, $K+$ current develops first at 72 hr postpupariation; then at 92 hr a delayed $K+$ current appears. The calcium current does not mature until eclosion of the adult from the pupal case. In spinal cord neurons of Xenopus embryos, generation of an action potential is first related to a specific calcium conductance, then to sodium and calcium, and finally to sodium conductance only (Spitzer and Lamborghini, 1976). If various ion channels develop sequentially in taste bud membranes too, then related changes in neural responses to taste stimuli might be predicted.

An alternative or additional basis for changes in neurophysiological taste responses is that a separate and specific NaCl *fiber* system develops later than an acid or acid–salt fiber system. Recently, Frank and colleagues (Frank *et al.*, 1983) have presented data to support a sodium fiber system in rat. Extending the argument to development, large-diameter fibers that synapse with acid–salt and acid-

CHARLOTTE M.
MISTRETTA AND
ROBERT M. BRADLEY

sensitive receptors might predominate early in development. Since NH_4Cl and KCl have taste properties that are not strictly "salty" but are acidic–salty or bitter–salty (Doetsch and Erickson, 1970; McBurney and Shick, 1971; Smith and Frank, 1972; Smith and McBurney, 1969), these stimuli would be very effective for the acid–salt system. Smaller-diameter fibers might be present that synapse with $NaCl$ receptors; since the fibers are small diameter, they would not be sampled frequently with neurophysiological recording techniques. Later in development these fibers might increase in diameter, so that sodium responses would increase.

This alternative requires that the initial, smaller-diameter, Na system not only increases in size, but also actually comes to predominate or "outnumber" the acid–salt or acid-system in rat. Thus the adult order of most effective stimuli (Na $> NH_4$ $>$ K) would emerge. In adult sheep the Na–fiber system would have to at least equal the large-diameter, acid system to achieve adult response characteristics.

Focusing on fiber diameter alone also implies an extreme specificity in taste responses; yet single fibers in rat chorda tympani nerve at all ages generally respond to all four of the monochloride salts: $0.1\ M$ NaCl, LiCl, NH_4Cl, and KCl (Hill *et al.*, 1982). However the response to NaCl and LiCl is of much lower frequency in very young rats than in older animals. We suggest that development of receptor components more simply predicts the developmental taste response changes, although fiber diameter changes do occur (Ferrell *et al.*, 1983) and obviously contribute to maturation of response characteristics, such as latency.

V. SUMMARY

For two species, rat and sheep, there are now extensive data on neurophysiological taste responses during development. The data include investigations of peripheral nerve responses from taste buds on the anterior and posterior tongue (in fungiform and circumvallate papillae) and from second-order taste neurons in the medulla. Very recently, reports are appearing on responses from higher central nervous system levels (pons). In addition, light and electron microscopic investigations of the taste bud are providing morphological correlates for neurophysiological changes. Thus our understanding of the developing gustatory sense has expanded rapidly in the past few years.

Several studies demonstrate that neurophysiological taste responses change during development, at peripheral and central nervous system levels. Response changes are observed for several salts, citric and hydrochloric acids, sucrose, and sodium saccharin.

In both rat and sheep similar developmental changes take place in salt taste responses. Most remarkable is the gradual emergence of increasing response magnitudes to NaCl compared to other monochloride salts. This developmental alteration occurs prenatally and continues postnatally in sheep; it is observed after birth in rats, since structural development of the rat taste bud is a postnatal phenomenon. Therefore taste responses alter in the two species against very different environmental backgrounds. In sheep, fetal taste buds are bathed in and stimulated by amniotic fluid, tracheal and salivary secretions. Postnatally, saliva, maternal milk, and available solid food provide changing backgrounds and stimuli for the gustatory system. In rats the fetal environment can have little to do with development of the sense of taste, since immature taste buds only appear near the end of gestation. Therefore in both species changing neural responses to salts probably occur due to intrinsic, preprogrammed, developmental factors rather than external environmental stimuli. However, the changing neural substrate that we have described provides opportunity for interactions among external chemical stimuli, the developing gustatory sense, and the postingestional consequences of tasting and swallowing; such interactions may play a role in establishing behavioral taste responses.

The reported developmental changes in salt taste responses demonstrate that different receptor components function in initiating responses to NH_4Cl versus KCl versus NaCl and LiCl. Furthermore, these components have different developmental sequences, since salt taste responses do not change in parallel as a function of age. Especially intriguing is the apparent, gradual development of receptor components that are particularly responsive to NaCl. There is no obvious adaptive value for a relatively late maturation of NaCl sensitivity in rats and sheep; this delayed development is probably dictated by changing receptor membrane properties.

Behavioral correlates for developing neurophysiological taste responses are appearing in studies of rats and humans. Furthermore, descriptions of neurophysiological plasticity after dietary exposure to specific chemical stimuli are emerging that clarify the basis for large changes in taste preferences and aversions throughout the life cycle.

Acknowledgment

Preparation of this manuscript was supported by National Science Foundation Grant BNS 8311497 to C.M.M. and R.M.B.

References

Ballard, F. J. Carbohydrate metabolism and the regulation of blood glucose. In U. Stawe and A. A. Weech (Eds.), *Perinatal physiology*. New York: Plenum Press, 1978.

Bartoshuk, L. M. Gustatory system. In R. B. Masterton (Ed.), *Handbook of Behavioral Neurobiology*. Vol. 1. *Sensory integration*. New York: Plenum Press, 1978.

Beauchamp, G. K., and Maller, O. The development of flavor preferences in humans: A review. In M. K. Kare and O. Maller (Eds.), *The Chemical senses and nutrition*. New York: Academic Press, 1977.

Beauchamp, G. K., and Moran, M. Dietary experience and sweet taste preference in human infants. *Journal for Intake Research*, 1092. *3*, 139–152.

Beidler, L. M. Taste receptor stimulation. *Progress in biophysics and biophysical chemistry*. London: Pergamon Press, 1961.

Beidler, L. M., and Gross, G. W. The nature of taste receptor sites. In W. D. Neff (Ed.), *Contributions to sensory physiology*. Vol. 5. New York: Academic Press, 1971.

Beidler, L. M., and Smallman, R. L. Renewal of cells within taste buds. *Journal of Cell Biology,* 1965, *27,* 263–272.

Bradley, R. M. Development of the taste bud and gustatory papillae in human fetuses. In J. F. Bosma (Ed.), *Third symposium on oral sensation and perception: The mouth of the infant.* Springfield, Ill.: Thomas, 1972.

Bradley, R. M. and Mistretta, C. M. The morphological and functional development of fetal gustatory receptors. In N. Emmelin and Y. Zotterman (Eds.), *Oral physiology,* Oxford: Pergamon Press, 1972.

Bradley, R. M., and Mistretta, C. M. The gustatory sense in foetal sheep during the last third of gestation. *Journal of Physiology,* 1973, *231,* 271–282.

Bradley, R. M., and Mistretta, C. M. Fetal sensory receptors. *Physiological Reviews,* 1975, *55,* 352–382.

Bradley, R. M., and Mistretta, C. M. Developmental changes in neurophysiological taste responses from the medulla in sheep. *Brain Research,* 1980, *191,* 21–34.

Bradley, R. M., and Stern, I. B. The development of the human taste bud during the foetal period. *Journal of Anatomy,* 1967, 101, 743–752.

Bradley, R. M., Hill, D. L., and Mistretta, C. M. Salt taste responses in rats depleted of NaCl during early development. *Society for Neuroscience Abstracts,* 1982, *8,* 754.

Bradley, R. M., Hill, D. L., and Mistretta, C. M. Effect of pre- and postnatal salt deprivation on taste responses. *Chemical Senses,* 1984, *8,* 246.

Brand, J. G., and Bayley, D. L. Peripheral mechanisms in salty taste reception. In M. R. Kare, M. J. Fregly, and P. A. Bernard (Eds.), *Biological and Behavioral Aspects of Salt Intake.* New York: Academic Press, 1980.

Contreras, R. J. and Kosten, T. Prenatal and early postnatal sodium chloride intake modifies the solution preferences of adult rats. *Journal of Nutrition,* 1983, *113,* 1051–1062.

Cowart, B. J. Development of taste perception in humans: Sensitivity and preference throughout the life span. *Psychological Bulletin,* 1981, *90,* 43–73.

Crook, C. K. Taste perception in the newborn infant. *Infant Behavior and Development,* 1978, *1,* 52–69.

Crook, C. K., and Lipsitt, L. P. Neonatal nutritive sucking: Effects of taste stimulation upon sucking rhythm and heart rate. *Child Development,* 1976, *47,* 518–522.

Delay, R. J., Kinnamon, J. C., and Roper, S. An HVEM Autoradiography study of cell turnover in the mouse vallate taste bud. *Abstracts of the Association for Chemoreception Sciences,* VI Meeting, 1984, No. 37.

DeSnoo, K. Das trinkende kind im uterus. *Monatsschrift für Geburtshilfe und Gynaekologie,* 1937, *105,* 88–97.

Desor, J. A., Maller, O., and Turner, R. E. Taste in acceptance of sugars by human infants. *Journal of Comparative and Physiological Psychology,* 1973, *84,* 496–501.

Doetsch, G. S., and Erickson, R. P. Synaptic processing of taste-quality information in the nucleus tractus solitarius of the rat. *Journal of Neurophysiology,* 1970, *33,* 490–507.

Farbman, A. I. Electron microscope study of the developing taste bud in rat fungiform papilla. *Developmental Biology,* 1965, *11,* 110–135.

Ferrell, F., Chole, R., and Tsuetaki, T. Developmental time course of myelination in chorda tympani nerve of postnatal rat. *Abstracts of the Association for Chemoreception Sciences,* V Meeting, 1983, No. 36.

Ferrell, M. F., Mistretta, C. M., and Bradley, R. M. Development of chorda tympani responses in rat. *Journal of Comparative Neurology,* 1981, *198,* 37–44.

Finberg, L., Kiley, J., and Luttrell, C. N. Mass accidental poisoning in infancy. *Journal of the American Medical Association,* 1963, *184,* 121–124.

Frank, M. E., Contreras, R. J., and Hettinger, T. P. Nerve fibers sensitive to ionic taste stimuli in chorda tympani of the rat. *Journal of Neurophysiology,* 1983, *50,* 941–960.

Gancrhow, J. R., and Matzner, H. Development of sucrose preference in rabbit pups. *Chemical Senses and Flavour,* 1979, *4,* 241–248.

Gottfried, D. S., Mistretta, C. M., Bradley, R. M., and Hill, D. L. Quantitative study of morphological development of rat taste bud. *Abstracts of the Association for Chemoreception Sciences,* VI Meeting, 1984, No. 57.

Hahn, P. Lipids. In U. Stawe and A. A. Weech (Eds.), *Perinatal physiology,* New York: Plenum Press, 1978.

Hatjis, C. G. and McLaughlin, M. K. Identification and ontogenesis of beta-adrenergic receptors in fetal and neonatal rabbit myocardium. *Journal of Developmental Physiology,* 1982, *4,* 327–338.

Heck, G. L., Mierson, S., and DeSimone, J. A. Salt taste transduction occurs through an amiloride-sensitive sodium transport pathway. *Science,* 1984, *223,* 403–405.

Hill, D. L. Development of pontine parabrachial nuclei taste responses in rat. *Society for Neuroscience Abstracts,* 1983, *9,* 378.

Hill, D. L., and Almli, R. C. Ontogeny of chorda tympani nerve responses to gustatory stimuli in the rat. *Brain Research,* 1980, *197,* 27–38.

Hill, D. L. Bradley, R. M., and Mistretta, C. M. Development of taste responses in rat nucleus of solitary tract. *Journal of Neurophysiology,* 1983, *50,* 879–895.

Hill, D. L., Mistretta, C. M., and Bradley, R. M. Developmental changes in taste response characteristics of rat single chorda tympani fibers. *Journal of Neuroscience,* 1982, *2,* 782–790.

Kashiwayanagi, M., Miyake, M., and Kurihara, K. Voltage-dependent Ca^{2+} channel and Na^+ channel in frog taste cells. *American Journal of Physiology,* 1983, *244,* C82–C88.

Keohe, P., and Blass, E. M., Gustatory determinants of suckling in albino rats 5–20 days of age. *Developmental Psychobiology,* 1985, in press.

Kinnamon, J. C., Delay, R., and Roper, S. Ultrastructure of taste cells and synapses in mouse vallate taste buds. *Abstracts of the Association for Chemoreception Sciences,* V Meeting, 1983, No. 72.

Maller, O., and Desor, J. A. Effects of taste on ingestion by human newborns. In J. F. Bosma (Ed.), *Fourth symposium on oral sensation and perception: Development of the fetus and infant.* DHEW Publication No. NIH 73-546, Washington, D.C.: U.S. Government Printing Office, 1973.

Marlin, N. A. Early exposure to sugars influences the sugar preference of the adult rat. *Physiology and Behavior,* 1983, *31,* 619–623.

McBurney, D. H., and Shick, T. R. Taste and water taste of twenty-six compounds for man. *Perception and Psychophysics,* 1971, *10,* 249–252.

Midkiff, E. E., and Bernstein, I. L. The influence of age and experience on salt preference of the rat. *Developmental Psychobiology,* 1983, *16,* 385–394.

Miller, A. J., McKoon, M., Pinneau, M., and Silverstein, R. Postnatal synaptic development of the nucleus tractus solitarius (NTS) of the rat. *Developmental Brain Research,* 1983, *8,* 205–213.

Mistretta, C. M. Topographical and histological study of the developing rat tongue, palate and taste buds. In J. F. Bosma (Ed.), *Third symposium on oral sensation and perception: The mouth of the infant,* Springfield, Ill.: Thomas, 1972.

Mistretta, C. M. Neurophysiological and anatomical aspects of taste development. In R. Aslin, J. R. Alberts, and M. R. Peterson (Eds.), *Development of perception: Psychobiological persepectives.* Vol. 1. New York: Academic Press, 1981.

Mistretta, C. M. and Bradley, R. M. Effects of early sensory experience on brain and behavioral development. In G. Gottlieb (Ed.), *Studies on the development of behavior and the nervous system. Vol 4: Early influences.* New York: Academic Press, 1978a.

Mistretta, C. M., and Bradley, R. M. Taste responses in sheep medulla: Changes during development. *Science,* 1978b, *202,* 535–537.

Mistretta, C. M., and Bradley, R. M. Neural basis of developing salt taste sensation: Response changes in fetal, postnatal and adult sheep. *Journal of Comparative Neurology,* 1983a, *215,* 199–210.

Mistretta, C. M., and Bradley, R. M. Developmental changes in taste responses from glossopharyngeal nerve in sheep and comparisons with chorda tympani responses. *Developmental Brain Research,* 1983b, *11,* 107–117.

Moe, K. Paradoxical sensitivity and aversion of 3 day old rat pups to NaCl, quinine and ammonium chloride. *Abstracts of the Association for Chemoreception Sciences,* VI Meeting, 1984, No. 95.

Moe, K., and Epstein, A. The ontogeny of sodium preference in rats. *Abstracts of the Association for Chemoreception Sciences,* V Meeting, 1983, No. 97.

Mook, D. Oral and postingestinal determinants of the intake of various solutions in rats with esophogeal fistulas. *Journal of Comparative and Physiological Psychology,* 1963, *56,* 645–659.

Moran, T. H., Schwartz, G. J. and Blass, E. M. Organized behavioral responses to lateral hypothalamic electrical stimulation in infant rats. *Journal of Neuroscience,* 1983, *3,* 10–19.

Mower, G. D., Christen, W. G., and Caplan, C. J. Very brief visual experience eleminates plasticity in the cat visual cortex. *Science,* 1983, *221,* 178–180.

Nowlis, G. H. Taste elicited tongue movements in human newborn infants: An approach to palatability. In J. F. Bosma (Ed.), *Fourth symposium on oral sensation and perception: Development of the fetus and infant.* DHEW Publication No. NIH 73-546, Washington, D.C.: U.S. Government Printing Office, 1973.

Nowlis, G. H., and Kessen, W. Human newborns differentiate differing concentrations of sucrose and glucose. *Science,* 1976, *191,* 865–866.

Peiper, A. Sinnesempfindungen des kindes vor seiner geburt. *Monatsschrift Kinderheilk,* 1925, *29,* 236–241.

Ramaley, J. A. Biological clocks and puberty onset. *Federation Proceedings,* 1980, *39,* 2355–2359.

Rozin, P. Psychobiological and cultural determinants of food choice. In T. Silverstone (Ed.), *Appetite and food intake.* Berlin: Abakon Verlags-Gesellschaft, 1976.

Salkoff, L., and Wyman, R. Ion channels in drosophilia muscle. *Trends in Neuroscience,* 1983, *6,* 128–133.

Sanes, D. H., and Constantine-Paton, M. Altered activity patterns during development reduce neural tuning. *Science,* 1983, *221,* 1183–1185.

Smith, D. V., and Frank, M. Cross adaptation between salts in the chorda tympani nerve of the rat. *Physiology and Behavior,* 1972, *8,* 213–220.

Smith, D. V., and McBurney, D. H. Gustatory cross-adaptation: Does a single mechanism code the salty taste? *Journal of Experimental Psychology,* 1969, *80,* 101–105.

Spitzer, N. C., and Lamborghini, J. E. The development of the action potential mechanism of amphibian neurons isolated in culture. *Proceedings of the National Academy of Science,* 1976, *73,* 1641–1645.

Steiner, J. E. Human facial expressions in response to taste and smell stimulation. *Advances in Child Development and Behavior,* 1979, *13,* 257–295.

Weiffenbach, J. M. The development of sweet preference. In J. H. Shaw and G. G. Roussos (Eds.), *Sweeteners and dental caries.* London: IRL Press, 1978.

Weiffenback, J. M., Daniel, P. A., and Cowart, B. J. Saltiness in developmental perspective. In M. R. Kare, M. J. Fregley, and R. A. Bernard (Eds.), *Biological and behavioral aspects of salt intake.* New York: Academic Press, 1980.

Weiffenbach, J. M., and Tach, B. T. Elicited tongue movements; Touch and taste in the mouth of the neonate. In J. F. Bosma (Ed.), *Fourth symposium on oral sensation and perception: Development of the Fetus and Infant.* DHEW Publication No. NIH 73-546, Washington, D.C.: U.S. Government Printing Office, 1973.

Wurtman, J. J. and Wurtman, R. J. Sucrose consumption early in life fails to modify the appetite of adult rats for sweet foods. *Science,* 1979, *205,* 321–322.

Yamada, T. Chorda tympani responses to gustatory stimuli in developing rats. *Japanese Journal of Physiology,* 1980, *30,* 631–634.

Zahm, D. S. and Munger, B. L. Fetal development of primate chemosensory corpuscles. I. Synaptic relationships in late gestation. *Journal of Comparative Neurology,* 1983, *213,* 146–162.

The Tunable Seer
Activity-Dependent Development of Vision

HELMUT V. B. HIRSCH

I. EXPERIENCE-DEPENDENT DEVELOPMENTAL PROGRAMS

> Species diversity and species constancy of behavior depend upon predictable patterns of development. In some species the course of development is relatively inflexible; learning and experience appear to play minor roles. . . . In other species, the structure and function of the brain make possible a strong reliance on learning and experience. . . . The greatest conceptual problem in the development of behavior is in understanding the interaction of the two fundamental determinants of the course of development, inheritance and experience. [Brown, 1975, p. 607]

For species of long-lived individuals, experience-dependent development provides flexibility for meeting a variety of environmental demands. This flexibility is accomplished by developmental programs that allow interactions with the environment to determine nervous system construction. Experience-dependent developmental programs therefore provide a means for (1) adapting to changing environmental conditions and (2) increasing the range of variables that guide nervous system development. Yet experience-dependent development of the nervous system is a double-edged sword. A nervous system that requires experience to develop is not only adaptable, but also vulnerable. Adequate experience must occur for the nervous system to develop successfully.

This is particularly well illustrated in the mammalian visual system, which requires both retinal and extraretinal stimulation for normal postnatal develop-

HELMUT V. B. HIRSCH   Center for Neurobiology, State University of New York at Albany, Albany, New York 12222.

ment. Most individual neurons in the mammalian visual cortex respond preferentially to lines, bars, or edges of a particular orientation (Hubel and Wiesel, 1959, 1962). This selectivity is thought to depend largely on the cell's connections with other visual system neurons (e.g., Hubel and Wiesel, 1959, 1962; Gilbert, 1983). Therefore development and maintenance of these interconnections are essential to visual system function and, as Wiesel and Hubel (1963, 1965) first demonstrated, require appropriate visual stimulation during early postnatal life. Obstructing the normal patterned visual input to one or both eyes has dramatic and deleterious effects on postnatal development of orientation selectivity.

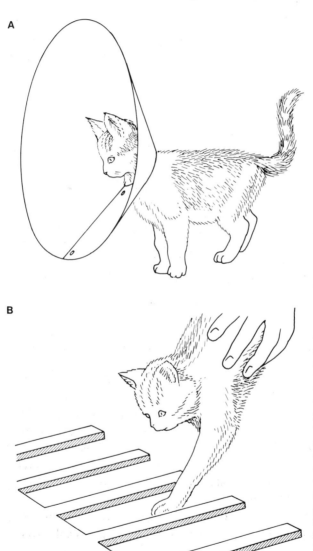

Fig. 1. (A) Kittens were reared with sight of their limbs blocked by an opaque neck ruff whenever they were in a lighted environment. (B) To test their ability to use vision for guiding placement of their limbs, the kittens were lowered toward the edge of a horizontal surface. They extended their forelimbs, but unlike normally reared kittens, they were not capable of guiding their paws accurately to the solid portions of an interrupted surface. This fractionalization of the visually controlled placing response reveals that the guided reach requires an integration of sensorimotor systems not necessary for the elicited extension response. These results show that if kittens are to be able to use vision to guide placement of their limbs, they must be able to observe their limbs while they are growing up. (From *Science,* October 11, 1967, *158,* 390–392. Copyright 1967 by the AAAS. Reproduced with permission from Hein and Held, 1967.)

The growing mammalian visual system requires more than just retinal stimulation for normal perceptual development. In order to use visual stimuli to guide motor activity, the animal must also have the opportunity to move about actively (Held and Hein, 1963), to see the consequences of its actions (Held and Bossom, 1961; Held and Hein, 1963), and to obtain nonvisual inputs such as feedback from eye movements (Hein *et al.*, 1979). For instance, kittens that cannot observe their limbs as they ambulate (Figure 1A) do not develop the ability to use vision to guide paw placement (Figure 1B) (Hein and Held, 1967). Furthermore, immobilization of one eye blocks development of normal sensorimotor coordination (Hein *et al.*, 1979) but does not affect performance of such coordination once established. A fuller understanding of visual system development therefore requires completing at least two related tasks.

The first task is to discover how development of connections within the visual system is affected by neuronal activity. I will show that part of the neuronal activity important in development is generated from within the visual system, the rest by stimulation from the environment. Internally generated patterns of neuronal activity may be determined by genetically guided developmental programs, whereas stimulation-dependent neuronal activity is determined by the characteristics of the organism's environment. A single mechanism, possibly that proposed by Hebb (1949), may determine how these two forms of activity influence connections among neurons. Neuronal activity can therefore serve as a "coin of the realm," enabling integration of genetic and environmental information.

The second task is to identify extraretinal factors, for example, those that affect visual system development. Work on this has barely begun and I can only enumerate some of the findings. Because some of these factors appear to affect both physiological and behavioral development, their study may help us to better understand the behavioral significance of experience-dependent development of the visual system.

II. The Retino-Geniculo-Cortical Pathway in Adult Cats

The most extensive data on normal development of the mammalian visual system and on the effects of neuronal activity on visual system ontogeny are available for cats and will be the focus of this chapter. I will begin with a brief description of the visual system in normal adult cats, limiting my discussion to the well-studied pathway from retina through thalamus to primary visual cortex (area 17). Next I will describe the changes occurring in the cat visual system during normal postnatal development. With that as a starting point I will examine the effects of experience on development of cortical cell orientation selectivity. Several recent reviews provide readers with additional material (Barlow, 1975; Blakemore, 1974; Fregnac and Imbert, 1984; Grobstein and Chow, 1975; Hirsch and Leventhal, 1978a,b; Movshon and Van Sluyters, 1981; Pettigrew, 1978; Sherman and Spear, 1982).

HELMUT V. B.
HIRSCH

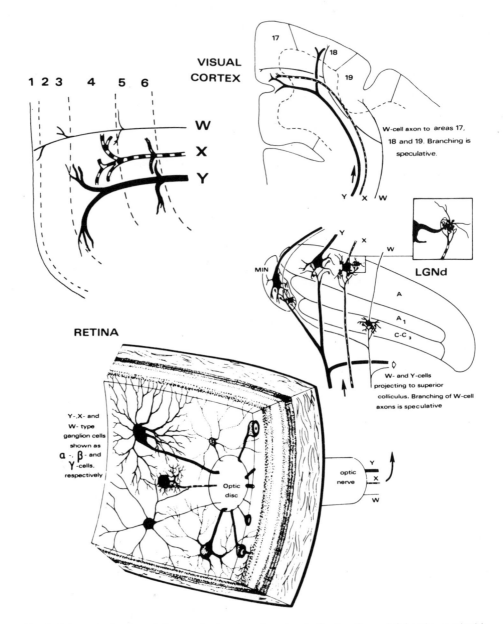

Fig. 2. Schematic drawing of the cat visual system. A segment of retina (lower left-hand corner) with the different physiologically defined classes of retinal ganglion cells (Y-, X-, and W-cells) represented by the corresponding morphologically defined types (α-, β-, and γ-cells). The axons of these cells go to a relay nucleus, the dorsal lateral geniculate nucleus or lgn, in the thalamus (middle). The axons of the Y-, X-, and W-cells contact different subgroups of lgn relay cells. These relay cells are known, respectively, as Y-, X-, and W-cells and in turn send their axons to the visual cortex. The physiologically defined classes of lgn relay cells are represented by the corresponding morphologically defined cell types. The Y-, X-, and W-cells in the lgn differ in the distribution of their axons to areas 17, 18, and 19 of the visual cortex (upper right-hand corner). All three cell types project to area 17; X-cells are predominant here and area 17 may be their only cortical target. Area 18 appears to be dominated by its Y-cell input, but a distinct W-cell input has also been described. Area 19 receives its principal input from W-type

The retina is a three-dimensional structure with levels of processing starting with rod and cone receptors, and ending with the output neurons, the ganglion cells. The sheet of ganglion cells contains a specialized central region, the area centralis, subserving high-resolution form vision (cf. Rodieck, 1973; Stone *et al.,* 1982). One way in which this functional specialization is achieved is by having the highest density of retinal ganglion cells in the area centralis. Outside the area centralis, density of retinal ganglion cells first decreases abruptly over a short distance and then gradually to the margins of the retina.

The population of retinal ganglion cells is not homogeneous, but consists of a number of different cell types. A first subdivision of retinal ganglion cells is based on the response to stimulation of the receptive field center; some cells respond at "light on" ("on-center" cells), others at "light-off" ("off-center" cells) (Kuffler, 1953). (Receptive fields of some retinal ganglion cells cannot be divided into a "center" and "surround" and are not included in this classification.)

A second, independent subdivision into three groups is based on physiological response properties of the cells, differences in their retinal distribution, and differences in their central projections (Figure 2). Each of these three physiologically characterized cell groups is associated with a morphologically identified ganglion cell type in the retina (Boycott and Wässle, 1974; Saito, 1983; Stanford and Sherman, 1984; for reviews see Hirsch and Leventhal, 1978b; Stone *et al.,* 1979; Sherman and Spear, 1982; Gilbert 1983; Rodieck, 1973, 1979; Stone, 1983). The three classes are (1) the X-cells (Enroth-Cugell and Robson, 1966; Stone and Fukuda, 1974) or sustained-cells (Ikeda and Wright, 1972; Cleland *et al.,* 1973) or Type II cells (Fukada, 1971); (2) the Y-cells (Enroth-Cugell and Robson, 1966; Stone and Fukuda, 1974) or transient cells (Cleland *et al.,* 1971, 1973; Ikeda and Wright, 1972) or Type I cells (Fukada, 1971); and (3) the W-cells (Stone and Hoffmann, 1972; Stone and Fukuda, 1974).

Retinal organization appears to have an important impact on other parts of the visual system. First, the specialization for high-resolution vision is elaborated in other parts of the visual system where a disproportionate number of cells are involved in the area centralis representation. This has suggested that "as a first approximation . . . regional variation in visual field representation seems to be derived from the regional variations in ganglion cell density found in the retina. The visual centers of the forebrain bear a strong impact of the topography of the retina, especially of the area centralis" (Stone *et al.,* 1982).

---

relay cells. The insert in the upper left-hand corner shows that axons of Y-, X-, and W-cells terminate in different layers of area 17. This pattern of termination is consistent with the idea that Y-, X-, and W-cell axons activate different subgroups of cells in area 17, at least at the level of their synaptic input to area 17. (Adapted with permission from Stone *et al., Brain Research Reviews,* 1979, *1,* 345–394, Figs. 1 and 2.)

Second, correlated firing of retinal ganglion cells may affect development and/or maintenance of connections among cells within the visual pathway. There are correlations in firing only among neighboring ganglion cells: if both cells are of the same type (e.g., both on-center or both off-center cells), then they tend to fire at the same time; but if they are of different types (e.g., one on-center, the other off-center), then they tend not to fire at the same time (Mastronarde, 1983a). There are also complex interactions among X- and Y-type ganglion cells (Mastronarde, 1983b,c). As will be discussed later, correlated activity of retinal ganglion cells is likely to influence the connectivity of their axons and thereby affect visual system development.

B. Parallel Pathways in the Visual System

The differences among cell types in the retina are preserved in the optic nerve and correspond to differences among the cell types onto which they synapse in the major thalamic relay nucleus to visual cortex, the dorsal lateral geniculate nucleus or lgn (Figure 2). The response properties of many lgn cells reflect those of the retinal ganglion cells that provide their direct, excitatory input. Relay cells in the lgn may receive direct, monosynaptic inputs from either X-, Y- or W-cells in the retina (Cleland et al., 1971; Hoffmann and Stone, 1971; Hoffmann et al., 1972; Mason, 1975; Wilson and Stone, 1975) and the resulting classes of relay cells display, with minor variations, response properties and distributions similar to those of the retinal ganglion cells. Each physiologically characterized relay cell group is associated with a distinct morphologically defined cell type in the lgn (Friedlander et al., 1981; Stanford et al., 1983). In summary, both physiological and morphological evidence indicates that information from the three types of retinal ganglion cells remains largely segregated in the lgn.

Relay cells in the lgn send their axons to various cortical visual areas (Figure 2). Some of the response properties of the cortical cells reflect the characteristics of the lgn cells that provide their direct, excitatory input. Each cortical visual area receives its own "blend" of the various classes of lgn afferents and this accounts for at least some of the differences among the areas (Dreher et al., 1980). For example, area 17 is the primary cortical target of lgn X-cells. Consistent with the properties of X-cells in retina or lgn, the cells in area 17 that receive direct, excitatory input from lgn X-cells have small receptive fields and respond well only to slowly moving stimuli (Dreher et al., 1980; Bullier and Henry, 1979b,c; Mustari et al., 1982). Area 17 also contains cells that receive direct, excitatory input from lgn Y-cells, and these tend to have large receptive fields and respond to rapidly moving stimuli (Dreher et al., 1980; Mustari et al., 1982). Finally, area 17 also contains cells that receive direct, excitatory input from lgn W-cells, and these tend to have large receptive fields and respond only to slowly moving stimuli (Dreher et al., 1980). Thus, the afferent streams originating in the retina are still segregated at the level of the first-order cells in area 17.

Excitatory connections within area 17 maintain the segregation of the different afferent streams; few higher-order cells are excited by more than one type of afferent (Singer *et al.*, 1975; Dreher *et al.*, 1980; Henry *et al.*, 1983). As a result, some response properties of higher-order cells still reflect those of the afferent stream that is the source of their input (Dreher *et al.*, 1980; Henry *et al.*, 1983).

In contrast, inhibitory interneurons provide connections between cells in the various afferent streams. For example, area 17 cells activated by lgn X-cells are often inhibited by other area 17 cells that receive their input from lgn Y-cells (Singer *et al.*, 1975; Tsumoto, 1978). Such inhibitory contacts allow activity in one afferent pathway to suppress activity in others. Since Y-cell axons conduct impulses much more rapidly than do X-cell axons, they can inhibit cortical targets of the latter before they have even started responding to an input. One possible function for such inhibition may be to suppress input during saccadic eye movements (e.g., Stark *et al.*, 1969; Matin, 1982).

C. Orientation Selectivity of Area 17 Cells

Hubel and Wiesel (1959, 1962) first showed that cells in the mammalian visual cortex are selective for stimulus orientation (Figure 3A). They found that cells within a column or slab of cortical tissue perpendicular to the cortical surface have similar preferred orientations. Furthermore, they found that the preferred orientation as represented in these orientation columns varied in a systematic fashion across the cortex (Hubel and Wiesel, 1963a).

Within a single column there are differences among cells as a function of cortical depth (e.g., Gilbert, 1977; Hubel and Wiesel, 1959, 1962; Leventhal and Hirsch, 1978). The first laminar difference to be discovered was the differential distribution of two cell types in area 17: the simple cells, and the complex cells (Hubel and Wiesel, 1959; 1962). Simple cell-receptive fields could be subdivided into distinct, mutually antagonistic "on" and "off" summating regions. The responses of a simple cell were thought to be determined by the position of these "on" and "off" regions. Simple cells are most common in layers IV and VI, where the majority of lgn afferents terminate (Figure 2).

Complex cell-receptive fields could not be subdivided into discrete "on" and "off" areas. In many cases these cells did not respond to flashed spots or gave "on–off" responses throughout the receptive field. The responses of complex cells were thought to be determined by the excitatory input received from nearby simple cells (Hubel and Wiesel, 1959, 1962). Complex cells are most common in cortical layers II, III, and V, which are adjacent to layers receiving most of the direct input from the lgn.

Subsequent work has shown that it is possible to subdivide both simple and complex cell groups further. Many simple cells apparently receive monosynaptic input from lgn X-cells (Dreher *et al.*, 1980; Mustari *et al.*, 1982; cf. Stone *et al.*, 1979), while some receive direct excitatory input from lgn Y-cells (Mustari *et al.*, 1982). The former have very small receptive fields and respond well to slowly but not to rapidly moving stimuli; the latter have somewhat larger receptive fields and respond well to a range of stimulus velocities (Dreher *et al.*, 1980; Mustari *et al.*, 1982). Many complex cells receive excitatory input from other cortical cells (Hubel

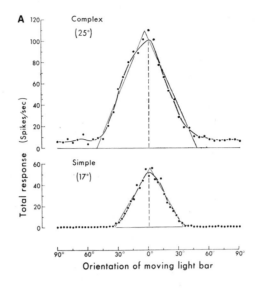

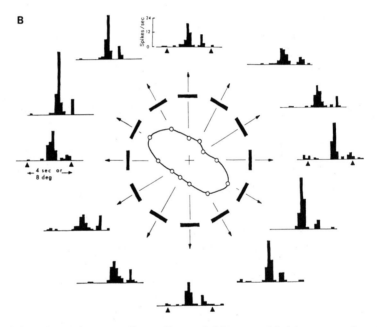

Fig. 3. (A) Orientation-tuning curve of two cells recorded from area 17. A long narrow bar of light was moved across the receptive field of the cell; the orientation of the bar varied from trial to trial. Data points indicate the number of action potentials elicited by the moving bar as a function of the orientation of the bar. Data points have been fitted with straight regression lines and with a smoothed curve made using a Gaussian weighting function. The upper curve is for a complex cell, the lower curve is for a simple cell. The optimal orientation for these two cells has been arbitrarily set to 0 deg. Note how rapidly the response of these area 17 cells falls off as the stimulus orientation deviates from the optimum orientation. Firing of the complex cell falls to half of maximum when the stimulus orientation is 25 deg from optimum; firing of the simple cell falls to half of maximum when the stimulus orientation is 17 deg from optimum. (Adapted with permission from Henry *et al.*, 1974.) (B) Orientation bias of a cell in the lgn. A black bar was moved across the receptive field of the cell; the orientation of the bar was varied from trial to trial. Histograms at the end of each line indicate the averaged response of the cell.

and Wiesel, 1959, 1962; Henry *et al.*, 1983) as well as a direct input from either lgn *X*- or *Y*-cells (Dreher *et al.*, 1980; Henry *et al.*, 1983), while others receive all their excitatory input from other area 17 cells (Henry *et al.*, 1983). As is the case for simple cells, differences in the source of the afferent input to complex cells are associated with differences in receptive field size and in the range of stimulus velocities to which the cell responds (Dreher *et al.*, 1980; Henry *et al.*, 1983). In short, there is variation among both simple and complex cells; a part of this variation arises because more than one afferent stream provides input to cells in each category (cf. Hagerty *et al.*, 1982).

D. Distribution of Preferred Orientations in Retina, Lgn, and Area 17

Cells in area 17 typically respond to only a limited range of stimulus orientations and thus they have narrow orientation tuning (Figure 3A; Henry *et al.*, 1974). In contrast, retinal ganglion cells and lgn cells usually respond to all stimulus orientations. However, the response of many lgn cells (Figure 3B; Daniels *et al.*, 1977; Creutzfeldt and Nothdurft, 1978; Lee *et al.*, 1979; Vidyasagar and Urbas, 1982) and retinal ganglion cells (Levick and Thibos, 1980, 1982) does vary with stimulus orientation so that they have an orientation bias.

Neighboring cells in the retina (Levick and Thibos, 1982), probably in the lgn (Vidyasagar and Urbas, 1982) and in area 17 (Hubel and Wiesel, 1963a), tend to have similar preferred orientations. Furthermore, for retinal ganglion cells (Levick and Thibos, 1982), for area 17 simple cells (Leventhal, 1983), and probably for lgn cells (Vidyasagar and Urbas, 1982), the distribution of preferred orientations of cells varies as a function of the retinal location of their receptive fields. Except for cells that have receptive fields near the centralis, the distribution of preferred orientations is like the spokes of a wheel with the axis at the area centralis (Figure 4).

In summary, there is a radial organization of orientation preferences at all levels from the retina to the first-order cells in area 17; at each stage the area centralis forms the center for this pattern. The orientation bias of retinal and lgn cells may provide a framework upon which to base the orientation selectivity of area 17 cells (Leventhal, 1983). This would require that lgn afferents to area 17 be segregated according to their orientation preferences. In that case, the afferent projection to area 17 preserves both the identity of the various classes of retinal ganglion cells and their orientation preferences. Correlated firing of retinal ganglion cells

---

Note two response peaks in each histogram, one representing a center-and-surround response as the black bar enters the center of the receptive field and leaves one side of the surround; the other smaller peak representing the response as the black bar leaves the other side of the surround. These histograms were used to construct the polar plot in the center which represents the response of the unit for each of the 12 directions at which the stimulus moved across the receptive field; only the first, larger peak was used in construction of the polar plot. Note that the lgn cell responds to all stimulus orientations, but there is a bias along the 150–330 deg axis. (From Daniels *et al.*, *Experimental Brain Research*, 1977, *29*, 155–172. Reproduced with permission.)

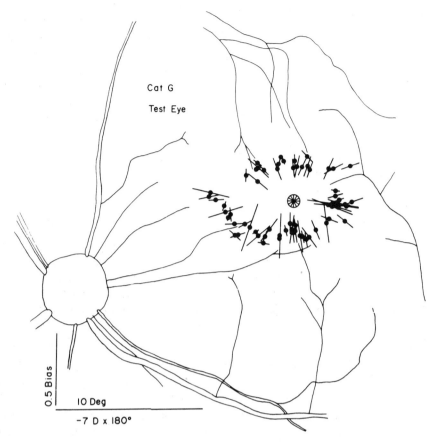

Fig. 4. Tracing of a fundus photograph showing the distribution of the orientation bias of 64 retinal ganglion cells. Dots indicate the location of the receptive field for each cell; lines through the dots indicate the orientation preference for each cell and the length of each line is proportional to the magnitude of the bias of the cell. The wheel symbol indicates the location of the area centralis. Note that the lines indicating the preferred orientation for each cell point toward the area centralis, showing the radial organization of orientation preferences of retinal ganglion cells. These data are taken from a cat raised wearing a lens that produced an astigmatism, but the distribution of orientation preferences is the same as in normally reared animals, showing that the distribution of orientation preferences of retinal ganglion cells is not affected by abnormal early visual experience. (From *Vision Research*, Vol. 22, Thibos and Levick. Copyright 1982 by Pergamon Press, Ltd. Reproduced with permission.)

may play an important role in development and/or maintenance of this segregation.

III. Development of the Retino-Geniculo-Cortical Pathway in Normal Cats

The cat visual system is relatively immature at birth. I have divided early postnatal development into three stages: (1) the first three weeks, during which most changes in the visual system are likely to be determined intrinsically; (2) the second three weeks, during which there are dramatic, experience-dependent changes, many of which may involve development of excitatory connectivity; and (3) the next 6–10 weeks, during which experience-dependent changes continue, albeit at a reduced rate; these late changes may involve development of inhibitory connectivity.

Adult cats are skillful predators, effective in using vision to guide their behavior. They begin to develop the necessary skills starting around the end of the first stage of postnatal life, but the bulk appear over the course of the second stage. During the third stage there is a qualitative change in behavior, marked by an upsurge in active efforts to seek out visual stimulation and explore the environment.

A. OPTICS AND ALIGNMENT OF THE EYES

Some of the change in visual function during early postnatal life may be caused by improvements in visual optics and/or of neural function. The kitten's eye is cloudy at birth because remnants of the embryological vasculature are still present. By the third postnatal week, however, this network has receded considerably and by the fourth postnatal week, maturation is essentially complete (Freeman and Lai, 1978). During this time the eye is also growing longer (Thorn et al., 1976; Olson and Freeman, 1980) and the cornea flatter (Freeman et al., 1978).

Direct measurement of the optical transfer function of the kitten eye shows that inputs are attenuated by about one log unit, because of the poor optical quality and because of the low efficiency of the tapetal surface (Bonds and Freeman, 1978). Optics, however, are probably not the limiting factor for visual acuity during the first month of postnatal life (Bonds, 1979; Bonds and Freeman, 1978); improvements in acuity during this period are more likely to reflect neuronal rather than optical changes.

In very young kittens the pupillary axes are divergent and this may mean that the optic axes of the two eyes are also not aligned, preventing the animal from obtaining concordant binocular stimulation (Sherman, 1972; Olson and Freeman, 1978). Available data on the alignment of the optic axes are somewhat contradictory. Pettigrew (1974) found that the optic axes are divergent in young, paralyzed kittens. Olson and Freeman (1978), however, report that the optic axes are aligned or even slightly convergent in alert kittens and they find no reliable changes in eye alignment with increasing age of the animal. Thus it is likely kittens can receive concordant stimulation as early as the second postnatal week.

B. RETINAL DEVELOPMENT

1. STAGE 1. Projections from retina to lgn and superior colliculus develop prenatally (Anker, 1977; Anker and Cragg, 1974; Shatz, 1983). The number of optic axons peaks 10–20 days before birth, at which time there are many more axons

than in the adult. This number decreases to adult levels by about 2 weeks after birth (Ng and Stone, 1982).

Development of a specialized region of the retina, later to become the area centralis, also begins prenatally around embryonic day 50, that is, during the final quarter of gestation, and continues postnatally. Those retinal ganglion cells that subserve central vision mature first, followed by the retinal ganglion cells within an elliptical region about the area centralis (in adult cats this is a specialized region called the horizontal visual streak), and then by the retinal ganglion cells in the remaining portions of the retina (Rapaport and Stone, 1982). In central retina genesis of cells is complete at birth, but in peripheral retina it continues until the end of the third postnatal week (Johns et al., 1979). Myelination of the optic axons begins during the last few prenatal days and increases rapidly after birth, reaching 90% at 30 days postnatal (Moore et al., 1976; Ng and Stone, 1982).

Retinal ganglion cell morphology changes dramatically over the course of the first postnatal stage. During the early part of this period cells appear quite immature (Donovan, 1966; Vogel, 1978), but by 2–3 weeks postnatal many are essentially mature (Donovan, 1966; Vogel, 1978; but cf. Tucker, 1978). By 3 weeks of age at least the beta ganglion cells, which are likely to correspond to the physiologically identified X-cells (Boycott and Wässle, 1974; Saito, 1983; Stanford and Sherman, 1984), have reached adult size near the area centralis, but not in more pheripheral regions of retina (Rusoff and Dubin, 1978). Despite their mature soma size, retinal ganglion cell terminals in the lgn are still very immature at 3 weeks postnatal (Friedlander et al., 1982; see also Mason, 1982a,b).

As retinal ganglion cells mature morphologically, their physiological properties develop. Some electrical activity can be recorded from the retina around the end of the first postnatal week (Zetterstrom, 1956). There is little information about the response properties of single retinal ganglion cells at this early stage of development. Single cells, however, have been recorded in 3-week-old kittens. Some cells in central (Hamasaki and Flynn, 1977) but not in peripheral retina (Rusoff and Dubin, 1977) have receptive field centers comparable in retinal extent to those in adult cats as well as center–surround organization. It is possible to classify some cells in central retina (Hamasaki and Sutija, 1979; but cf. Friedlander et al., 1982), but not in peripheral retina (Rusoff and Dubin, 1977), as X- or Y-cells. Even in central retina many cells still display mixed X- and Y-cell properties (Hamasaki and Sutija, 1979). Relatively more retinal ganglion cells that have characteristics of Y-cells are found at 3 weeks postnatal than in adult animals (Hamasaki and Sutija, 1979). Finally, cells in central retina have lower spontaneous activity levels, longer response latency to visual stimulation, and lower temporal resolution, and require higher intensity stimulation than do cells in adults (Hamasaki and Flynn, 1977). Conduction velocity of their axons is also lower than in adult cats (Friedlander et al., 1982).

The retina is therefore maturing both physiologically and morphologically during the first phase of postnatal life, and central retina matures earlier than peripheral retina. There is physiological evidence that Y-cells may be the first to mature (Hamasaki and Sutija, 1979), but this is difficult to reconcile with the anatomical evidence that beta cells, which are likely to correspond to X-cells, may be the first to mature in central retina (Rusoff and Dubin, 1978). Little is known about the status of W-cells in the young kitten retina, perhaps because their small somata hinder recording. It is also not known when retinal ganglion cells develop the orientation bias that has been reported for retinal ganglion cells in adult cats (e.g., Levick and Thibos, 1982).

2. STAGE 2. The retina matures considerably between the third and sixth postnatal week. By the fourth postnatal week some ganglion cells in peripheral retina are mature enough to be classified as X- or Y-cells (Rusoff and Dubin, 1977). The numbers of these cells increase over the next few weeks (Rusoff and Dubin, 1977; see also Friedlander *et al.*, 1982). There are also increases in the number of cells in central retina that can be identified by their receptive field centers as X- or Y-cells, although some continue to have mixed properties (Hamasaki and Sutija, 1979). The proportion of X-cells reaches adult levels by 5–6 weeks postnatal (Hamasaki and Sutija, 1979). Receptive field center size reaches adult levels by 7 weeks postnatal, although at that age receptive field surrounds are larger and possibly less effective than in adult cats; this may help explain why cells are hyperexcitable at this stage (Rusoff and Dubin, 1977). Conduction velocities of the ganglion cell axons are still somewhat slower than in the adult (Friedlander *et al.*, 1982). Finally, although Y-cell axons have achieved adultlike patterns of innervation of the lgn by the fifth postnatal week, in absolute terms the terminal arbors of these cells in the upper layers of the lgn (the A laminae) are smaller than those in the adult, and in the lower layers (the C laminae) they are poorly developed compared with those of the adult (Friedlander *et al.*, 1982). Receptive field centers of retinal ganglion cells are therefore quite mature by the end of the sixth or seventh postnatal week, and as a result, cells may be assigned to groups comparable to those found in adult cats. Some cells, however, still have mixed X- and Y-cell responses (Hamasaki and Sutija, 1979). Furthermore, development of receptive field surrounds lags behind development of the centers.

3. STAGE 3. During the course of the third postnatal stage, retinal ganglion cells continue to mature so that by 12 weeks postnatal, cells with mixed X- and Y-cell responses are no longer found, and response properties of the ganglion cells are adultlike (Hamasaki and Sutija, 1979). During this final phase of retinal development, there are changes in the balance between center and surround responses as the surrounds mature.

4. SUMMARY. There is an initial, prenatal overproduction of retinal ganglion cells, followed by a reduction to adult levels during the first postnatal stage. Cells in the area centralis mature first; some of them may have relatively mature morphology and physiology by the end of the first postnatal stage. In contrast, cells in more peripheral parts of the retina remain immature during the first postnatal stage; they develop mature, receptive field centers during the second postnatal stage. Although receptive field centers are generally mature by the end of the second postnatal stage, complete development of surrounds is delayed until the third

postnatal stage. Maturation of inhibitory connectivity in the retina may thus be the final stage in ontogeny of this structure. Since "fine tuning" of the balance between excitatory and inhibitory inputs to a cell is often influenced by neuronal activity (cf. Murphey and Hirsch, 1982), it is possible there is some experience-dependent optimization of retinal ganglion cell properties during the third postnatal stage.

An overproduction of young neurons, such as occurs in the retina, is common in the developing nervous system (for reviews see Jacobson, 1978; Lund, 1978). It may help determine the final shape of many of the cell groups or nuclei in the nervous system (e.g., Saunders, 1966), provide a means for "error correction" (cf. Jacobson, 1970), and provide an opportunity for modifying the development of the nervous system in accordance with conditions prevailing in the organism's environment (Jacobson, 1973; Hirsch and Jacobson, 1975). An overproduction of neuronal elements, followed by a "pruning away" of the excess might help to ensure a match between the operational characteristics of the surviving neurons and the functional requirements of the animal as they are determined by its environment and past history (cf. Roux, 1881; Cowan, 1973; Prestige, 1974; Hollyday and Hamburger, 1976; Jacobson, 1978).

C. LGN DEVELOPMENT

1. STAGE 1. Although there are afferents to the lgn before birth (Anker, 1977; Shatz 1983; Shatz and Kirkwood, 1984), its basic structure develops during the first postnatal stage. The cell layers characteristic of the adult lgn are maturing during the first postnatal week and do not become clearly established until 14 days postnatal (Kalil, 1978b). Furthermore, cells that are of fairly uniform size at birth develop a variety of soma sizes by the end of the first postnatal week (Hickey, 1980; Kalil, 1978b). Over the next 3 weeks there is an increase in the proportion of medium-sized cells. Although available information on the time course of maturational changes in the lgn is not as complete as in the retina, it appears that, unlike in retina, cells subserving central vision do not develop more rapidly than those subserving peripheral vision. If anything, cells representing the periphery acquire the adult distribution of cell sizes before more central cells (Kalil, 1978b).

The overall changes in lgn structure during the first postnatal stage are accompanied by maturation of the morphology of individual cells (Mason, 1983). There is a phase of rapid dendritic extension during the second postnatal week; this is associated with the development of spines along the dendritic processes. Maturation does not proceed far enough during the first postnatal stage to make it possible to distinguish the cell types in the lgn morphologically (Figure 5; Mason, 1983).

The synaptic organization of the lgn is also very immature at birth; most lgn synapses are formed during the first postnatal month (Cragg, 1975). These synapses can be divided into two main groups, asymmetric and symmetric, which may differ in their function; asymmetric synapses are thought to be excitatory, sym-

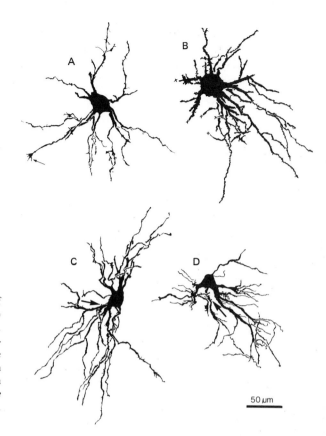

Fig. 5. Camera lucida drawings of cells in the lgn of 3-week-old (A, B) and 4-week-old (C, D) kittens. Note the growth cone on a dendrite (small arrow), and the rounded grapelike appendage on (C) and (D) (large arrows). (From C. A. Mason, *J. of Comparative Neurology*, 1983, *217*, 458–469. Reproduced with permission.)

50 μm

metric ones inhibitory. From birth to about 20 days postnatal, most of the synapses that form are asymmetric; the proportion of symmetric synapses remains low (Winfield *et al.*, 1980; cf. Winfield and Powell, 1980). This would suggest that during early postnatal life, lgn maturation primarily involves development of excitatory connectivity.

Accompanying the morphological maturation of the lgn are changes in its physiology. Neuronal activity can be recorded in the lgn as early as two days postnatal. The mean discharge frequency of the cells increases during the first postnatal month and by 3 weeks postnatal is modulated by changes in state (sleep–waking) in much the same way as in adults (Adrien and Roffwarg, 1974).

Cells recorded in the lgn before 3 weeks postnatal have some characteristics of lgn cells in adult cats, yet in other ways they are still immature (Norman *et al.*, 1977; Daniels *et al.*, 1978; Ikeda and Tremain, 1978; Friedlander, 1982). Lgn cells may be classified as sustained or transient (Ikeda and Tremain, 1978) but attempts to use other response properties to identify precursors of X- and Y-cells have met with limited success (Norman *et al.*, 1977; Daniels *et al.*, 1978; Friedlander, 1982).

By 3 weeks postnatal, there are cells in the A laminae of the lgn that display orientation biases comparable to those in adult cats (Albus *et al.*, 1983). In addition, at 3–4 weeks postnatal relatively mature direction- or motion-selective cells are present in the C laminae of the lgn (Daniels *et al.,* 1978; Friedlander, 1982). These same cells have mature morphology and sizes of their cell bodies fall within the range characteristic of *W*-cells in adults (Friedlander, 1982).

During the first postnatal stage the lgn thus develops much of its adult structure. The distribution of soma sizes begins to approach that seen in the mature lgn. Individual cells begin to mature, but the cell types present in the adult cannot be readily distinguished either morphologically or physiologically. Finally, dendritic spines, which are not characteristic of lgn cells in normal animals, begin to develop as dendrites start growing.

2. STAGE 2. The soma size of lgn cells continues to increase to near adult levels during the second postnatal month (Kalil, 1978b). Furthermore, there is an increase in the relative numbers of large-sized cells. By 8 weeks postnatal, the proportions of different-sized somas in the lgn have reached adult levels, at least in the peripheral and monocular part of lamina A (Kalil, 1978b).

Morphology of individual lgn cells also continues to develop during the second postnatal stage (Mason, 1983). Between the third postnatal week and the sixth to eighth postnatal week, cells develop the shape and dendritic extent characteristic of mature lgn neurons (Figure 6; compare with Figure 5) (Mason, 1983). However, the cells still differ from adult lgn cells in one important way: they continue to have spines (Mason, 1983). These spines, which are present from the time that dendritic elongation begins, during the first postnatal stage, disappear from the soma and proximal dendrites between the third and sixth to eighth postnatal week while remaining on the peripheral dendrites (Mason 1983).

Synaptic development in the lgn also continues after the third postnatal week. Specifically, the porportion of symmetric synapses increases dramatically to about 30% of the total shortly after the beginning of the fourth postnatal week (Winfield *et al.,* 1980). To the extent that symmetric synapses may be identified with inhibitory connections, this suggests that there is a burst in development of inhibitory connectivity during this phase of postnatal life (Winfield *et al.,* 1980).

During the period that individual lgn cells acquire distinctive dendritic morphology, their response properites become more adultlike (Norman *et al.,* 1977; Daniels *et al.,* 1978; Ikeda and Tremain, 1978) making it easier to distinguish X-, Y-, and W-type cells physiologically (Daniels *et al.,* 1978). Surround responses have developed but they are still weaker than in adult cats (Daniels *et al.,* 1978; Ikeda and Tremain, 1978).

3. STAGE 3. Although lgn cells have acquired their adult size and morphology by the sixth to eighth postnatal week (Kalil, 1978b; Mason, 1983), several important developments occur after that time. These include a loss of the spines from the peripheral dendrites of the lgn cells (Mason, 1983), an increase in lgn volume to adult values (Kalil, 1978b), and development of an adult complement of large cells in the paracentral portion of lamina A (Kalil, 1978b). Furthermore, the proportion of symmetrical synapses continues to increase (at about one-tenth its rate of increase during the second postnatal stage) to the adult level of 45%, which is reached by about 15 weeks postnatal (Winfield *et al.,* 1980).

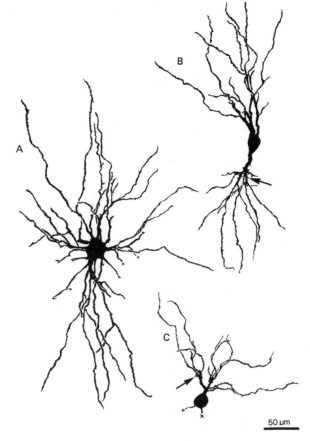

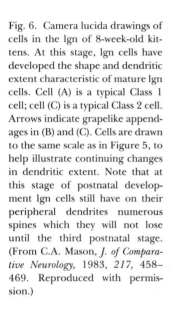

Fig. 6. Camera lucida drawings of cells in the lgn of 8-week-old kittens. At this stage, lgn cells have developed the shape and dendritic extent characteristic of mature lgn cells. Cell (A) is a typical Class 1 cell; cell (C) is a typical Class 2 cell. Arrows indicate grapelike appendages in (B) and (C). Cells are drawn to the same scale as in Figure 5, to help illustrate continuing changes in dendritic extent. Note that at this stage of postnatal development lgn cells still have on their peripheral dendrites numerous spines which they will not lose until the third postnatal stage. (From C.A. Mason, *J. of Comparative Neurology*, 1983, *217*, 458–469. Reproduced with permission.)

The physiological properties of lgn cells continue to mature after the sixth postnatal week; most of the changes represent continued development of inhibitory surrounds (Ikeda and Tremain, 1978). Consistent with this, lgn cells continue to increase their sensitivity to higher spatial frequences and the spatial resolving power of lgn cells improves and reaches adult levels by 16 weeks postnatal (Ikeda and Tremain, 1978). Finally, temporal inhibition also increases between the sixth and sixteenth postnatal week (Ikeda and Tremain, 1978).

Morphological and physiological changes in the lgn thus continue after the first 6 weeks of postnatal life. The loss of dendritic spines is likely to be associated with a reduction in the proportion of asymmetric synapses (Winfield *et al.*, 1980). These morphological changes may be correlated with physiological changes in the balance of excitatory and inhibitory processes, thereby improving both spatial and temporal resolving power of lgn cells (Ikeda and Tremain, 1978).

4. SUMMARY. The appearance of the basic structural characteristics of the adult lgn during the first postnatal stage is accompanied by physiological and morphological maturation of some lgn cells. Maturation does not proceed far enough during that stage for the cell types characteristic of the adult to be distinguished

readily; this level of development is not reached in the lgn until the second post-natal stage. Most lgn synapses develop postnatally. Many of those formed during the first postnatal stage may be excitatory; production of inhibitory synapses is apparently delayed until the second and third postnatal stages. Consistent with this, development of receptive field centers precedes development of surrounds.

In many ways development in the lgn parallels that in the retina, with retinal development possibly leading the way. However, while conditions present in the area centralis may provide a focal point for orchestrating retinal development, they do not appear to dictate the time course of development in the lgn. As will be discussed below, retinal organization does, however, affect development of con-nections between retinal ganglion cell axons and lgn relay cells.

While overproduction of dendritic spines followed by a subsequent elimina-tion or decrease has been documented for a number of regions in the central ner-vous system, the significance of this process is not clear. This phenomenon may reflect not only a loss of synapses but a redistribution of synapses (cf. Mates and Lund, 1983a,b,c). As we shall see, sensory stimulation plays a role in lgn develop-ment; perhaps changes in dendritic spines represent the structural basis for some of this experience-dependent development (cf. Valverde, 1967).

D. DEVELOPMENT OF AREA 17

1. STAGE 1. Afferents from the lgn reach area 17 some 12 days before birth (Anker, 1977). Their distribution within area 17 changes significantly over the course of the first postnatal stage. During the first postnatal week lgn afferents project more heavily to the upper cortical layers (especially layer I) than in adult cats (Anker, 1977; Anker and Cragg, 1974; Laemle et al., 1972; LeVay et al., 1978; Kato et al., 1983). At this stage the projection to layer IV may be wider than that in adult cats, so that there is an abnormally dense projection to layer V (LeVay et al., 1978). In addition, during the first 2 postnatal weeks afferents from the two eyes have extensive overlap in layer IV (LeVay et al., 1978). The adult distribution of lgn afferents in area 17, in which the highest density of projections is restricted to layer IV and afferents from the two eyes are segregated into ocular dominance "patches," appears by the end of the third postnatal week (Laemle et al., 1972; LeVay et al., 1978; Kato et al., 1983).

During the time that the distribution of the lgn afferents is maturing, most area 17 synapses are being formed. A few synapses are present as early as 3 weeks before birth, but at birth less than 1% of the adult number of synapses are found on an average neuron (Cragg, 1975). Synaptic density increases during post-natal life, especially during the first 40 postnatal days (Cragg, 1975; Winfield, 1981).

As more synapses develop in area 17, response properties of individual cells begin to mature. Area 17 cells may be activated by light stimulation by the sixth postnatal day (Huttenlocker, 1967), but at this stage few, if any, display the ori-entation selectivity characteristic of area 17 cells in older animals (Pettigrew, 1974; Fregnac and Imbert, 1978; but cf. Hubel and Wiesel, 1963b; Beckmann and Albus,

1982; Albus and Wolf, 1984). During the second postnatal week responses to visual stimuli are usually erratic, cells fatigue rapidly with repeated stimulation, and the maintained firing rates of the cells are very low (Hubel and Wiesel, 1963b; McCall, 1983). Moving stimuli are especially effective (Pettigrew, 1974; Blakemore and Van Sluyters, 1975; Bonds, 1979; McCall, 1983), one or two directions often being preferred to the others (Barlow and Pettigrew, 1971; Pettigrew, 1974; Blakemore and Van Sluyters, 1975). In adult cats such direction selectivity may be associated with neurons receiving direct excitatory input from lgn W-type cells (Dreher *et al.*, 1980). There is, however, no direct evidence that early responses of area 17 cells are mediated by W-type cells (but see Kato *et al.*, 1983), although the heavy projection to layer I (Anker, 1977; Anker and Cragg, 1974; Laemle *et al.*, 1972; LeVay *et al.*, 1978; Kato *et al.*, 1983) may be consistent with this (cf. Leventhal, 1979).

Toward the end of the first postnatal stage some area 17 cells have matured sufficiently that they have selectivity for stimulus orientation comparable to that in adult cats (Hubel and Wiesel, 1963b; Blakemore and Van Sluyters, 1975; Buisseret and Imbert, 1976; Fregnac and Imbert, 1978; Bonds, 1979; Beckmann and Albus, 1982; McCall, 1983; Albus and Wolf, 1984). Most area 17 cells, however, are still immature at 2–3 weeks postnatal; they are either completely nonselective for orientation or display only a bias for one orientation (Pettigrew, 1974; Blakemore and Van Sluyters, 1975; Buisseret and Imbert, 1976; Fregnac and Imbert, 1978; Albus and Wolf, 1984). There is thus a subset of early maturing cells in retina, lgn, and area 17.

The cells that are orientation selective by the end of the first postnatal stage have several distinctive characteristics. One is that they are likely to be first-order neurons, activated monosynaptically by lgn afferents. The most direct evidence for this is from the latency to electrical stimulation of the afferent pathways; although transmission time to area 17 is very long in young kittens, Beckmann and Albus (1982) interpret the overall latencies of orientation-selective cells as being consistent with monosynaptic activation. Indirect evidence that the early maturing cells are first-order neurons derives from their laminar location and response properties. Orientation-selective cells in 2- or 3-week-old kittens tend to be concentrated in the deeper layers of area 17, especially in layers IV and VI where the bulk of the lgn afferents terminate (Blakemore and Van Sluyters, 1975; Albus and Wolf, 1984; but see also Tsumoto and Suda, 1982, who claim that the first cells to display orientation selectivity are located in layer V). In addition, the first orientation selective cells have receptive field characteristics (Blakemore and Van Sluyters, 1975; Beckmann and Albus, 1982; McCall, 1983; Albus and Wolf, 1984) associated with simple cells in area 17 of adult cats, and simple cells are good candidates for first-order neurons (Hubel and Wiesel, 1959, 1962, Singer *et al.*, 1975; Mustari *et al.*, 1982). The nonselective cells are more like complex cells (Blakemore and Van Sluyters, 1975; Buisseret and Imbert, 1976; Fregnac and Imbert, 1978) and thus many of them may be second-order cells (Hubel and Wiesel, 1959, 1962, Singer *et al.*, 1975; Mustari *et al.*, 1982; Henry *et al.*, 1983).

A second distinctive feature of the first orientation-selective cells is that their response properties are, in part, determined by inhibitory inputs from other area

17 cells. Iontophoretic application of a drug (*N*-methylbicuculine) that is an antagonist for a putative neurotransmitter, gamma-aminobutyric acid (GABA) (Sillito, 1975), reduces the sharpness of tuning of these orientation-selective cells (Sato and Tsumoto, 1984); the cells retain an orientation bias so it is likely that this is derived from their excitatory inputs. The immature cells are much less likely to receive GABA-mediated inhibitory input. Response properties of some of the orientation-biased cells are slightly altered by iontophoretic application of *N*-methylbicuculine, and responses of the nonselective cells are unchanged (Sato and Tsumoto, 1984). A part of the selectivity of the early-maturing cells may thus be produced by intracortical inhibitory connectivity; the immature cells may lack such inhibitory inputs.

There are two further distinctive features of the first area 17 cells that display orientation selectivity. First, the distribution of their orientation preferences is not uniform. Results of some studies show a bias towards horizontal and vertical (Fregnac and Imbert, 1978; McCall, 1983), while results of another study show only a slight bias towards horizontal (Albus and Wolf, 1984). [The peaks in the distribution of preferred orientation may vary with the retinal locus of the cells' receptive fields; see Leventhal (1983) for a possible explanation of this relationship.] Second, these orientation-selective cells are mostly monocular (Blakemore and Van Sluyters, 1975; Fregnac and Imbert, 1978; McCall, 1983) and are dominated by the contralateral eye (Fregnac and Imbert, 1978; but cf. Blakemore and Van Sluyters, 1975, and McCall, 1983, for failures to find a contralateral dominance among orientation-selective cells in young kittens). The nonselective cells tend to be binocular (Blakemore and Van Sluyters, 1975; Fregnac and Imbert, 1978; McCall, 1983).

2. STAGE 2. Synaptic density continues to increase during the second postnatal stage (Cragg, 1975; Winfield, 1981). During the same period the proportion of orientation-selection cells increases dramatically, reaching adult levels at some point between the fourth postnatal week (Hubel and Wiesel, 1963b; Blakemore and Van Sluyters, 1975; Sherk and Stryker, 1976; Bonds, 1979, Albus and Wolf, 1984) and the end of the fifth postnatal week (Fregnac and Imbert, 1978; Pettigrew, 1974). Furthermore, the "tuning width" of cortical cells continues to improve, that is, to become narrower, once the proportion of cells that are selective reaches adult levels (Figure 7) (Bonds, 1979). Finally, selectivity for the spatial frequency of an optimally oriented sine wave grating gradually increases to adult levels between the third and eighth postnatal week (Derrington and Fuchs, 1981). In short, during the fourth, fifth, and possibly sixth postnatal week there is a dramatic increase in the proportion of area 17 cells that display mature selectivity. After this time, as in adult cats, the great majority of the cells display such selectivity.

3. STAGE 3. During the third postnatal stage synaptic density overshoots adult levels only to drop down again by the end of this period. Total synaptic density reaches a maximum at about 10 weeks postnatal, and then declines to adult levels by about 15 weeks postnatal (Cragg, 1975; Winfield, 1981). During this period the density of asymmetric synapses remains relatively constant while the density of sym-

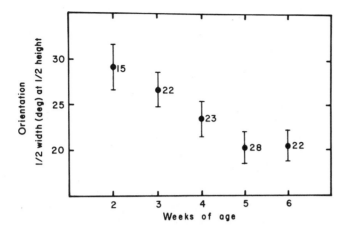

Fig. 7. Orientation tuning of orientation-selective cells in area 17 as a function of age. The ordinate represents one-half the width of the orientation-tuning curve at that level for which firing of the cell has fallen to one-half of the response to the optimum orientation (see Figure 3A). The bars represent $+/-$ one standard error of the mean. Note that orientation-selective cells become more finely tuned for orientation over the course of the second postnatal stage. (From A. B. Bonds, *Developmental Neurobiology of Vision*, 1979, pp. 31–49. Reproduced with permission.)

metric synapses continues to climb (Winfield, 1981). Furthermore, the number of asymmetric synapses per neuron drops after the tenth postnatal week, while the number of symmetric synapses per neuron continues to increase after that time (Winfield, 1981). The relative increase in the number of symmetric synapses per neuron may mean that there is an overall increase in inhibitory connectivity in area 17 during this phase of postnatal development.

Very little is known about possible physiological changes that might accompany the changes in synaptic connectivity during the third postnatal stage. The proportion of orientation-selective cells in layers IV, V, and upper VI is close to adult levels at about 6 weeks postnatal, but the proportion of orientation-selective cells in layers II/III and lower VI may be only 50% of what it will be in the adult (Tsumoto and Suda, 1982). There is other, indirect evidence that maturation of the cells in layers II/III and lower VI lags behind maturation of cells in the other layers of area 17 (Tsumoto and Suda, 1982). These results suggest that development of orientation selectivity may continue after 6 weeks of postnatal life in those layers (II/III) of area 17 that provide the bulk of the efferent projection to other cortical visual areas (Gilbert and Kelly, 1975; see Gilbert, 1983, and Swadlow, 1983, for recent reviews). Selectivity for the spatial frequency of a stimulus continues to increase during this stage, reaching adult levels by the eighth postnatal week (Derrington and Fuchs, 1981); this is consistent with an increase in inhibitory input to area 17 cells during this stage of development.

4. SUMMARY. During the first postnatal stage lgn afferents to area 17 develop their adult distribution, large numbers of synapses are formed, and some cells

develop orientation selectivity. These selective cells are likely to be first-order neurons and to receive intracortical inhibitory inputs and they are concentrated in the deeper layers of area 17, especially layers IV and VI. Most of them are monocular, activated by the contralateral eye, and respond preferentially to lines near horizontal and vertical.

During the second postnatal stage synaptic development continues in area 17, and there is a dramatic increase in the proportion of orientation-selective cells. Many of the cells that acquire orientation selectivity during this period are likely to be located outside layers IV and VI and receive significant excitatory input from other area 17 neurons. This development of selectivity may reflect changes in both excitatory and inhibitory intracortical connectivity.

During the third postnatal stage synaptic density in area 17 overshoots adult levels and then drops down again; this may indicate that there is an overproduction of synapses during postnatal development followed by a "pruning away" of those that are not appropriate to the functional requirements of the animal. There is a relative increase in the proportion of symmetric synapses per neuron during the third postnatal stage, and so there may be an overall increase in inhibitory connectivity. Very little is known about development of the response properties of area 17 cells during the third postnatal stage. Available evidence suggests that cortico–cortical projection cells in the upper layers (II/III) may not develop adult levels of orientation selectivity until this stage of development. The improvement in selectivity for spatial frequency during this stage of development suggests that there is continued development of inhibitory connectivity in area 17.

IV. ACTIVITY-DEPENDENT DEVELOPMENT OF THE VISUAL SYSTEM

Neuronal activity influences postnatal development of connections in the visual system. During the first postnatal stage this activity is largely internally generated; during later stages it is generated externally by visual stimulation.

These conclusions are based on results from three lines of research. First, the importance of activity generated by the neurons themselves has been demonstrated by manipulating such activity pharmacologically, for example, by repeated intraocular injection of tetrodotoxin (TTX) to block retinal ganglion cell action potentials (e.g., Archer *et al.*, 1982; Meyer, 1982, 1983; Kalil *et al.*, 1983; Schmidt and Edwards, 1983; Schmidt *et al.*, 1983). Second, the importance of stimulation-dependent neuronal activity has been shown by depriving young animals of all patterned visual stimulation by raising them from birth in total darkness (e.g., Bonds, 1979; Cynader *et al.*, 1976; Leventhal and Hirsch, 1977, 1980; Mower *et al.*, 1981a,b). Third, the specific effects of stimulation-dependent neuronal activity on the course of postnatal ontogeny have been shown by raising animals in controlled visual environments (e.g., Blakemore and Cooper, 1970; Hirsch and Spinelli, 1970, 1971).

1. The Retina. Blocking retinal ganglion cell activity by intraocular injection of TTX during the first 5 to 8 postnatal weeks appears not to have gross effects on development of the retinal ganglion cells; once the TTX wears off, their physiology is normal (Archer *et al.*, 1982). The absence of retinal ganglion cell activity does, however, affect connections that retinal ganglion cell axons make in the lgn (Kalil *et al.*, 1983). In cats reared for 2 months with retinal ganglion cells silenced by TTX retinogeniculate terminals have many features in common with terminals in neonatal kittens (Kalil *et al.*, 1983).

There is limited information about the effects of dark-rearing on retinal development. Several studies have shown that long periods of deprivation may produce both biochemical (Rasch *et al.*, 1961) and morphological (Rasch *et al.*, 1961; Weiskrantz, 1958) changes in the retina (see also Chow, 1973; Gyllensten *et al.*, 1965, 1966; Hendrickson and Boothe, 1976); the effects of short periods of dark-rearing have not been studied in cats.

2. The Lgn. Lgn cells develop abnormal response properties if retinal activity is blocked for periods of 5–8 weeks starting within a few days of birth. These changes all point to a breakdown in the normal segregation of the different types of retinal ganglion cell afferents (Archer *et al.*, 1982). First, lgn cells normally receive excitatory input from either on- or off-center retinal ganglion cells, but not from both; in lgn layers connected to an eye in which activity was blocked, many cells receive excitatory input from both on- and off-center ganglion cells (Archer *et al.*, 1982). Second, most lgn cells normally receive excitatory input from either X- or Y-type ganglion cells; the proportion of cells receiving input from both X- and Y-cells is higher than normal in layers connected to the blocked eye (Archer *et al.*, 1982). Third, lgn cells are usually monocular, but after retinal ganglion cell activity has been blocked significant numbers of lgn cells are activated by either eye (Archer *et al.*, 1982).

Development of the lgn during the first postnatal stage appears not to be affected by dark-rearing; if the dark-rearing continues, subsequent development is retarded. Specifically, rearing cats in darkness during the first 3 weeks of postnatal life does not affect the rate of increase in lgn soma size (Figure 8) (Kalil, 1978a). If the dark-rearing continues beyond the third postnatal week until the eighth postnatal week, lgn soma size continues to increase, but not as fast or as much as in normally reared animals (Figure 8). Continuing the dark-rearing beyond 8 weeks leads to an actual reduction in soma size so that by 12 weeks postnatal the differences between normal and dark-reared animals reach a maximum. Finally, if the dark-rearing lasts more than 12 weeks, soma size again increases so that by 14 weeks postnatal there is little difference between soma size in normal and dark-reared animals (Figure 8; Kalil, 1978a; Kratz *et al.*, 1979).

While soma size is normal, the physiological properties of some lgn cells are abnormal after 4 or more months of dark-rearing (Kratz *et al.*, 1979; Kratz, 1982; Mower *et al.*, 1981a; Derrington and Hawken, 1981). Dark-rearing appears to pro-

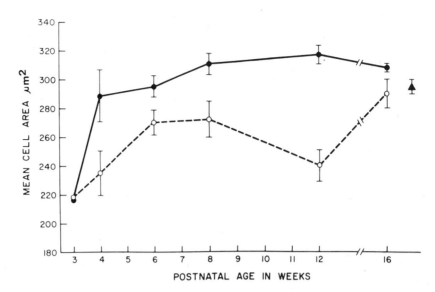

Fig. 8. Mean cross-sectional area (+/− S.E.) of cells in the lgn of normal cats (filled symbols) and of dark-reared cats (open symbols) at various postnatal ages. For comparison, the filled triangle indicates the mean cell size in cats that were dark-reared and then returned to a lighted environment. Note that by 16 weeks postnatal cell size in the lgn of dark-reared cats is essentially normal. (From R. E. Kalil, *J. of Comparative Neurology*, 1978, *178*, 451–468. Reproduced with permission from Kalil.)

duce some breakdown in the normal segregation of retinal ganglion cell afferents; many abnormal cells display characteristics of both X- and Y-cells (Kratz *et al.*, 1979; Kratz, 1982; Mower *et al.*, 1981a; Derrington and Hawken, 1981). Perhaps the largest group of these cells have conduction latencies characteristic of Y-cells, but lack the nonlinear response associated with Y-cells (Kratz, 1982; cf. Mower *et al.*, 1981a).

The bulk of retinal ganglion cell afferents must, however, still remain segregated after 4 months of dark-rearing. There are some normal lgn Y-cells, although somewhat fewer than in normal cats (Kratz *et al.*, 1979; Mower *et al.*, 1981a; Kratz, 1982), as well as normal lgn X-cells, possibly more than in normal cats (Kratz *et al.*, 1979; Kratz, 1982). Both the surviving Y- and X-cells have essentially normal spatial resolution (Mower *et al.*, 1981a) as well as normal spatial and temporal contrast sensitivity (Kratz, 1982).

3. SUMMARY. Neuronal activity, whether internally or externally generated, has little effect on retinal development. Blocking retinal activity with TTX, however, does have a drastic effect on lgn development, apparently disrupting the development and/or maintenance of the normal segregation among various classes of retinal afferents. Prolonged periods of dark-rearing produce some disruption in segregation of retinal afferents; lgn cells may display characteristics of both X- and Y-cells. Most lgn cells in dark-reared cats, however, appear to continue to receive their normal afferent input. There is some evidence that lgn Y-cells are affected more by deprivation than lgn X-cells (Kratz *et al.*, 1979; Kratz, 1982).

1. EFFECTS OF DARK-REARING. It is not known whether internally generated activity of the retinal ganglion cells is involved in the development of orientation-selective cells during the first postnatal stage; stimulation-dependent neuronal activity certainly is not. Development of selectivity during the first 2–3 postnatal weeks is not affected by dark-rearing (Bonds, 1979; Buisseret and Imbert, 1976; Fregnac and Imbert, 1978). Similarly, during this stage, development of spatial frequency selectivity is also not affected by dark-rearing (Derrington, 1984).

The extension of orientation selectivity to the bulk of area 17 cells does require visual stimulation (Pettigrew, 1974; Blakemore and Van Sluyters, 1975; Buisseret and Imbert, 1976; Fregnac and Imbert, 1978; Bonds, 1979); if kittens are kept in darkness beyond the third postnatal week, the proportion of orientation-selective cells remains at (Bonds, 1979), or possibly drops below (Blakemore and Van Sluyters, 1975; Buisseret and Imbert, 1976; Fregnac and Imbert, 1978), the 3-week level. (cf. Buisseret *et al.*, 1982 for a possible reason for the differences between the results of the earlier studies and the more recent one by Bonds, 1979). The development of selectivity for spatial frequency that normally occurs after the third postnatal week (Derrington and Fuchs, 1981) does not take place if kittens are dark-reared between the third and eighth postnatal week (Derrington, 1984).

Some orientation-selective cells retain (or possibly regain) orientation selectivity if dark-rearing is continued until the animals are 7–10 months of age (Leventhal and Hirsch, 1977, 1980); as a group these cells are not as finely tuned for orientation as their counterparts in normal adults (Leventhal and Hirsch, 1980). In both kittens and dark-reared adult cats orientation-selective cells have the smallest receptive fields and relatively low spontaneous activity, are concentrated in the lower layers of cortex, especially in and around layer IV, have a distribution of orientation preferences that is biased toward horizontal and vertical, and are activated monocularly through the contralateral eye. In normal cats these response properties are associated with direct, excitatory input from lgn X-cells (Dreher *et al.*, 1980; Leventhal and Hirsch, 1980; Mustari *et al.*, 1982).

The effects of prolonged dark-rearing thus suggest that first-order area 17 cells, especially those receiving excitatory input from lgn X-cells, have robust orientation preferences that they can develop and maintain in the absence of stimulation-dependent neuronal activity. Other cells, including many that receive their excitatory input from other area 17 cells and/or from lgn Y- and W-cells, require visual stimulation to develop and/or maintain orientation selectivity.

The effects of dark-rearing on morphological development of area 17 cells have not been extensively studied. Coleman and Riesen (1968) found that morphology of layer IV stellate cells, but not of layer V pyramidal cells, is affected by dark-rearing. After 6 months of dark-rearing, stellate cells have fewer and shorter dendrites that branch less often than in normally reared animals. The layer IV stellate cells are likely to be first-order neurons and thus would be expected to retain their orientation preferences despite the prolonged dark-rearing (Leventhal and Hirsch, 1980). The morphological changes observed by Coleman and Riesen

(1968) may thus not have involved structural correlates of the cell's preferred orientation (e.g., the elongation of the cell's dendritic field in a plane parallel to the pial surface; cf. Colonnier, 1964; Tieman and Hirsch, 1982, 1985). This interpretation is strengthened by our findings (to be presented below) that early postnatal exposure to lines of only one orientation does not change the distribution of dendritic field orientations of layer IV spiny stellate cells (Tieman and Hirsch, 1982).

There is some evidence that the course of synaptogenesis in area 17 is affected by stimulation-dependent neuronal activity. Bilateral lid suture, which reduces but does not eliminate patterned visual stimulation, slows the rate of formation of synapses during the first 70 days of life (Cragg, 1975; Winfield, 1981). The peak in synaptic density (in normally reared animals 70 days postnatal, Winfield, 1981) is delayed; in bilaterally lid-sutured animals synaptic density increases until 110 days and then declines, eventually reaching approximately normal adult levels (Winfield, 1981). Despite the eventual normality of synaptic density, the relative proportion of symmetric and asymmetric synapses remains abnormal. Long-term bilateral lid suture results in a slightly diminished population of symmetric synapses and this may offer an anatomical basis for the reduced levels of intracortical inhibition found in physiological recordings from area 17 of bilaterally sutured animals (e.g., Singer and Tretter, 1976; Watkins *et al.*, 1978).

Neuronal activity thus plays a major role in development of area 17. Although the inputs to area 17, especially those from lgn X-cells, are relatively normal in dark-reared cats, stimulation-dependent activity is needed if area 17 cells are to develop the capability to process these inputs. The bulk of area 17 cells, especially those receiving much or all of their excitatory input from other area 17 cells, fail to develop normal orientation and spatial frequency selectivity unless visual stimulation is provided. Because many of these cells send their axons outside area 17, normal development of the output from area 17 is dependent on visual stimulation.

2. EFFECTS OF SELECTIVE EXPOSURE. Visual experience has both nonspecific and specific effects that reflect the characteristics of the stimulation provided to the animal. For example, exposure to small spots enhances the responsiveness of area 17 cells (Pettigrew and Freeman, 1973; Van Sluyters and Blakemore, 1973; Blakemore and Van Sluyters, 1975), while exposure to elongated contours is necessary to increase the proportion of cells that are orientation selective (Blakemore and Van Sluyters, 1974). Among the most dramatic and well documented of these specific changes occur in kittens exposed only to lines of one orientation, a procedure referred to as stripe-rearing (Figure 9). The distribution of the preferred orientations of area 17 cells is biased toward the exposed orientation (Figure 10) (Blakemore and Cooper, 1970; Hirsch and Spinelli, 1970, 1971; Pettigrew *et al.*, 1973; Leventhal and Hirsch, 1975; Blakemore, 1976; Blasdel *et al.*, 1977; Stryker et al., 1978; Rauschecker and Singer, 1981; Gordon and Presson, 1982; Tieman and Hirsch, 1982; Hirsch *et al.*, 1983). This effect of stripe-rearing is strongest between the third and seventh postnatal week (Blakemore, 1974). Thus during the second and to a lesser extent during the third stage of postnatal development,

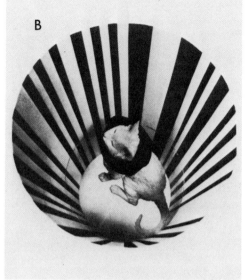

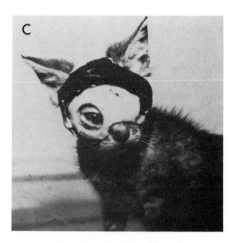

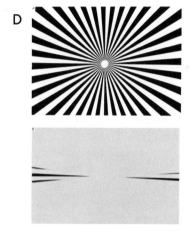

Fig. 9. Different techniques used for providing controlled postnatal visual stimulation. (A) Kitten wearing goggles used for stripe-rearing. Stimulus patterns are mounted on the inside surface of the black plastic sheet: one pattern in front of the left eye, a second in front of the right eye. A lens is mounted in the mask in front of each eye so that the patterns are at the focal plane of the lens. Light entering through the white diffusing plastic illuminates the patterns. (From Hirsch, *Cellular and Molecular Neurobiology*, 1985, *5*, 103–121. Reproduced with permission from Hirsch.) (B) Kitten inside a striped cylinder used to restrict visual exposure. The animal is placed onto a clear plastic shelf mounted in the center of the cylinder. Although the animal is free to move about and turn its head, the predominant orientation it sees is that of the lines on the inside of the cylinder. (From C. Blakemore, *The Neurosciences: Third Study Program*, MIT Press, 1974, pp. 105–113. Reproduced with permission from Blakemore.) (C) Kitten wearing a hood that positions a cylindrical lens in front of one eye; the other eye is covered by the hood. To illustrate the effect of the cylindrical lens, (D) shows a set of radiating lines (above) as seen through the cylindrical lens (below). (From Rauschecker and Singer, *J. of Physiology* (Lond.), 1981, *310*, 215–239. Reproduced with permission.)

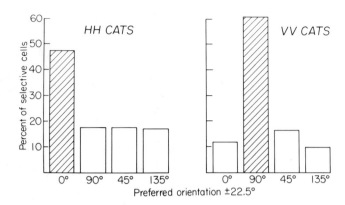

Fig. 10. Distribution of preferred orientations of orientation-selective cells recorded from area 17 of (A) six cats exposed to horizontal lines and (B) six cats exposed to vertical lines. Note that for each group of cats the distribution peaks at the exposed orientation. Furthermore, the exposed orientation lies within the 95% confidence limits for the mean of the distribution of the preferred orientations of the sample of cells recorded. (From Hirsch, *Cellular and Molecular Neurobiology*, 1985, *5*, 103–121. Reproduced with permission.)

exposure to elongated contours increases the proportion of orientation-selective cells; the orientation of the contours has a systematic effect on the distribution of perferred orientations.

3. Differences in Experience Sensitivity among Area 17 Cells. As one step toward discovering how visual experience affects development of cortical cell orientation selectivity, Hirsch *et al.*, (1983) sought to discover if area 17 cells differ among themselves in experience sensitivity. To do this they attempted to identify those area 17 cells that are likely to resist environmental manipulation, namely, the cells that develop and maintain orientation selectivity in dark-reared animals (Leventhal and Hirsch, 1980). These cells were identified physiologically by their receptive field size (small) and by the maximum velocity (slow) to which they respond. They were referred to as Small and Slow or SAS cells (Leventhal and Hirsch, 1980; Hirsch *et al.*, 1983).

Hirsch *et al.* (1983) found that the distribution of perferred orientations of SAS cells is not affected by stripe-rearing. In contrast, the distribution of the preferred orientations of the remaining cells shows a clear bias toward the exposed orientation (Figure 11). There are thus differences in experience sensitivity among physiologically characterized cell groups in area 17.

To look for corresponding differences in experience sensitivity among morphologically identified cell groups, Tieman and Hirsch (1982) examined basal dendritic morphology of layer III pyramidal cells and of layer IV spiny stellate cells in normal and stripe-reared cats. In normally reared cats, the dendritic fields of both layer III pyramidal cells and layer IV stellate cells were elongated in the plane of the map, and the distributions of the orientations of the long axis of the fields were uniform. In stripe-reared cats, however, the distributions of the orientations of layer III pyramidal cells, but not of layer IV stellate cells, were shifted (Figures 12 and 13). In cats viewing vertical lines, the dendritic fields of the layer III pyramidal cells were oriented perpendicular to the representation of the vertical meridian, and in cats reared viewing horizontal lines the dendritic fields were oriented parallel to the representation of the vertical meridian.

These results (see also Coleman *et al.,* 1981) suggest that the layer IV stellate cells are affected less by stimulation-induced neuronal activity than other cell types, that is, the layer III pyramidal cells. The effects of dark-rearing on layer IV stellate cells (Coleman and Riesen, 1968), however, suggest that visual exposure may have some nonspecific effects on their morphology; perhaps these changes reduce orientation tuning without abolishing it (see Hirsch *et al.,* 1983, for additional evidence for nonspecific, experience-dependent changes in orientation tuning of SAS cells).

There are accordingly differences in experience sensitivity among both physiologically and morphologically identified cell groups in area 17. On the basis of the physiological and morphological effects of stripe-rearing, as well as on available data on the distribution of lgn afferents to cells in area 17, Tieman and Hirsch (1982) suggested that the layer IV stellate cells are major contributors to the population of physiologically identified SAS cells and that the layer III pyramidal cells are included in the population of physiologically characterized cell groups whose

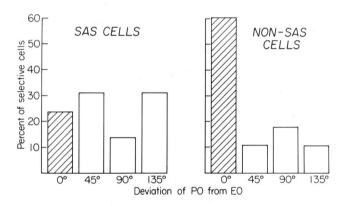

Fig. 11. Distribution of the preferred orientations of orientation-selective cells recorded from area 17 of stripe-reared cats. Data for cells with small receptive field and low cutoff velocity, SAS cells, are shown in (A); data for all other cells are shown in (B). For these histograms, preferred orientations were computed as deviations from the exposed orientation as follows. Cells in the bins labeled 0 deg have a preferred orientation within 22.5 deg of the exposed orientation; cells in the bins labeled 45 deg have a preferred orientation within 22.5 deg of the exposed orientation + 45 deg; cells in the bins labeled 90 deg have a preferred orientation within 22.5 deg of the exposed orientation + 90 deg; cells in the bins labeled 135 deg have a preferred orientation within 22.5 deg of the exposed orientation + 135 degrees. Note that for non-SAS cells (B) the largest bin consists of cells with preferred orientations within 22.5 deg of the exposed orientation. The distribution of preferred orientations for non-SAS cells is thus biased toward the exposed orientation. For SAS cells (A) there is no such relationship between the number of cells in a bin and the differences between the preferred and exposed orientation. (From Hirsch, *Cellular and Molecular Neurobiology,* 1985, *5,* 103–121. Reproduced with permission.)

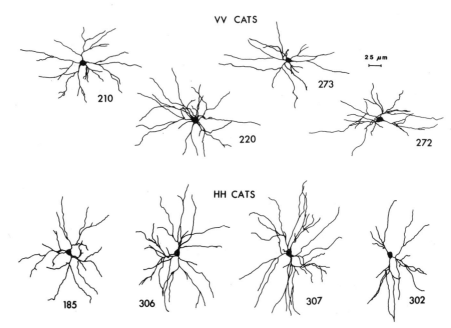

Fig. 12. Camera lucida drawings of basal dendrites of individual layer III pyramidal cells from stripe-reared cats exposed to vertical lines with both eyes (VV cats) and from stripe-reared cats exposed to horizontal lines with both eyes (HH cats). The drawings have been rotated so that dorsal is to the right and the representation of the vertical meridian would be vertical in the drawing. These examples were chosen to illustrate the relative elongation and dendritic field orientation typical of cats from the same rearing condition and atypical of cats from the other rearing conditions. Note that dendritic fields of layer III pyramidal cells from the cats exposed to vertical lines are oriented horizontally while dendritic fields of layer III pyramidal cells from cats exposed to horizontal lines are oriented vertically. Individual cells from normally reared control animals are not shown since the variability in orientation and elongation of dendritic fields of layer III pyramidal cells (see Figure 15) makes selection of a small "representative" sample impossible. (From Tieman and Hirsch, *J. of Comparative Neurology*, 1982, *211*, 353–362. Reproduced with permission from Tieman and Hirsch.)

members have a high cutoff velocity and are likely to be activated by way of lgn Y-cells (Dreher *et al.*, 1980; Leventhal and Hirsch, 1980; Mustari *et al.*, 1982; Henry *et al.*, 1983).

C. FACTORS CONTROLLING DIFFERENCES IN EXPERIENCE SENSITIVITY OF CORTICAL CELLS

Several factors appear to influence the degree to which the orientation selectivity of area 17 cells is affected by early visual exposure. Among these are (1) afferent input, that is, whether they receive excitatory input from X-, Y-, or W-cells, and (2) ordinal position, that is, whether they are first-order cells, activated primarily by lgn afferents, or higher-order cells receiving significant excitatory input from other area 17 cells.

1. AFFERENT INPUT. In dark-reared cats, cells in the afferent stream arising from retinal X-cells appear to develop normally in both the lgn (Kratz *et al.*, 1979;

Kratz, 1982) and area 17 (Leventhal and Hirsch, 1980). Furthermore, in stripe-reared cats the distribution of the preferred orientations of cells with response properties normally associated with direct, excitatory input from lgn X-cells (Dreher *et al.*, 1980; Mustari *et al.*, 1982) is not biased toward the exposed orientation (Hirsch *et al.*, 1983). In contrast, in dark-reared cats, cells in the afferent stream arising from retinal Y-cells may fail to develop normally both in the lgn (Kratz *et al.*, 1979; Kratz, 1982), and in area 17 (Leventhal and Hirsch, 1980). Furthermore, stripe-rearing produces a bias in the distribution of preferred orientations of cells with response properties associated with an excitatory input from lgn Y-cells (Hirsch *et al.*, 1983). Accordingly, there are differences in experience sensitivity among target cells of the various afferent streams that begin in the retina.

2. ORDINAL POSITION. Several lines of evidence indicate that a cell's ordinal position may affect its experience sensitivity. First, the area 17 cells that do not require visual experience to develop orientation selectivity are likely to be first-order neurons, activated monosynaptically from the lgn (Beckmann and Albus, 1982; cf. also Blakemore and Van Sluyters, 1975; Fregnac and Imbert, 1978; McCall, 1983; Albus and Wolf, 1984), whereas cells that require visual stimulation to develop orientation selectivity (Blakemore and Van Sluyters, 1975; Buisseret and Imbert, 1976; Fregnac and Imbert, 1978; Bonds, 1979; cf. also Leventhal and Hirsch, 1977, 1980) are most common in the extragranular layers (Blakemore and Van Sluyters, 1975) where they probably receive significant excitatory input from other area 17 cells (Hubel and Wiesel, 1962; Henry *et al.*, 1979; Bullier and Henry, 1979a; Henry *et al.*, 1983). Second, the effects of stripe-rearing also suggest that cells in the granular and extragranular layers may differ in experience sensitivity;

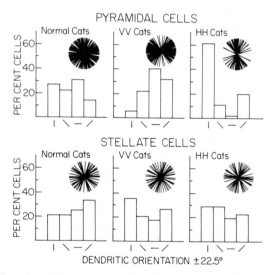

Fig. 13. Dendritic field orientations of layer III pyramidal cells and layer IV stellate cells from area 17 of normal cats, cats exposed to vertical lines (VV cats) and cats exposed to horizontal lines (HH cats). The data from all the cats in each condition are grouped together. The histograms plot the percentage of cells having dendritic field orientations within 22.5 deg of vertical, 135 deg, horizontal, or 45 deg. The inserts are the polar plots of these distributions. The dendritic fields of the pyramidal cells from the normal cats were uniformly distributed, whereas most of the pyramidal cells from the cats exposed to vertical lines had dendritic fields within 22.5 deg of horizontal, and most of the pyramidal cells from cats exposed to horizontal lines had dendritic field orientations within 22.5 deg of vertical. In contrast, the dendritic field orientations of layer IV stellate cells were uniformly distributed in all three groups. (From Tieman and Hirsch, *J. of Comparative Neurology*, 1982, *211*, 353–362. Reproduced with permission.)

the effects of stripe-rearing are most pronounced in extragranular layers (Singer *et al.,* 1981), and the layer III pyramidal cells whose dendritic morphology is affected by stripe-rearing have somas located just outside layer IV while the spiny stellate cells whose morphology is not affected by stripe-rearing have their somas in layer IV (Tieman and Hirsch, 1982). Visual experience is thus likely to play an important role in postnatal development of orientation selectivity of cells in extragranular layers of area 17.

3. MECHANISMS FOR GENERATING ORIENTATION SELECTIVITY. Correlations among ordinal position, type of afferent input, and experience sensitivity may be related to differences in the mechanisms by which orientation selectivity of the various area 17 cell types is generated; several mechanisms may be involved in the genesis of orientation selectivity for a particular cell type, and these may differ from one cell type to another. Let us consider first those area 17 cells that do not require visual stimulation for development or maintenance of orientation selectivity.

a. The Intrinsically Selective Cells. Since the sensory input to area 17 is organized topographically, dendritic field shape may determine receptive field shape. Elongation of the dendritic fields (in a plane parallel to the pial surface) of area 17 cells whose primary activation comes from lgn afferents could generate elongated receptive fields and hence a preference for one stimulus orientation (Colonnier, 1964; cf. also Young, 1960; Hammond, 1974; Kirk *et al.,* 1983; Leventhal and Schall, 1983; Mariani, 1983; Peichl and Wässle, 1983; Bacon and Murphy, 1984). An orientation bias generated in this fashion would depend on the excitatory inputs to the cell. Since the intrinsically orientation-selective cells in area 17 retain an orientation bias when GABA-mediated inhibitory inputs are blocked by iontophoretic application of *N*-methylbicuculine (Sato and Tsumoto, 1984), it is possible that the orientation preferences of these cells are generated by the elongation of their dendritic fields (cf. Tieman and Hirsch, 1985).

Several factors could enhance an orientation bias produced in this fashion and thus produce the degree of selectivity that is characteristic of intrinsically selective area 17 cells (Blakemore and Van Sluyters, 1975; Buisseret and Imbert, 1976; Fregnac and Imbert, 1978; Bonds, 1979). First, the orientation bias of cells in retina and lgn could amplify the orientation bias generated by elongated dendritic fields of area 17 cells (cf. Leventhal, 1983; Tieman and Hirsch, 1985). Second, a set of lateral inhibitory connections among these neurons could accentuate their orientation biases (cf. Blakemore and Tobin, 1972; Hess *et al.,* 1975; Fries *et al.,* 1977; Nelson and Frost, 1978; Sillito, 1975, 1979; Sillito *et al.,* 1980; Toyama *et al.,* 1974; Morrone *et al.,* 1982). The effects of iontophoretic application of *N*-methylbicuculine on orientation selective cells in very young kittens suggest that inhibitory inputs are involved in generating their selectivity (Sato and Tsumoto, 1984).

Stimulation-dependent neuronal activity apparently has little role in the development of either the excitatory or inhibitory connectivity responsible for generating the orientation selectivity of early maturing area 17 cells. Internally generated neuronal activity could be involved in establishing the selectivity of these cells, perhaps by producing a segregation among orientation-biased lgn afferents ending on an individual area 17 cell. Such a segregation would help to generate the orientation preferences of the area 17 cell (e.g., Leventhal, 1983). This would imply that there is correlated firing of retinal ganglion cells that have a similar orientation preference.

   b. The Experience-Sensitive Cells.   A different set of mechanisms may be involved in generating the orientation selectivity of late-maturing area 17 cells. Many of these are higher-order cells, receiving much or all of their excitatory input from other area 17 cells. They may rely on first-order neurons to generate orientation selectivity; experience-dependent development of intracortical excitatory connections could then play a role in transferring this selectivity to the second-order cells (Tieman and Hirsch, 1985). The effects of stripe-rearing on the shape of the basal dendrites of layer III pyramidal cells suggest that there are such experience-dependent changes in intracortical excitatory connectivity. Tieman and Hirsch (1985) suggested that the layer III pyramidal cells are orientation selective because they receive excitatory input from a selected population of cortical cells (Figure 14), and that development of these connections may be affected by exposure to con-

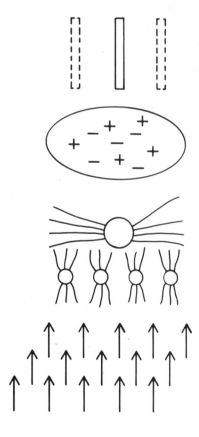

Fig. 14. A model relating the orientation of a cell's dendritic field to its preferred orientation. The arrows at the bottom represent the retinotopic array of incoming geniculocortical axons, the cell with horizontal dendrites represents a layer III pyramidal cell, and the line at the top represents the cell's preferred orientation. Between the cell and the line is a schematic drawing of the cell's receptive field. Note that the geniculocortical axons influence the layer III pyramidal cell indirectly, through the layer IV stellate cells, which are shown as small, vertical cells. Illustrated here are four stellate cells that prefer vertical lines. These cells are scattered along a line that is horizontal in the plane of the map, and hence their receptive fields are also scattered along a horizontal line in the retina. The layer III pyramidal cell, which is one of Hubel and Wiesel's (1962) complex cells, optimizes the inputs from these stellate cells by having a dendritic field oriented horizontally in the map. Its receptive field is the sum of the receptive fields of the stellate cells, and has mixed "on" and "off" responses. The pyramidal cell shares with the stellate cells a preference for vertical, and responds to a vertical line anywhere within its receptive field. This model suggests that development of intracortical excitatory connectivity is affected by early visual experience. (Adapted with permission from Tieman and Hirsch, *Models of Visual Cortex,* [Rose and Dobson, eds.]. Copyright 1985 by Wiley.)

tours. Inhibitory interconnections may serve to amplify orientation selectivity further (Sillito, 1975; 1979; Sillito *et al.*, 1978; Leventhal and Hirsch, 1980; Derrington, 1984). Thus development of both excitatory and inhibitory intracortical connectivity may be experience dependent.

4. Activity-Dependent Synaptic Changes. Hebb (1949) proposed a simple mechanism by which neuronal activity may affect connectivity. He suggested that the synaptic connections made by presynaptic inputs active just before the postsynaptic cell fires are strengthened, while the synaptic connections of other, inactive inputs become weaker. Neuronal activity, whether internally generated or produced by sensory stimulation, can thus affect synaptic connections. For example, the correlated firing of neighboring retinal ganglion cells of the same type enables them to maintain synaptic contacts onto a common postsynaptic cell, while axons whose firing is negatively correlated (e.g., those of neighboring retinal ganglion cells of differing types) are not likely to do so. The apparent ease with which nervous system development can incorporate and integrate information from genetic and environmental sources may result because information from both genetic and environmental sources is translated into patterns of neuronal activity that affect neuronal connectivity through a common mechanism.

V. Extraretinal Factors in Visual System Development

We have seen that the experience-dependent changes that occur during the second and third stages of postnatal development enable the visual system to incorporate information from the environment into the growing nervous system. Recent results indicate that several nonretinal factors can influence this process.

The first of these is the monoamine system of the brain stem, which has long been suspected of having a role in cortical development (Kety, 1970; Pettigrew, 1978). Consistent with this role, the specific neurotoxin 6-hydroxydopamine may somewhat retard development of cells in area 17 in very young kittens (Kasamatsu and Pettigrew, 1976). Furthermore, 6-hydroxydopamine appears to protect area 17 cells in kittens from the effects of monocular deprivation, possibly by decreasing noradrenaline levels in the cortex (Kasamatsu and Pettigrew, 1976, 1979; Kasamatsu *et al.*, 1981; Paradiso *et al.*, 1983; cf. also Daw *et al.*, 1985). Indeed, perfusing exogenous noradrenaline (Kasamatsu *et al.*, 1979, 1981) restores sensitivity to monocular deprivation even in adult cats.

A second set of extraretinal factors affecting cortical development are the projections from the mesencephalic reticular formation and from the medial thalamic nuclei. Lesions in these areas (Singer, 1982) appear to impede both normal maturation of the responses of area 17 cells, as well as much of the change in ocular dominance normally resulting from monocular deprivation. The importance of input from these areas is also shown by stimulating them electrically (Singer and

Rauschecker, 1982). In anesthetized, paralyzed kittens visual stimulation of only one eye does not produce changes in ocular dominance (cf. also Freeman and Bonds, 1979), unless the mesencephalic reticular formation or the medial thalamic nuclei are stimulated electrically at the same time (Singer and Rauschecker, 1982). Interestingly, simple cells are least dependent on the projections from the mesenchephalic reticular formation and the medial thalamic nuclei; they display the most normal responsiveness and the greatest shift in ocular dominance in animals with lesions disturbing the projections from the mesencephalic reticular formation or the medial thalamic nuclei.

A third important extraretinal factor involved in modulation of experience sensitivity is proprioceptive feedback from the extraocular muscles (cf. Imbert and Fregnac, 1983). In dark-reared kittens, 6 hr of exposure to a normal environment increases the proportion of orientation-selective cells to essentially normal levels (Imbert and Buisseret, 1975; Buisseret *et al.*, 1978, 1982). This extension of orientation selectivity to the bulk of area 17 cells does not take place in kittens in which oculomotor activity is blocked because the animal is paralyzed and anesthetized (Buisseret *et al.*, 1978) or in kittens in which proprioceptive afferents from the extraocular muscles are blocked because the ophthalmic branch of the trigeminal nerve has been cut (Buisseret and Gary-Bobo, 1979; Trotter *et al.*, 1981). In addition, bilateral section of the ophthalmic branch of the trigeminal nerve impedes experience-dependent development of orientation selectivity of area 17 cells, and prevents most of the changes in ocular dominance that would otherwise result from monocular deprivation and experimental strabismus (Buisseret and Singer, 1983). It is interesting to note again that simple cells, which in these animals are usually monocular and have a preferred orientation near horizontal or vertical, are the most mature and show the most pronounced changes in ocular dominance (Buisseret and Singer, 1983). Blocking the proprioceptive feedback from the extraocular muscles therefore appears to interfere with the experience-dependent developmental changes that characterize area 17 cells during the second and third stage of postnatal life. The pattern of results is consistent with the possibility that much of the disruption involves development of intracortical connectivity, rather than development of connectivity between lgn afferents and first-order cells in area 17 (Singer, 1982).

In summary, several extraretinal factors appear to contribute importantly to experience-dependent changes in area 17. These factors include projections from the monoaminergic cells in the brainstem, projections from the midbrain reticular activating system and the medial thalamic nuclei, as well as proprioceptive afferents from the extraocular muscles. The importance of intact extraocular muscles in development of normal sensorimotor coordination (Hein *et al.*, 1979) has been documented. Identification of the extraretinal factors affecting experience-dependent development of the visual system may thus provide some insights into the behavioral significance of these changes. This will be considered in the following section.

HELMUT V. B.
HIRSCH

The extraretinal factors that influence experience-dependent development of the visual system may function to ensure that under normal conditions stimulation-induced changes are made in accordance with behaviorally relevant sensory stimulation. If so, then as Singer (1982) suggests,

> the developing kitten has the option to screen the continuous stream of sensory signals and to accept only selected activity patterns for modifying cortical connectivity. Thus developmental plasticity appears to share an important property with general learning. In both cases signals leave traces only when the animal pays attention to them and uses them for the control of behavior. [p. 221]

In this section I will review the available data on the behavioral significance of experience-dependent changes in visual system connectivity. I will begin with a brief description of behavioral development in normal kittens, and then describe behavioral changes resulting from stripe-rearing. In considering the behavior of a developing organism it is important to remember that it may subserve two distinct functions: (1) to permit the young animal to deal with the problem of surviving in its current environmental niche, that is, while it is still developing, and (2) to enable it to acquire the skills that it will need to survive as an adult (Blass, 1980; Galef, 1981; Oppenheim, 1981; Martin, 1982). Experience-dependent changes in the visual system may be important for development of behaviors involved in either or both of these functions.

A. Development of Visually Guided Behavior in the Kitten

1. Stage 1. A kitten makes little use of vision during the first weeks of postnatal life. As early as the end of the first week of postnatal life a kitten removed from the nest or "home cage" will use olfactory and thermal stimuli to guide its return to the nest (Freeman and Rosenblatt, 1978), but it will not use visual stimuli until around the fourteenth to eighteenth postnatal day (Rosenblatt et al., 1964). At about this age kittens are able to orient to visual targets (Norton, 1974) and track them by either head or eye movements (Warkentin and Smith, 1937; Hubel and Wiesel, 1963b; Norton, 1974; but cf. Vital-Durand and Jeannerod, 1974, and Sherman, 1972, who do not report seeing visual tracking until at least the fourth week of life).

Kittens become increasingly active over the course of the first three postnatal weeks. There is an overall increase in motor activity between 9 and 14 days postnatal (Levine et al. 1980), but kittens do not yet venture forth from the home cage (Rosenblatt et al., 1964). Consistent with this reluctance to leave the nest, locomotor activity measured in an open field actually decreases slightly over the course of the first three weeks postnatal (Levine et al., 1980). Thus, the kitten's earliest use of vision appears to be in guiding its return to the nest; development of the ability to orient to visual targets may be a prerequisite for using vision in this fashion.

2. STAGE 2. There is a dramatic increase in the prevalence of visually guided behaviors over the course of the second postnatal stage. Around the fourth postnatal week kittens first begin to avoid the deep side of a visual cliff (Warkentin and Smith, 1937; Walk, 1968; Karmel *et al.*, 1970; Norton, 1974; Van Hof-Van Duin, 1976; Villablanca and Olmstead, 1979; Martin, 1982; but cf. Fox, 1970, who reports that reaction to the visual cliff develops near the end of the third postnatal week), they are able to avoid obstacles in their path (Norton, 1974; Van Hof-Van Duin, 1976) and they extend their paws to meet an edge when lowered toward it (Warkentin and Smith, 1937; Karmel *et al.*, 1970; Fox, 1970; Sherman, 1972; Norton, 1974; Van Hof-Van Duin, 1976; Villablanca and Olmstead, 1979). Four-week-old kittens show an optokinetic nystagmus to stimuli that move rapidly (over 20 deg/sec) (Van Hof-Van Duin, 1976; but cf. Warkentin and Smith, 1937, who report seeing optokinetic nystagmus as early as the second or third postnatal week). When tested with only one eye, 4-week-old kittens follow only temporal-to-nasal movement. Responses to slowly moving stimuli (6–12 deg/sec) do not appear until 5–6 weeks postnatal (Van Hof-Van Duin, 1976).

Visually guided behaviors improve during the fifth postnatal week. For example, when lowered toward a serrated edge kittens now use visual cues to guide the placement of their paws so as to land on the solid portions of the edge (Norton, 1974; Van Hof-Van Duin, 1976). Furthermore, at this age kittens display a willingness to jump down from a platform, although the response still lacks adultlike "precision" and "certainty" (Van Hof-Van Duin, 1976). There is an increase in the range of stimulus directions to which an optokinetic response is made: when tested monocularly, 5-week-old kittens follow movement of a rapidly moving stimulus proceeding in either temporal-to-nasal or nasal-to-temporal directions (Van Hof-Van Duin, 1978). One or two additional weeks are required for comparable responses to slow stimulus motion to appear (Van Hof-Van Duin, 1978).

As kittens develop these visual capabilities, they also become increasingly active. During the fourth postnatal week kittens first begin to leave their nest (Rosenblatt *et al.*, 1964). Furthermore, locomotor activity in an open field increases markedly during the fourth, fifth, and sixth postnatal weeks (Frederickson and Frederickson, 1979b; Levine *et al.*, 1980). In addition to the quantitative changes in locomotor activity there are changes in the pattern of locomotion in the environment. Between the fourth and fifth postnatal weeks there are changes in the extent to which animals will go back and forth between the same portions of the open field (Frederickson and Frederickson, 1979b), and around the fifth postnatal week kittens will begin to spontaneously alternate which arm of a T-maze they enter (Frederickson and Frederickson, 1979a). Finally, near the end of the second postnatal stage kittens are able to learn to make differential responses to visual stimuli; starting at around 6–7 weeks postnatal they can learn to discriminate between complex visual patterns (Wilkinson and Dodwell, 1980). Whether and how kittens make use of the ability to acquire visual discriminations at this stage of development is not known.

The changes in sensory and locomotor behavior during the second postnatal stage are accompanied by important changes in the kitten's lifestyle. First, kittens begin to engage in a group of behaviors such as stalking, chasing, and pouncing that involve either siblings or the mother as play partners; these behaviors are referred to collectively as "social play" (Rosenblatt and Schneirla, 1962; West, 1974; Moelk, 1979; Martin, 1982). Social play starts to become apparent around the fourth postnatal week and increases in frequency over the next few weeks (West, 1974; 1979; Kolb and Nonneman, 1975; Barrett and Bateson, 1978; Baerends–van Roon and Baerends, 1979; Villablanca and Olmstead, 1979; Martin, 1982). Next, over the course of the second postnatal stage the mother cat begins to wean her kittens. This is achieved by a gradual increase in how much difficulty the kittens have in obtaining maternal care (Martin, 1982).

In summary, sensory and locomotor behaviors characteristic of the adult cat begin to appear during the course of the second postnatal stage. Kittens begin to venture forth from the nest, but much of their activity involves interactions with siblings or the mother and it generally takes place in the vicinity of the nest (Martin, 1982).

3. STAGE 3. The third postnatal stage is characterized by the emergence of behaviors that suggest the kitten is now beginning to seek interactions with stimuli and objects outside the home cage. Perhaps the most dramatic evidence for this is that by the eighth or ninth postnatal week kittens begin to make active efforts to obtain visual stimulation. They will perform an operant task (lever pressing) for which the only reward is to view a visual stimulus. Although younger kittens can perform this task, they do so very infrequently, even if housed in the dark (Dodwell et al., 1976; Timney et al., 1979). Suddenly, at 8 or 9 weeks postnatal, the frequency and duration of lever pressing to obtain visual stimulation increases dramatically (Figure 15). This increase takes place in kittens that have been housed in the dark since birth, and in kittens raised in a lighted environment who are placed into the dark at the end of the first postnatal month (Timney et al., 1979). In short, if visual stimulation is not provided automatically by the environment, an 8- or 9-week-old kitten will work to obtain it.

Further evidence for an increase in active exploration of the environment during the course of the third postnatal stage comes from studies of behavior in an open field. Between the sixth and ninth postnatal week the overall amount of locomotor activity in an open field remains fairly constant (Levine et al., 1980), but then during the third and fourth postnatal month activity in the open field increases (Levine et al., 1980). Finally, during the third postnatal stage there are changes in the kitten's play behavior. Between the seventh and eighth postnatal week (Barrett and Bateson, 1978) there is a sharp increase in a group of behaviors that include mouthing, patting, poking, batting, and leaping over almost any suitable toy; these behaviors are collectively referred to as "object play" (West, 1977; Moelk, 1979; Martin, 1982). The behavioral changes that characterize the third postnatal stage thus point to an increase in the kitten's interaction with the environment outside the nest.

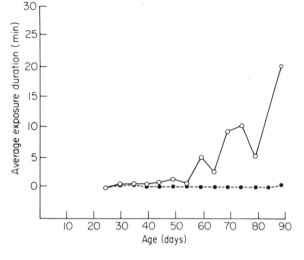

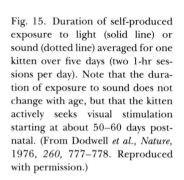

Fig. 15. Duration of self-produced exposure to light (solid line) or sound (dotted line) averaged for one kitten over five days (two 1-hr sessions per day). Note that the duration of exposure to sound does not change with age, but that the kitten actively seeks visual stimulation starting at about 50–60 days postnatal. (From Dodwell *et al.*, *Nature*, 1976, *260*, 777–778. Reproduced with permission.)

4. SUMMARY. Each of the three stages of postnatal development is characterized by a pattern of developmental changes in the visual system. Furthermore, each stage is characterized by significant changes in the animal's behavior. Neuronal activity plays a part in many of the visual system changes, especially during the second postnatal stage when stimulation-dependent neuronal activity can influence the development of at least some area 17 cells. In the following sections I will consider some of the available evidence on the functional significance of developmental changes in the cat visual system.

B. CORRELATING BEHAVIORAL AND VISUAL SYSTEM DEVELOPMENT

One way of trying to establish a link between visual system and behavioral development is to search for correlations between the time course of development of specific cell types and an animal's visual capabilities. An example of such a correlation is that between the third and sixth postnatal week the spatial resolving power of sustained cells in the lgn matures at the same rate (Figure 16a) (Ikeda and Tremain, 1978) as does visual acuity measured by using evoked potentials to measure contrast sensitivity (Figure 16b) (Freeman and Marg, 1975). Furthermore, the relative improvement in lgn cell acuity parallels the relative improvement in acuity measured behaviorally (Figure 16b) (Mitchell *et al.*, 1976). This suggests that neuronal changes in the lgn may be a limiting factor in the development of the kitten's ability to resolve fine details in a stimulus.

Correlations between physiological and behavioral development, of course, do not provide evidence for a causal link between the two. Furthermore, even for this very simple behavioral task, maturation of single cells cannot by itself account for the full development of the animal's behavioral capability. Before the seventh post-

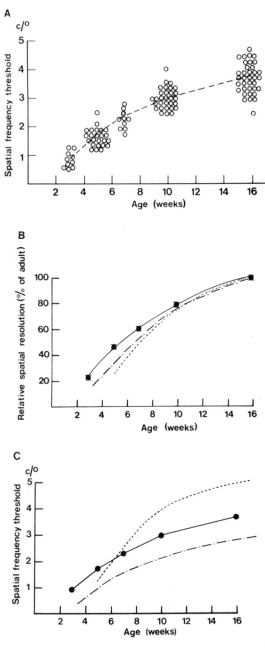

Fig. 16. (A) Age-related changes in spatial frequency threshold (the highest spatial frequency of a sinusoidal grating to which a cell responds with a modulation of its firing rate) of sustained cells in the lgn. Note that the highest frequency which sustained cells can respond to increases over the course of all three postnatal stages. (B) Comparison of relative changes in visual acuity as a function of age as determined by three different methods. To compare the relative visual acuity, the three curves were replotted so that visual acuity at 16 weeks postnatal is equal to 100%. The solid curve has been redrawn from (A) above. The dashed curve presents acuity determined from cortical evoked potentials (replotted from Freeman and Marg, 1975). The dotted curve presents acuity determined behaviorally (replotted from

natal week, the levels of visual acuity determined behaviorally are lower than those predicted from spatial resolution of lgn cells, while after 7 weeks postnatal acuity measured behaviorally is better than acuity measured for individual lgn cells (Figure 16c) (Ikeda and Tremain, 1978). This suggests that around the seventh postnatal week the visual system develops the capability to integrate information from many cells and therefore can resolve details beyond the resolution of any one cell. Identification of correlations between visual system development and behavioral development is thus just a starting point for study of the functional role of developmental changes in the visual system.

C. EFFECTS OF STRIPE-REARING ON VISUAL DEVELOPMENT

To provide more direct evidence for the functional significance of some of the developmental changes in the visual system, we can take advantage of the fact that the course of development of some cells (e.g., the bulk of area 17 cells) is influenced by stimulation-induced neuronal activity. We can manipulate an animal's early visual environment and look for correspondences between the resulting changes in the visual system and in the animal's behavioral capabilities. If behavioral capabilities and the visual system are both affected by the particular stimuli present during rearing, then we may have some confidence in postulating a functional relationship between the visual system changes and the behavioral deficits (Hirsch, 1972).

Stripe-rearing has several behavioral effects that are related in a systematic, specific fashion to the stimulus orientations presented during the rearing. These include deficits in sensorimotor coordination (Blakemore and Cooper, 1970; Muir and Mitchell, 1975), orientation-specific reductions in visual acuity (Muir and Mitchell 1973, 1975; Blasdel et al., 1977; Fiorentini and Maffei, 1978; Thibos and Levick, 1982; but see also Wark and Peck, 1982), orientation-specific deficits in the ability to discriminate differences in stimulus orientation (Hirsch, 1972; Wark and Peck, 1982), and orientation-specific changes in the stimulus to which the animals attend (Hirsch, 1972).

Stripe-rearing produces dramatic changes in sensorimotor coordination. Immediately after the rearing the animals fail to respond to a moving rod (Blake-

---

Mitchell et al., 1976). Note that the relative improvement in acuity determined by the three measures is very similar. (C) Comparison of changes in visual acuity as a function of age as determined by three different methods. The same data as are plotted above in (B), except that now the actual values for acuity are indicated. Note that the curves for threshold of sustained lgn cells and for acuity measured from cortical evoked potentials are approximately parallel. Note that after 7 weeks postnatal, acuity measured behaviorally exceeds both the acuity measured from cortical evoked potentials and acuity determined on the basis of the thresholds of sustained cells in the lgn. This suggests that the animals may integrate information from more than one cell. (From Ikeda and Tremain, *Experimental Brain Research*, 1978, *31*, 193–206. Reproduced with permission.)

more and Cooper, 1970; Muir and Mitchell, 1975). After several days of normal visual exposure stripe-reared cats acquire the ability to follow visually a moving rod of the same orientation as the lines presented during rearing; they continue to fail to respond to a moving rod oriented at right angles to the lines presented during the rearing (Blakemore and Cooper, 1970; Muir and Mitchell, 1975). In some animals this deficit may last for up to two months (Muir and Mitchell, 1975). Eventually, the orientation-specific sensorimotor deficit does largely disappear; after several months of normal visual exposure it is difficult, if not impossible, to identify an animal's exposure history from casual inspection of its visual behavior (Muir and Mitchell, 1975). Since the distribution of the orientation preferences of area 17 cells continues to reflect the bias present during the rearing (Spinelli *et al.*, 1972; Pettigrew *et al.*, 1973; Leventhal and Hirsch, 1975), it is unlikely that the orientation-specific sensorimotor deficits are related in any simple fashion to the changes observed in area 17.

Other orientation-specific deficits, though perhaps not as dramatic, persist despite many months of normal visual exposure. First, there are orientation-specific deficits in the ability of stripe-reared cats to detect gratings; cats exposed to horizontal lines are better at detecting gratings consisting of horizontal lines than they are at detecting gratings consisting of vertical lines, and vice versa (Muir and Mitchell, 1973, 1975; Blasdel *et al.*, 1977; Thibos and Levick, 1982, but cf. Wark and Peck, 1982, who failed to find any evidence for such deficits in grating acuity). Second, there are orientation-specific deficits in the ability of stripe-reared cats to discriminate orientation differences. Cats exposed to horizontal lines are better at discriminating horizontal from nearly horizontal than they are at discriminating vertical from nearly vertical, and vice versa (Figure 17) (Hirsch, 1972).

Stripe-rearing, however, does not appear to affect the capability of the animals to make simple form discriminations. For example, stripe-rearing so that one eye is exposed only to horizontal lines while the other eye is exposed only to vertical lines (Hirsch and Spinelli, 1970, 1971) has relatively little effect on interocular transfer (Hirsch, 1972). Animals that were trained using only the eye exposed to vertical lines were able to perform the discrimination essentially immediately when tested with the other eye that had been exposed only to horizontal lines (Hirsch, 1972). There must therefore be considerable similarity in the processing of inputs by populations of area 17 cells that have very different distributions of preferred orientations.

Stripe-rearing thus has specific effects on the distribution of the preferred orientations of area 17 cells and on the ability to resolve fine details in a visual stimulus. This implicates experience-sensitive area 17 cells in cortical processing of fine details in a visual stimulus. Consistent with this, lesions of area 17 result in moderate deficits in grating acuity and in severe deficits in discriminating whether or not lines are parallel (Berkley and Sprague, 1979).

Stripe-rearing apparently does not affect the basic capacity for form discriminations. In part, this may reflect the fact that much of the basic structure of area

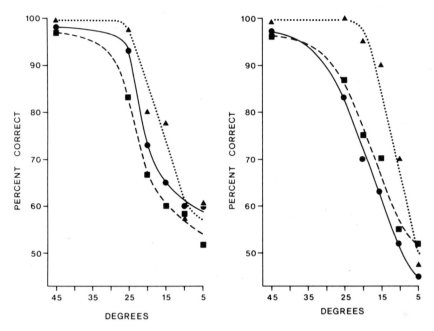

Fig. 17. Orientation discrimination thresholds for two normal cats and for four stripe-reared cats each exposed to horizontal lines with one eye and to vertical lines with the other eye. Mean percentage correct scores are plotted as a function of the orientation differences between the positive and negative stimuli. For the vertical orientation threshold (VOT—left-hand side) the positive stimulus was always oriented vertically and the negative stimuli were oriented at 45, 25, 20, 15, 10, and 5 deg relative to the vertical axis. For the horizontal orientation threshold (HOT—right-hand side) the positive stimulus was always oriented horizontally and the negative stimuli were oriented at 45, 25, 20, 15, 10, and 5 deg relative to the horizontal axis. Data for stripe-reared cats tested using only the eye exposed to vertical lines (VE) are represented by filled circles; data for stripe-reared cats tested using only the eye exposed to horizontal lines (HE) are represented by filled squares. Data for the normal cats are represented by filled triangles (normal cats were tested with one eye at a time, but the data for the two eyes were then averaged). Sigmoid curves were fitted by hand to the data points: a solid curve for VE data, a dashed curve for HE data, and a dotted curve for data from normal cats. Note that overall the normal animals performed better than stripe-reared cats. Furthermore, the orientation of the lines to which stripe-reared cats were exposed affected their performance: stripe-reared cats were better at discriminating orientation differences about vertical when tested with their VE, while they were better at discriminating orientation differences about horizontal when tested with their HE. Exposure to a restricted range of stimulus orientations thus results in an overall reduction in the ability to discriminate orientation differences, even when the orientation of the test stimuli is similar to that of the stimuli presented during rearing. This deficit may be related to the presence of large numbers of nonselective cells in area 17 of cats reared in this fashion. In addition to this nonspecific deficit, there is an orientation-dependent difference in performance so that stripe-reared cats are better at discriminating small orientation differences about the exposed orientation than they are at discriminating orientation differences between lines orthogonal to the exposed orientation. The orientation-dependent difference in performance may be related to differences in the distribution of the orientation preferences of area 17 cells activated by the VE and by the HE. (From Hirsch, *Experimental Brain Research*, 1972, *15*, 405–423. Reproduced with permission.)

17 is not affected by stripe-rearing; stripe-rearing does not modify the distribution of the preferred orientations of intrinsically selective cells (Hirsch *et al.*, 1983), nor does it affect the distribution of orientation columns in layer IV of area 17 (Singer *et al.*, 1981). In addition, other parts of the visual system may be able to mediate form discrimination; lesions restricted to area 17 do not abolish this ability in cats (e.g., Berkley and Sprague, 1979). Thus, as Leventhal and Hirsch (1975) demonstrated, there is a basic, invariant cortical framework; superimposed on this are experience-sensitive cells so that the visual system can be tuned to process fine details of patterns present in the animal's visual environment.

D. Experience-Dependent Changes in "Attention"

Since most visually guided behaviors involve responses to suprathreshold stimuli, one may ask whether the distribution of preferred orientations of area 17 cells can affect the choice of stimuli used to guide behavior. One means of asking this question is to permit the animal to choose one of several cues in making a discrimination. Hirsch (1972) found some evidence that stripe-reared cats tend to use cues that match the stimuli presented during rearing. A cat trained to approach a black outline square chooses the two vertical sides when tested with the eye that had seen only vertical lines, and it chooses the two horizontal lines at the top and bottom of the outline square when tested with the other eye, which had been exposed only to horizontal lines. Though very preliminary, these results point to a functional role for experience-dependent changes in the visual system.

One means for studying the effects of early experience on the choice and utilization of cues would be to assess visual preferences of stripe-reared cats with stimulus-seeking procedures similar to those described above (Dodwell et al., 1976; Emerson and Timney, 1977; Timney *et al.*, 1979). Because an 8- or 9-week-old kitten will work to obtain visual stimulation, it may be possible to determine whether postnatal visual exposure serves the function of guiding an animal's choices in the kinds of stimuli that it seeks to obtain. Studies of rats raised in controlled visual environments (Tees *et al.*, 1980) suggest that stimulus choices are affected by a bias in the animal's early visual exposure. The results obtained suggest that deprivation has relatively limited effects on the capacity to discriminate even very small differences between stimuli (e.g., Tees, 1968, 1979); early visual experience does, however, affect preferences among stimuli that are readily discriminated (Tees *et al.*, 1980). In summary, for the rat attentional, not discriminative, processes are experience dependent (Tees *et al.*, 1980). Complete interpretation of these data on rats must await physiological studies of the effects that a biased visual environment has on visual system development in this species. These data, as well as the preliminary results obtained using stripe-reared cats (Hirsch, 1972), hold out great hope that research on the effects of early visual experience on attention to visual cues will provide a better understanding of the functional role of experience-dependent visual system development.

In summary, it is possible that the dominant features in a kitten's visual environment may help fine-tune the response properties of area 17 cells and thereby optimize the ability of the animal to use these features. Furthermore, visual experience may determine which features an animal attends to and thus have a profound bias on its interactions with the environment. Behavioral tests of stripe-reared cats can provide an important means for learning more about perceptual development: the use of stripe-rearing as a tool in this fashion remains as a challenge to the adventurous researcher.

VII. SUMMARY: THE TUNABLE SEER

One major difference among the three stages of development described above is in the importance of visual experience for normal development. During the first 3 postnatal weeks developmental changes may require internally generated neuronal activity, but not stimulation-dependent neuronal activity. In sharp contrast, between the third and sixth postnatal week many developmental changes are experience dependent. After the sixth postnatal week, developmental changes may also be experience dependent, but there have been few direct tests of this.

A. DEVELOPMENTAL CHANGES

1. STAGE 1. By the end of the first postnatal stage, intrinsic developmental programs have established a framework that will be used to guide the course of subsequent experience-dependent development. Key elements in this framework are precociously mature cells in the retina, lgn, and area 17. The orientation-selective cells in area 17 are predominantly first-order neurons; they are concentrated in layers IV and VI. Most of them are activated monocularly. Many of them almost certainly receive direct input from lgn X-cells; whether some of them also receive direct input from lgn Y-cells is an open question.

Several mechanisms may generate the orientation selectivity of the intrinsically selective area 17 cells. Elongation of their dendritic fields (in a plane parallel to the pial surface) could generate a preference for one orientation. This bias may be amplified by the orientation bias of cells in the lgn (and the retina); adjacent cells in the lgn (and retina) tend to have similar preferred orientations so an area 17 cell that has a small receptive field is likely to receive input from a population of afferents with similar orientation preferences. This bias could be sharpened further by a system of lateral inhibitory connections among area 17 cells. The intrinsically orientation-selective cells in area 17 may provide a starting point for experience-dependent development of orientation columns.

The orientation preferences of the intrinsically selective cells are very robust; they are largely unaffected by profound changes in sensory input such as that resulting from dark- or stripe-rearing. Because of their durability, the intrinsically selective cells may ensure that experience-dependent development of orientation selectivity during the next 3 weeks of life proceeds in an orderly fashion; they may provide the framework needed to establish the orientation selectivity of the bulk of area 17 cells.

2. STAGE 2. Once the framework of intrinsically selective cells has been completed, experience-dependent changes may begin. Between the third and sixth postnatal weeks, the bulk of the cells in retina, lgn, and area 17 develop many of the physiological and morphological properties characteristic of the adult visual system. In area 17, orientation selectivity spreads to cells located outside layers IV and VI, and there is now a more uniform distribution of preferred orientations. Cells develop adult levels of selectivity for spatial frequency. Some of the cells in layers II/III and lower VI may lag behind the rest and complete their development during the third postnatal stage.

3. STAGE 3. After the rapid pace of the developmental changes during the second postnatal stage, there is a gradual, asymptotic approach to the final adult state. There is some physiological and morphological evidence in both lgn and area 17 for extensive changes in inhibitory connectivity after the sixth postnatal week. Furthermore, orientation selectivity of some cells in the upper layers (II/III) and in the lowest layer (lower VI) is completed during the third postnatal stage; final development of the cells that project from layers II/III of area 17 to other cortical visual areas is thus delayed until this stage of development. Since much of the selectivity characteristic of neurons in the adult animal depends on the balance between excitatory and inhibitory input (Murphey and Hirsch, 1982), it is likely that the third stage of postnatal life is a time for fine-tuning of neurons, and of the interconnections between them.

B. EXPERIENCE-DEPENDENT CHANGES IN AREA 17

Most of the developmental changes occurring in area 17 after the third postnatal week are experience dependent; they are delayed or blocked by depriving the animal of normal patterned visual stimulation. Extension of orientation selectivity to the bulk of area 17 cells does not occur if animals are deprived of all patterned visual stimulation. Normal development of selectivity for spatial frequency is also blocked by depriving the animal of patterned visual stimulation. Furthermore, exposure to a limited range of stimulus orientations can lead to an increase in the proportion of orientation-selective cells, and the range of orientation preferences that the cells acquire is constrained by the range of orientations to which the animal is exposed. Selective visual exposure has this effect without apparently changing the physiology or morphology of intrinsically selective cells; it may have its effect by influencing the connections between the intrinsically selective cells and higher-order neurons in area 17. These results suggest that the experience-dependent changes that take place after the third postnatal week provide the ability to incorporate specific information present in the animal's early visual environment.

Most of the behavioral skills that develop either near the end of the first stage of postnatal development or during the second stage involve fairly simple responses to environmental stimulation; animals develop the skills to avoid obstacles, guide limb placement, and navigate within familiar surroundings. Available data suggest that during this stage of development major stages in the postnatal maturation of other portions of the nervous system, especially the hippocampus, are being completed (cf. Frederickson and Frederickson, 1979a). For example, the relationship between hippocampal theta activity and behavior is not fully mature until kittens are 6–8 weeks old (e.g., Passouant, 1975).

It is only after this time, that is, during the third stage of postnatal development, that kittens actively seek to obtain visual stimulation, that they play with objects in their environment, and generally gain familiarity with their world. Through these efforts, kittens may be obtaining visual stimulation that is necessary to consolidate changes that occurred during the earlier stages of development. Furthermore, they may be obtaining stimulation needed to fine-tune the responses of cells in the visual system and to elaborate connections between cells in the various cortical visual areas. Finally, they may be developing the internal representation of their environment that they will need as mature adults to guide their behavior. The appearance of stimulus seeking at this stage of postnatal development may thus be correlated with the maturation of neural structures, such as the hippocampus, that are thought to be involved in the acquisition and storage of information about the environment (cf. Altman *et al.*, 1973).

The onset and time course of behavioral development may be altered by events in the kitten's environment. For example, the time at which weaning begins can affect subsequent frequency of certain behavior patterns. If lactation is blocked for 24 hours when kittens are 4 weeks old, there is an increase in the frequency of "social play" near the end of the second postnatal stage (Martin, 1982); if lactation is blocked for 24 hr when kittens are 5 weeks old, there is an increase in the frequency of "object play" during the third postnatal stage (Bateson *et al.*, 1981). In both cases the behavioral effects of the early weaning appear several weeks after lactation is blocked and are thus not likely to be a direct effect of the interruption in lactation.

The effects of sensory stimulation on visual system development are particularly evident during the second postnatal stage, yet there is no evidence that kittens will actively seek visual stimulation at this stage of development (Dodwell *et al.*, 1976; Timney *et al.*, 1979). Perhaps behavioral effects of these experience-dependent changes are delayed in much the same manner as are behavioral effects of disruptions in lactation. If so, then a possible functional role for those experience-dependent changes in the visual system that reflect the particular stimulus patterns present in the environment could be to affect behavioral responses to those patterns during the course of subsequent stages of postnatal development (cf. Tees *et al.*, 1980). The temporary orientation-specific deficit in sensorimotor coordination that has been observed in stripe-reared cats may be one reflection of a delayed, experience-sensitive bias in the animal's interaction with its environment. Once stripe-reared kittens are returned to a normal visual environment they have to compensate for such an extreme bias by gradually acquiring the ability to respond to all orientations. Any bias that occurs in a normally reared animal would be less extreme, but could still influence the kinds of stimuli to which the animal chooses to respond.

In summary, by the end of the first stage of postnatal life, an intrinsic framework has been established in the visual system. Subsequently, there is a series of developmental changes, many of which may be dependent on visual stimulation. By having stimulation affect both fine-tuning of cells in the visual system and the acquisition of sensorimotor coordination, it is possible to ensure that the animal will subsequently use the dominant visual features of its environment to guide its behavior.

D. Integration of Genetic and Environmental Information

Internally generated neuronal activity can be used to represent information about the outcome of genetically guided developmental programs in one part of the visual system (e.g., the retina); in this form the information can be transmitted to other parts of the visual system (e.g., lgn or area 17). Neuronal activity can also be used to represent information contained in the animal's environment within the same parts of the visual system. Because there is a common mechanism for translating activity into changes in connectivity, possibly much like that proposed by Hebb (1949), it is possible to integrate genetic and environmental information into the structure of the nervous system.

E. Implications of Experience-Dependent Development

In concluding, I suggest that proper development of visual system function requires that inputs be rich enough to allow fine-tuning of sensory processing; this will maximize the system's response to the dominant features of the environment. Once sensory processing has been optimized in this fashion, it can serve as a means of helping the growing animal focus its "attention" on significant features in the environment. Restricting an animal's early visual environment would thus limit its subsequent behavioral choices. To the extent that similar experience-dependent changes are a part of normal development in primates, these conclusions imply that restricting a person's early environment will influence the range choices he or she can make once actively guided behaviors begin to develop.

> But under that root [of Yggdrasil] which faces [the world of] the frost giants there is the well of Mimir in which wit and wisdom are hidden. . . . Thither came Othin and asked for a draught from the well, but got it not before giving his one eye as a pledge. . . . [*Poetic Edda,* Snorri Sturleson]

Several colleagues read earlier versions of this chapter and provided helpful criticisms and comments: H. Ghiradella, A. G. Leventhal, C. Mason, J. D. Schall, S. B. Tieman, D. G. Tieman, and N. Tumosa. A. Appleby, P. Caruccio, and E. Hirsch helped in the preparation of the manuscript. R. Loos and R. Speck prepared some of the figures. H. Ghiradella provided the literary allusions. E. Blass encouraged me to venture beyond the narrow walls of my own discipline. Support provided by NSF Grant BNS 8217540.

References

Adrien, J., and Roffwarg, H. P. The development of unit activity in the lateral geniculate nucleus of the kitten. *Experimental Neurology*, 1974, *43*, 261–275.

Albus, K., and Wolf, W. Early post-natal development of neuronal function in the kitten's visual cortex: A laminar analysis. *Journal of Physiology (London)*, 1984, *348*, 153–185.

Albus, K., Wolf, W., and Beckmann, R. Orientation bias in the response of kitten LGNd neurons to moving light bars. *Developmental Brain Research* 1983 *6*, 308–313.

Altman, J., Brunner, R. L., and Bayer, S. A. The hippocampus and behavioral maturation. *Behavioral Biology*, 1973, *8*, 557–594.

Anker, R. L. The prenatal development of some of the visual pathways in the cat. *Journal of Comparative Neurology*, 1977, *173*, 185–204.

Anker, R. L., and Cragg, B. G. Development of the extrinsic connections of the visual cortex in the cat. *Journal of Comparative Neurology*, 1974, *154*, 29–42.

Archer, S. M., Dubin, M. W., and Stark, L. A. Abnormal development of kitten retino-geniculate connectivity in the absence of action potentials. *Science*, 1982, *217*, 743–745.

Bacon, J., and Murphey, R. K. Receptive fields of cricket giant interneurones are related to their dendritic structure. *Journal of Physiology (London)*, 1984, *352*, 601–623.

Baerends-van Roon, J. M., and Baerends, G. P. *The morphogenesis of the behaviour of the domestic cat.* Amsterdam: North-Holland, 1979.

Barlow, H. B. Visual experience and cortical development. *Nature*, 1975, *258*, 199–203.

Barlow, H. B., and Pettigrew, J. D. Lack of specificity of neurones in the visual cortex of young kittens. *Journal of Physiology (London)*, 1971, *218*, 98–100.

Barrett, P., and Bateson, P. The development of play in cats. *Behaviour* 1978, *66*, 106–120.

Bateson, P., Martin, P., and Young, M. Effects of interrupting cat mothers' lactation with bromocriptine on the subsequent play of their kittens. *Physiology and Behavior*, 1981, *27*, 841–845.

Beckmann, R., and Albus, K. The geniculocortical system in the early postnatal kitten: An electrophysiological investigation. *Experimental Brain Research*, 1982, *47*, 49–56.

Berkley, M. A., and Sprague, J. M. Striate cortex and visual acuity functions in the cat. *Journal of Comparative Neurology*, 1979, *187*, 679–702.

Blakemore, C. Developmental factors in the formation of feature extracting neurons. In F. O. Schmidt and F. G. Worden (Eds.), *The neurosciences: Third study program*. Cambridge, Mass.: MIT Press, 1974, pp. 105–113.

Blakemore, C. The conditions required for the maintenance of binocularity in the kitten's visual cortex. *Journal of Physiology (London)*, 1976, *261*, 423–444.

Blakemore, C., and Cooper, G. F. Development of the brain depends on the visual environment. *Nature*, 1970, *228*, 477–478.

Blakemore, C., and Tobin, E. A. Lateral inhibition between orientation detectors in the cat's visual cortex. *Experimental Brain Research* 1972, *15*, 439–440.

Blakemore, C., and Van Sluyters, R. C. Innate and environmental factors in the development of the kitten's visual cortex. *Journal of Physiology (London)*, 1975, *248*, 663–716.

Blasdel, G. G., Mitchell, D. E., Muir, D. W., and Pettigrew, J. D. A physiological and behavioural study in cats of the effect of early visual experience with contours of a single orientation. *Journal of Physiology (London)*, 1977, *265*, 615–636.

Blass, E. M. The ontogenesis of suckling, a goal-directed behavior. In R. F. Thompson, L. H. Hicks, and V. B. Shvyrkov (Eds.), *Neural mechanisms of goal directed behavior and learning*. New York: Academic Press, 1980, pp. 461–470.

Bonds, A. B. Development of orientation tuning in the visual cortex of kittens. In R. D. Freeman (Ed.), *Developmental neurobiology of vision*. New York: Plenum Press, 1979, pp. 31–49.

Bonds, A. B., and Freeman, R. D. Development of optical quality in the kitten eye. *Vision Research*, 1978, *18*, 391–398.

Boycott, B. B., and Wässle, H. The morphological types of ganglion cells of the domestic cat's retina. *Journal of Physiology (London)*, 1974, *240*, 397–419.

Brown, J. L. *The evolution of behavior*. New York: Norton, 1975.

Buisseret, P., and Imbert, M. Visual cortical cells: Their developmental properties in normal and dark reared kittens. *Journal of Physiology (London)*, 1976, *255*, 511–525.

Buisseret, P., and Gary-Bobo, E. Development of visual cortical orientation specificity after dark-rearing: Role of extraocular proprioception. *Neuroscience Letters*, 1979, *13*, 259–263.

Buisseret, P., Gary-Bobo, E., and Imbert, M. Ocular motility and recovery of orientational properties of visual cortical neurones in dark-reared kittens. *Nature*, 1978, *272*, 816–817.

Buisseret, P., Gary-Bobo, E., and Imbert, M. Plasticity in the kitten's visual cortex: Effects of the suppression of visual experience upon the orientational properties of visual cortical cells. *Developmental Brain Research*, 1982, *4*, 417–426.

Buisseret, P., and Singer, W. Proprioceptive signals from extraocular muscles gate experience-dependent modifications of receptive fields in the kitten visual cortex. *Experimental Brain Research*, 1983, *51*, 443–450.

Bullier, J., and Henry, G. H. Ordinal position of neurons in cat striate cortex. *Journal of Neurophysiology*, 1979a, *42*, 1251–1263.

Bullier, J., and Henry, G. H. Neural path taken by afferent streams in striate cortex of the cat. *Journal of Neurophysiology*, 1979b, *42*, 1264–1270.

Bullier, J., and Henry, G. H. Laminar distribution of first-order neurons and afferent terminals in cat striate cortex. *Journal of Neurophysiology*, 1979c, *42*, 1271–1281.

Chow, K. L. Neuronal changes in the visual system following visual deprivation. In R. Jung (Ed.), *Handbook of sensory physiology*. Vol. VII/3A. New York: Springer-Verlag, 1973, pp. 599–627.

Cleland, B. G., Dubin, M. W., and Levick, W. R. Sustained and transient neurones in the cat's retina and lateral geniculate nucleus. *Journal of Physiology (London)*, 1971, *217*, 473–496.

Cleland, B. G., Levick, W. R., and Sanderson, K. J. Properties of sustained and transient ganglion cells in the cat retina. *Journal of Physiology (London)*, 1973, *228*, 649–680.

Coleman, P. D., and Riesen, A. H. Environmental effects on cortical dendritic fields. I. Rearing in the dark. *Journal of Anatomy*, 1968, *102*, 363–374.

Coleman, P. D., Flood, D. G., Whitehead, M. C., and Emerson, R. C. Spatial sampling by dendritic trees in visual cortex. *Brain Research*, 1981, *214*, 1–21.

Colonnier, M. The tangential organization of the visual cortex. *Journal of Anatomy*, 1964, *98*, 327–344.

Cowan, W. M. Neuronal death as a regulative mechanism in the control of cell number in the nervous system. In M. Rockstein (Ed.), *Development and aging in the nervous system*. New York: Academic Press, 1973, pp. 19–41.

Cragg, B. G. The development of synapses in the visual system of the cat. *Journal of Comparative Neurology*, 1975, *160*, 147–166.

Creutzfeldt, O. D., and Nothdurft, H. C. Representation of complex visual stimuli in the brain. *Naturwissenschaften*, 1978, *65*, 307–318.

Cynader, M., Berman, N., and Hein, A. Recovery of function in cat visual cortex following prolonged deprivation. *Experimental Brain Research*, 1976, *25*, 139–156.

Daniels, J. D., Norman, J. L., and Pettigrew, J. D. Biases for oriented moving bars in lateral geniculate nucleus neurons of normal and stripe-reared cats. *Experimental Brain Research*, 1977, *29*, 155–172.

Daniels, J. D., Pettigrew, J. D., and Norman, J. L. Development of single-neuron responses in kitten's lateral geniculate nucleus. *Journal of Neurophysiology*, 1978, *41*, 1373–1393.

Daw, N. W., Videen, T. O., Robertson, T., and Rader, R. K. An evaluation of the hypothesis that noradrenaline affects plasticity in the developing visual cortex. In A. Fein and J. S. Levine (Eds.), *The visual system,* New York: Alan R. Liss, 1985, pp. 133–144.

Derrington, A. M. Development of spatial frequency selectivity in striate cortex of vision-deprived cats. *Experimental Brain Research,* 1984, *55,* 431–437.

Derrington, A. M., and Fuchs, A. F. Development of spatial frequency selectivity in kitten striate cortex. *Journal of Physiology (London),* 1981, *316,* 1–10.

Derrington, A. M., and Hawken, M. J. Spatial and temporal properties of cat geniculate neurones after prolonged deprivation. *Journal of Physiology (London),* 1981, *314,* 107–120.

Dodwell, P. C., Timney, B. N., and Emerson, V. F. Development of visual stimulus-seeking in dark-reared kittens. *Nature,* 1976, *260,* 777–778.

Donovan, A. The postnatal development of the cat's retina. *Experimental Eye Research,* 1966, *5,* 249–254.

Dreher, B., Leventhal, A. G., and Hale, P. T. Geniculate input to cat visual cortex: A comparison of area 19 with areas 17 and 18. *Journal of Neurophysiology,* 1980, *44,* 804–826.

Emerson, V. F., and Timney, B. Measurement of visual preferences in cats. *Perception,* 1977, *6,* 173–179.

Enroth-Cugell, C., and Robson, J. G. The contrast sensitivity of retinal ganglion cells of the cat. *Journal of Physiology (London),* 1966, *187,* 517–552.

Fiorentini, A., and Maffei, L. Selective impairment of contrast sensitivity in kittens exposed to periodic gratings. *Journal of Physiology (London),* 1978, *277,* 455–466.

Fox, M. W. Reflex development and behavioral organization. In W. A. Himwich (Ed.), *Developmental Neurobiology.* Springfield, Ill.: Thomas, 1970, pp. 553–580.

Frederickson, C. J., and Frederickson, M. H. Emergence of spontaneous alternation in the kitten. *Developmental Psychobiology,* 1979a, *12,* 615–621.

Frederickson, C. J., and Frederickson, M. H. Developmental changes in open-field behavior in the kitten. *Developmental Psychobiology,* 1979b, *12,* 623–628.

Freeman, D. N., and Marg, E. Visual acuity development coincides with the sensitive period in kittens. *Nature,* 1975, *254,* 614–615.

Freeman, N. C. G., and Rosenblatt, J. S. The interrelationship between thermal and olfactory stimulation in the development of home orientation in newborn kittens. *Developmental Psychobiology,* 1978, *11,* 437–457.

Freeman, R. D., and Bonds, A. B. Cortical plasticity in monocularly deprived immobilized kittens depends on eye movement. *Science,* 1979, *206,* 1093–1095.

Freeman, R. D., and Lai, C. E. Development of the optical surfaces of the kitten eye. *Vision Research,* 1978, *18,* 399–407.

Freeman, R. D., Wong, S., and Zezula, S. Optical development of the kitten cornea. *Vision Research,* 1978, *18,* 409–414.

Fregnac, Y., and Imbert, M. Early development of visual cortical cells in normal and dark-reared kittens: Relationship between orientation selectivity and ocular dominance. *Journal of Physiology (London),* 1978, *278,* 27–44.

Fregnac, Y., and Imbert, M. Development of neuronal selectivity in primary visual cortex of cat. *Physiological Reviews,* 1984, *64,* 325–434.

Friedlander, M. J. Structure of physiologically classified neurones in the kitten dorsal lateral geniculate nucleus. *Nature* 1982, *300,* 180–183.

Friedlander, M. J., Lin, C. S., Stanford, L. R., and Sherman, S. M. Morphology of functionally identified neurons in lateral geniculate nucleus of the cat. *Journal of Neurophysiology,* 1981, *46,* 80–129.

Friedlander, M. J., Martin, K. A. C., and Vahle-Hinz, C. The postnatal development of structure of physiologically identified retinal ganglion cell (r.g.c.) axons in the kitten. *Journal of Physiology (London),* 1982, *336,* 28–29P.

Fries, W., Albus, K., and Creutzfeldt, O. D. Effects of interacting visual patterns on single cell responses in cat's striate cortex. *Vision Research,* 1977, *17,* 1001–1008.

Fukada, Y. Receptive field organization of cat optic nerve fibers with special reference to conduction velocity. *Vision Research,* 1971, *11,* 209–226.

Galef, B. G., Jr. The ecology of weaning: Parasitism and the achievement of independence by altricial mammals. In D. J. Gubernick and P. H. Klopfer (Eds.), *Parental care in mammals.* New York: Plenum Press, 1981, pp. 211–242.

Gilbert, C. D. Laminar differences in receptive field properties of cells in cat primary visual cortex. *Journal of Physiology (London)*, 1977, *268*, 391–421.

Gilbert, C. D. Microcircuitry of the visual cortex. *Annual Review of Neuroscience*, 1983, *6*, 217–247.

Gilbert, C. D., and Kelly, J. P. The projections of cells in different layers of the cat's visual cortex. *Journal of Comparative Neurology*, 1975, *163*, 81–106.

Gordon, B., and Presson, J. Orientation deprivation in cat: What produces the abnormal cells? *Experimental Brain Research*, 1982, *46*, 144–146.

Grobstein, P., and Chow, K. L. Receptive field development and individual experience. *Science*, 1975, *190*, 352–358.

Gyllensten, L., Malmfors, T., and Norrlin, M. L. Effect of visual deprivation on the optic centers of growing and adult mice. *Journal of Comparative Neurology*, 1965, *124*, 149–160.

Gyllensten, L., Malmfors, T., and Norrlin-Grettve, M. L. Developmental and functional alterations in the fiber composition of the optic nerve in visually deprived mice. *Journal of Comparative Neurology*, 1966, *128*, 413–418.

Hagerty, C. M., Lees, F. C., Tieman, S. B., and Hirsch, H. V. B. Principal components analysis of cells in cat visual cortex. *Brain Research*, 1982, *251*, 45–53.

Hamasaki, D. I., and Flynn, J. T. Physiological properties of retinal ganglion cells of 3-week-old kittens. *Vision Research*, 1977, *17*, 275–284.

Hamasaki, D. I., and Sutija, V. G. Development of X- and Y-cells in kittens. *Experimental Brain Research*, 1979, *35*, 9–23.

Hammond, P. Cat retinal ganglion cells: Size and shape of receptive field centres. *Journal of Physiology (London)*, 1974, *242*, 99–118.

Hebb, D. O. *The Organization of behaviour*. New York: Wiley, 1949.

Hein, A., and Held, R. Dissociation of the visual placing response into elicited and guided components. *Science* 1967, *158*, 390–392.

Hein, A., Vital-Durand, F., Salinger, W., and Diamond, R. Eye movements initiate visual-motor development in the cat. *Science*, 1979, *204*, 1321–1322.

Held, R., and Bossom, J. Neonatal deprivation and adult rearrangement: Complementary techniques for analyzing plastic sensory-motor coordinations. *Journal of Comparative and Physiological Psychology*, 1961, *54*, 33–37.

Held, R., and Hein, A. Movement-produced stimulation in the development of visually guided behavior. *Journal of Comparative and Physiological Psychology*, 1963, *56*, 872–876.

Hendrickson, A., and Boothe, R. Morphology of the retina and dorsal lateral geniculate nucleus in dark-reared monkeys *(Macaca nemestrina)*. *Vision Research*, 1976, *16*, 517–521.

Henry, G. H., Dreher, B., and Bishop, P. O. Orientation specificity of cells in cat striate cortex. *Journal of Neurophysiology*, 1974, *37*, 1394–1409.

Henry, G. H., Harvey, A. R., and Lund, J. S. The afferent connections and laminar distribution of cells in the cat striate cortex. *Journal of Comparative Neurology*, 1979, *187*, 725–744.

Henry, G. H., Mustari, M. J., and Bullier, J. Different geniculate inputs to B and C cells of cat striate cortex. *Experimental Brain Research*, 1983, *52*, 179–189.

Hess, R., Negishi, K., and Creutzfeldt, O. The horizontal spread of intracortical inhibition in the visual cortex. *Experimental Brain Research*, 1975, *22*, 415–419.

Hickey, T. L. Development of the dorsal lateral geniculate nucleus in normal and visually deprived cats. *Journal of Comparative Neurology*, 1980, *189*, 467–481.

Hirsch, H. V. B. Visual perception in cats after environmental surgery. *Experimental Brain Research*, 1972, *15*, 405–423.

Hirsch, H. V. B. The role of visual experience in the development of cat striate cortex. *Cellular and Molecular Neurobiology*, 1985, *5*, 103–121.

Hirsch, H. V. B., and Jacobson, M. The perfectible brain: Principles of neuronal development. In M. Gazzaniga and C. Blakemore (Eds.), *Handbook of psychobiology*. New York: Academic Press, 1975, pp. 107–137.

Hirsch, H. V. B., and Leventhal, A. G. Cortical effects of early visual experience. In S. J. Cool and E. L. Smith, III (Eds.), *Frontiers in visual science*. New York: Springer-Verlag, 1978a, pp. 660–673.

Hirsch, H. V. B., and Leventhal, A. G. Functional modification of the developing visual system. In Marcus Jacobson (Ed.), *Handbook of sensory physiology*. Vol. IX. *Development of sensory systems*, New York: Springer-Verlag, 1978b, pp. 279–335.

Hirsch, H. V. B., and Spinelli, D. N. Visual experience modifies distribution of horizontally and vertically oriented receptive fields in cats. *Science*, 1970, *168*, 869–871.

Hirsch, H. V. B., and Spinelli, D. N. Modification of the distribution of receptive field orientation in cats by selective visual exposure during development. *Experimental Brain Research*, 1971, *12*, 509–527.

Hirsch, H. V. B., Leventhal, A. G., McCall, M. A., and Tieman, D. G. Effects of exposure to lines of one or two orientations on different cell types in striate cortex of cat. *Journal of Physiology (London)*, 1983, *337*, 241–255.

Hoffmann, K.-P., and Stone, J. Conduction velocity of afferents to cat visual cortex: A correlation with cortical receptive field properties. *Brain Research*, 1971, *32*, 460–466.

Hoffmann, K.-P., Stone, J., and Sherman, S. M. Relay of receptive-field properties in dorsal lateral geniculate nucleus of the cat. *Journal of Neurophysiology*, 1972, *35*, 518–531.

Hollyday, M., and Hamburger, V. Reduction of the naturally occurring motor neuron loss by enlargement of the periphery. *Journal of Comparative Neurology*, 1976, *170*, 311–320.

Hubel, D. H., and Wiesel, T. N. Receptive fields of single neurones in the cat's striate cortex. *Journal of Physiology (London)*, 1959, *148*, 574–591.

Hubel, D. H., and Wiesel, T. N. Receptive fields, binocular interaction and functional architecture in the cat's visual cortex. *Journal of Physiology (London)*, 1962, *160*, 106–154.

Hubel, D. H., and Wiesel, T. N. Shape and arrangement of columns in cat's striate cortex. *Journal of Physiology (London)*, 1963a, *165*, 559–568.

Hubel, D. H., and Wiesel, T. N. Receptive fields of cells in striate cortex of very young, visually inexperienced kittens. *Journal of Neurophysiology*, 1963b, *26*, 994–1002.

Huttenlocher, P. R. Development of cortical neuronal activity in the neonatal cat. *Experimental Neurology*, 1967, *17*, 247–262.

Ikeda, H., and Tremain, K. E. The development of spatial resolving power of lateral geniculate neurones in kittens. *Experimental Brain Research*, 1978, *31*, 193–206.

Ikeda, H., and Wright, M. J. Receptive field organization of "sustained" and "transient" retinal ganglion cells which subserve different functional roles. *Journal of Physiology (London)*, 1972, *227*, 769–800.

Imbert, M., and Buisseret, P. Receptive field characteristics and plastic properties of visual cortical cells in kittens reared with or without visual experience. *Experimental Brain Research*, 1975, *22*, 25–36.

Imbert, M., and Fregnac, Y. Specification of cortical neurons by visuomotor experience. In J. P. Changeux, J. Glowinski, M. Imbert, and F. E. Bloom (Eds.), *Molecular and cellular interactions underlying higher brain functions*. New York: Elsevier Science Publishers, 1983, pp. 427–436.

Jacobson, M. *Developmental neurobiology*. New York: Holt, Rinehart and Winston, 1970.

Jacobson, M. Genesis of neuronal specificity. In M. Rockstein (Ed.), *Development and aging in the nervous system*. New York: Academic Press, 1973, pp. 105–119.

Jacobson, M. *Developmental neurobiology*, 2nd ed. New York: Plenum Press, 1978.

Johns, P. R., Rusoff, A. C., and Dubin, M. W. Postnatal neurogenesis in the kitten retina. *Journal of Comparative Neurology*, 1979, *187*, 545–556.

Kalil, R. E. Dark rearing in the cat: Effects on visuomotor behavior and cell growth in the dorsal lateral geniculate nucleus. *Journal of Comparative Neurology*, 1978a, *178*, 451–468.

Kalil, R. E. Development of the dorsal lateral geniculate nucleus in the cat. *Journal of Comparative Neurology*, 1978b, *182*, 265–292.

Kalil, R. E., Dubin, M. W., Scott, G. L., and Stark, L. A. Effects of retinal ganglion cell blockade on the morphological development of retinogeniculate synapses in the cat. *Society for Neuroscience Abstracts*, 1983, *11.4*.

Karmel, B. Z., Miller, P. N., Dettweiler, L., and Anderson, G. Texture density and normal development of visual depth avoidance. *Developmental Psychobiology*, 1970, *3*, 73–90.

Kasamatsu, T., and Pettigrew, J. D. Depletion of brain catecholamines: Failure of ocular dominance shift after monocular occlusion in kittens. *Science*, 1976, *194*, 206–209.

Kasamatsu, T., and Pettigrew, J. D. Preservation of binocularity after monocular deprivation in the striate cortex of kittens treated with 6-hydroxydopamine. *Journal of Comparative Neurology*, 1979, *185*, 139–162.

Kasamatsu, T., Pettigrew, J. D., and Ary, M. Restoration of visual cortical plasticity by local microperfusion of norepinephrine. *Journal of Comparative Neurology*, 1979, *185*, 163–182.

Kasamatsu, T., Pettigrew, J. D., and Ary, M. Cortical recovery from effects of monocular deprivation: Acceleration with norepinephrine and suppression with 6-hydroxydopamine. *Journal of Neurophysiology*, 1981, *45*, 254–266.

Kato, N., Kawaguchi, S., Yamamoto, T., Samejima, A., and Miyata, H. Postnatal development of the geniculocortical projection in the cat: Electrophysiological and morphological studies. *Experimental Brain Research*, 1983, *51*, 65–72.

Kety, S. S. The biogenic amines in the central nervous system: Their possible roles in arousal, emotion and learning. In F. O. Schmitt (Ed.), *The neurosciences: second study program*. New York: Rockefeller University Press, 1970, pp. 324–336.

Kirk, M. D., Waldrop, B., and Glantz, R. M. A quantitative correlation of contour sensitivity with dendritic density in an identified visual neuron. *Brain Research*, 1983, *274*, 231–237.

Kolb, B., and Nonneman, A. J. The development of social responsiveness in kittens. *Animal Behaviour*, 1975, *23*, 368–374.

Kratz, K. E. Spatial and temporal sensitivity of lateral geniculate cells in dark-reared cats. *Brain Research*, 1982, *251*, 55–63.

Kratz, K. E., Sherman, S. M., and Kalil, R. Lateral geniculate nucleus in dark-reared cats: Loss of Y cells without changes in cell size. *Science*, 1979, *203*, 1353–1355.

Kuffler, S. W. Discharge patterns and functional organization of mammalian retina. *Journal of Neurophysiology*, 1953, *16*, 37–68.

Laemle, L., Benhamida, C., and Purpura, D. P. Laminar distribution of geniculo-cortical afferents in visual cortex of the postnatal kitten. *Brain Research*, 1972, *41*, 25–37.

Lee, B. B., Creutzfeldt, O. D., and Elepfandt, A. The responses of magno- and parvocellular cells of the monkey's lateral geniculate body to moving stimuli. *Experimental Brain Research*, 1979, *35*, 547–557.

LeVay, S., Stryker, M. P., and Shatz, C. J. Ocular dominance columns and their development in layer IV of the cat's visual cortex: A quantitative study. *Journal of Comparative Neurology*, 1978, *179*, 223–244.

Leventhal, A. G. Evidence that the different classes of relay cells of the cat's lateral geniculate nucleus terminate in different layers of the striate cortex. *Experimental Brain Research*, 1979, *37*, 349–372.

Leventhal, A. G. Relationship between preferred orientation and receptive field position of neurons in cat striate cortex. *Journal of Comparative Neurology*, 1983, *220*, 476–483.

Leventhal, A. G., and Hirsch, H. V. B. Cortical effect of early selective exposure to diagonal lines. *Science*, 1975, *190*, 902–904.

Leventhal, A. G., and Hirsch, H. V. B. Effects of early experience upon orientation sensitivity and binocularity of neurons in visual cortex of cats. *Proceedings of the National Academy of Science USA*, 1977, *74*, 1272–1276.

Leventhal, A. G., and Hirsch, H. V. B. Receptive-field properties of neurons in different laminae of visual cortex of the cat. *Journal of Neurophysiology*, 1978, *41*, 948–962.

Leventhal, A. G., and Hirsch, H. V. B. Receptive-field properties of different classes of neurons in visual cortex of normal and dark-reared cats. *Journal of Neurophysiology*, 1980, *43*, 1111–1132.

Leventhal, A. G., and Schall, J. D. Structural basis of orientation sensitivity of cat retinal ganglion cells. *Journal of Comparative Neurology*, 1983, *220*, 465–475.

Levick, W. R., and Thibos, L. N. Orientation bias of cat retinal ganglion cells. *Nature*, 1980, *286*, 389–390.

Levick, W. R., and Thibos, L. N. Analysis of orientation bias in cat retina. *Journal of Physiology (London)*. 1982, *329*, 243–261.

Levine, M. S., Hull, C. D., and Buchwald, N. A. Development of motor activity in kittens. *Developmental Psychobiology*, 1980, *13*, 357–371.

Lund, R. D. *Development and plasticity of the brain*. New York: Oxford University Press, 1978.

Mariani, A. P. A morphological basis for verticality detectors in the pigeon retina: Asymmetric amacrine cells. *Naturwissenschaften*, 1983, *70*, 368–369.

Martin, P. H. Weaning and behavioural development in the cat. Ph.D. Thesis, Christ's College, Cambridge, England, 1982.

Mason, C. A. Development of terminal arbors of retino-geniculate axons in the kitten. I. Light microscopical observations. *Neuroscience*, 1982a, *7*, 541–559.

Mason, C. A. Development of terminal arbors of retino-geniculate axons in the kitten. II. Electron microscopical observations. *Neuroscience*, 1982b, *7*, 561–582.

Mason, C. A. Postnatal maturation of neurons in the cat's lateral geniculate nucleus. *Journal of Comparative Neurology*, 1983, *217*, 458–469.

Mason, R. Cell properties in the medial interlaminar nucleus of the cat's lateral geniculate complex in relation to the transient/sustained classification. *Experimental Brain Research*, 1975, *22*, 327–329.

Mastronarde, D. N. Correlated firing of cat retinal ganglion cells. I. Spontaneously active inputs to X- and Y-cells. *Journal of Neurophysiology*, 1983a, *49*, 303–324.

Mastronarde, D. N. Correlated firing of cat retinal ganglion cells. II. Responses of X- and Y-cells to single quantal events. *Journal of Neurophysiology*, 1983b, *49*, 325–349.

Mastronarde, D. N. Interactions between ganglion cells in cat retina. *Journal of Neurophysiology*, 1983c, *49*, 350–365.

Mates, S. L., and Lund, J. S. Neuronal composition and development in lamina 4C of monkey striate cortex. *Journal of Comparative Neurology*, 1983a, *221*, 60–90.

Mates, S. L., and Lund, J. S. Spine formation and maturation of type 1 synapses on spiny stellate neurons in primate visual cortex. *Journal of Comparative Neurology*, 1983b, *221*, 91–97.

Mates, S. L., and Lund, J. S. Developmental changes in the relationship between type 2 synapses and spiny neurons in the monkey visual cortex. *Journal of Comparative Neurology*, 1983c, *221*, 98–105.

Matin, E. Saccadic suppression and the dual mechanism theory of direction constancy. *Vision Research*, 1982, *22*, 335–336.

McCall, M. A. The relationship between ocular dominance and other response properties of cortical cells in normal kittens and in monocularly deprived kittens. Ph.D. Thesis, SUNY Albany, 1983.

Meyer, R. L. Tetrodotoxin blocks the formation of ocular dominance columns in goldfish. *Science*, 1982, *218*, 589–591.

Meyer, R. L. Tetrodotoxin inhibits the formation of refined retinotopography in goldfish. *Developmental Brain Research*, 1983, *6*, 293–298.

Mitchell, D. E., Giffin, F., Wilkinson, F., Anderson, P., and Smith, M. L. Visual resolution in young kittens. *Vision Research*, 1976, *16*, 363–366.

Moelk, M. The development of friendly approach behavior in the cat: A study of kitten–mother relations and the cognitive development of the kitten from birth to eight weeks. *Advances in the Study of Behaviour*, 1979, *10*, 163–224.

Moore, C. L., Kalil, R., and Richards, W. Development of myelination in optic tract of the cat. *Journal of Comparative Neurology*, 1976, *165*, 125–136.

Morrone, M. C., Burr, D. C., and Maffei, L. Functional implications of cross-orientation inhibition of cortical visual cells. I. Neurophysiological evidence. *Proceedings of the Royal Society of London B*, 1982, *216*, 335–354.

Movshon, J. A., and Van Sluyters, R. C. Visual neural development. *Annual Review of Psychology*, 1981, *32*, 477–522.

Mower, G. D., Burchfiel, J. L., and Duffy, F. H. The effects of dark-rearing on the development and plasticity of the lateral geniculate nucleus. *Developmental Brain Research*, 1981a, *1*, 418–424.

Mower, G. D., Berry, D., Burchfiel, J. L., and Duffy, F. H. Comparison of the effects of dark rearing and binocular suture on development and plasticity of cat visual cortex. *Brain Research*, 1981b, *220*, 255–267.

Muir, D. W., and Mitchell, D. E. Visual resolution and experience: Acuity deficits in cats following early selective visual deprivation. *Science*, 1973, *180*, 420–422.

Muir, D. W., and Mitchell, D. E. Behavioral deficits in cats following early selected visual exposure to contours of a single orientation. *Brain Research*, 1975, *85*, 459–477.

Murphey, R. K., and Hirsch, H. V. B. From cat to cricket: The genesis of response selectivity on interneurons. In R. K. Hunt (Ed.), *Current topics in developmental biology*. (Vol. 17). New York: Academic Press, 1982, pp. 241–256.

Mustari, M. J., Bullier, J., and Henry, G. H. Comparison of response properties of three types of monosynaptic S-cell in cat striate cortex. *Journal of Neurophysiology*, 1982, *47*, 439–454.

Nelson, J. I., and Frost, B. J. Orientation-selective inhibition from beyond the classic visual receptive field. *Brain Research*, 1978, *139*, 359–365.

Ng, A. Y. K., and Stone, J. The optic nerve of the cat: Appearance and loss of axons during normal development. *Developmental Brain Research*, 1982, *5*, 263–271.

Norman, J. L., Pettigrew, J. D., and Daniels, J. D. Early development of X-cells in kitten lateral geniculate nucleus. *Science*, 1977, *198*, 202–204.

Norton, T. T. Receptive-field properties of superior colliculus cells and development of visual behavior in kittens. *Journal of Neurophysiology,* 1974, *37,* 674–690.

Olson, C. R., and Freeman, R. D. Eye alignment in kittens. *Journal of Neurophysiology,* 1978, *41,* 848–859.

Olson, C. R., and Freeman, R. D. Rescaling of the retinal map of visual space during growth of the kitten's eye. *Brain Research,* 1980, *186,* 55–65.

Oppenheim, R. W. Ontogenetic adaptations and retrogressive processes in the development of the nervous system and behaviour: A neuroembryological perspective. In K. J. Connolly and H. F. R. Prechtl (Eds.), *Maturation and development: Biological and psychological perspectives.* Philadelphia: Lippincott, 1981, pp. 73–109.

Paradiso, M. A., Bear, M. F., and Daniels, J. D. Effects of intracortical infusion of 6-hydroxydopamine on the response of kitten visual cortex to monocular deprivation. *Experimental Brain Research,* 1983, *51,* 413,–422.

Passouant, P. Ontogenesis and phylogenesis of sleep. In P. Passouant (Ed.), *Handbook of EEG and clinical neurology.* Vol. 7A: *EEG and sleep.* Amsterdam: Elsevier, 1975, pp. 23–24.

Peichl, L., and Wässle, H. The structural correlate of receptive field centre of α ganglion cells in the cat retina. *Journal of Physiology (London),* 1983, *341,* 309–324.

Pettigrew, J. D. The effect of visual experience on the development of stimulus specificity by kitten cortical neurones. *Journal of Physiology (London),* 1974, *237,* 49–74.

Pettigrew, J. D. The paradox of the critical period for striate cortex. In C. W. Cotman (Ed.), *Neuronal plasticity.* New York: Raven Press, 1978, pp. 311–330.

Pettigrew, J. D., and Freeman, R. D. Visual experience without lines: Effect on developing cortical neurons. *Science,* 1973, *182,* 599–601.

Pettigrew, J. D., Olson, C., and Hirsch, H. V. B. Cortical effect of selective visual experience: degeneration or reorganization? *Brain Research,* 1973, *51,* 345–351.

Prestige, M. C. Axon and cell numbers in the developing nervous system. *Brain Medical Bulletin,* 1974, *30,* 107–111.

Rapaport, D. H., and Stone, J. The site of commencement of maturation in mammalian retina: Observations in the cat. *Developmental Brain Research,* 1982, *5,* 273–279.

Rasch, E., Swift, H., Riesen, A. H., and Chow, K. L. Altered structure and composition of retinal cells in dark-reared mammals. *Experimental Cell Research,* 1961, *25,* 348–363.

Rauschecker, J. P., and Singer, W. The effects of early visual experience on the cat's visual cortex and their possible explanation by Hebb synapses. *Journal of Physiology (London),* 1981, *310,* 215–239.

Rodieck, R. W. *The vertebrate retina.* San Francisco: W. H. Freeman, 1973.

Rodieck, R. W. Visual pathways. *Annual Review of Neuroscience,* 1979, *2,* 193–225.

Rosenblatt, J. S., and Schneirla, T. C. The behaviour of cats. In E. S. E. Hafez (Ed.), *The behaviour of domestic animals.* London: Bailliere, Tindall and Cox, 1962, pp. 453–488.

Rosenblatt, J. S., Turkewitz, G., and Schneirla, T. C. Development of home orientation in newly born kittens. *Transactions of New York Academy of Sciences,* 1964, *31,* 231–250.

Roux, W. *Der Kampf der Theile im Organismus,* Leipzig: Wilhelm Engelmann Verlag, 1881.

Rusoff, A. C., and Dubin, M. W. Development of receptive-field properties of retinal ganglion cells in kittens. *Journal of Neurophysiology,* 1977, *40,* 1188–1198.

Rusoff, A. C., and Dubin, M. W. Kitten ganglion cells: Dendritic field size at 3 weeks of age and correlation with receptive field size. *Investigative Ophthalmology and Visual Science,* 1978, *17,* 819–821.

Saito, H. Morphology of physiologically identified X-, Y-, and W-type retinal ganglion cells of the cat. *Journal of Comparative Neurology,* 1983, *221,* 279–288.

Sato, H., and Tsumoto, T. GABAergic inhibition already operates on a group of neurons in the kitten visual cortex at the time of eye opening. *Developmental Brain Research,* 1984, *12,* 311–315.

Saunders, J. W., Jr. Cell death in embryonic systems. *Science,* 1966, *154,* 604–612.

Schmidt, J. T., and Edwards, D. L. Activity sharpens the map during the regeneration of the retinotectal projection in goldfish. *Brain Research,* 1983, *269,* 29–39.

Schmidt, J. T., Edwards, D. L., and Stuermer, C. The re-establishment of synaptic transmission by regenerating optic axons in goldfish: Time course and effects of blocking activity by intraocular injection of tetrodotoxin. *Brain Research,* 1983, *269,* 15–27.

Shatz, C. J. The prenatal development of the cat's retinogeniculate pathway. *Journal of Neuroscience,* 1983, *3,* 482–499.

Shatz, C. J., and Kirkwood, P. A. Prenatal development of functional connections in the cat's retino-geniculate pathway. *Journal of Neuroscience,* 1984, *4,* 1378–1397.

Sherk, H., and Stryker, M. P. Quantitative study of cortical orientation selectivity in visually inexperienced kitten. *Journal of Neurophysiology,* 1976, *39,* 63–70.

Sherman, S. M. Development of interocular alignment in cats. *Brain Research,* 1972, *37,* 187–203.

Sherman, S. M., and Spear, P. D. Organization of visual pathways in normal and visually deprived cats. *Physiological Reviews,* 1982, *62,* 738–855.

Sillito, A. M. The contribution of inhibitory mechanisms to the receptive field properties of neurones in the striate cortex of the cat. *Journal of Physiology (London),* 1975, *250,* 305–329.

Sillito, A. M. Inhibitory mechanisms influencing complex cell orientation selectivity and their modification at high resting discharge levels. *Journal of Physiology (London),* 1979, 289, 33–53.

Sillito, A. M., Kemp, J. A., Milson, J. A., and Berardi, N. A re-evaluation of the mechanisms underlying simple cell orientation selectivity. *Brain Research,* 1980, *194,* 517–520.

Singer, W. Central core control of developmental plasticity in the kitten visual cortex: I. Diencephalic lesions. *Experimental Brain Research,* 1982, *47,* 209–222.

Singer, W., and Rauschecker, J. P. Central core control of developmental plasticity in the kitten visual cortex: II. Electrical activation of mesencephalic and diencephalic projections. *Experimental Brain Research,* 1982, *47,* 223–233.

Singer, W., and Tretter, F. Receptive-field properties and neuronal connectivity in striate and parastriate cortex of contour-deprived cats. *Journal of Neurophysiology,* 1976, *39,* 613–630.

Singer, W., Freeman, B., and Rauschecker, J. Restriction of visual experience to a single orientation affects the organization of orientation columns in cat visual cortex. *Experimental Brain Research,* 1981, *41,* 199–215.

Singer, W., Tretter, F., and Cynader, M. Organization of cat striate cortex: A correlation of receptive-field properties with afferent and efferent connections. *Journal of Neurophysiology,* 1975, *38,* 1080–1098.

Spinelli, D. N., Hirsch, H. V. B., Phelps, R. W., and Metzler, J. Visual experience as a determinant of the response characteristics of cortical receptive fields in cats. *Experimental Brain Research* 1972, *15,* 289–304.

Stanford, L. R., and Sherman, S. M. Structure/function relationships of retinal ganglion cells in the cat. *Brain Research* 1984, *297,* 381–386.

Stanford, L. R., Friedlander, M. J., and Sherman, S. M. Morphological and physiological properties of geniculate W-cells of the cat: A comparison with X- and Y-cells. *Journal of Neurophysiology,* 1983, *50,* 582–608.

Stark, L., Michael, J. A., and Zuber, B. L. Saccadic suppression: A product of the saccadic anticipatory signal. In C. R. Evans and T. B. Mulholland (Eds.), *Attention in neurophysiology.* London: Butterworths, 1969, pp. 281–303.

Stone, J. *Parallel processing in the visual system.* New York: Plenum Press, 1983.

Stone, J., and Fukuda, Y. Properties of cat retinal ganglion cells: A comparison of W-cells with X- and Y-cells. *Journal of Neurophysiology,* 1974, *37,* 722–748.

Stone, J., and Hoffmann, K.-P. Very slow-conducting ganglion cells in the cat's retina: a major, new functional type? *Brain Research,* 1972, *43,* 610–616.

Stone, J., Dreher, B., and Leventhal, A. Hierarchical and parallel mechanisms in the organization of visual cortex. *Brain Research Reviews,* 1979, *1,* 345–394.

Stone, J., Rapaport, D. H., Williams, R. W., and Chalupa, L. Uniformity of cell distribution in the ganglion cell layer of prenatal cat retina: Implications for mechanisms of retinal development. *Developmental Brain Research,* 1982, *2,* 231–242.

Stryker, M. P., Sherk, H., Leventhal, A. G., and Hirsch, H. V. B. Physiological consequences for the cat's visual cortex of effectively restricting early visual experience with oriented contours. *Journal of Neurophysiology,* 1978, *41,* 896–909.

Swadlow, H. A. Efferent systems of primary visual cortex: A review of structure and function. *Brain Research Reviews,* 1983, *6,* 1–24.

Tees, R. C. Effect of early restriction on later form discrimination in the rat. *Canadian Journal of Psychology,* 1968, *22,* 294–298.

Tees, R. C. The effect of visual deprivation on pattern recognition in the rat. *Developmental Psychobiology,* 1979, *12,* 485–497.

Tees, R. C., Midgley, G., and Bruinsma, Y. Effect of controlled rearing on the development of stimulus-seeking behavior in rats. *Journal of Comparative and Physiological Psychology*, 1980, *94*, 1003–1018.

Thibos, L. N., and Levick, W. R. Astigmatic visual deprivation in cat: Behavioral, optical and retino-physiological consequences. *Vision Research*, 1982, *22*, 43–53.

Thorn, F., Gollender, M., and Erickson, P. The development of the kitten's visual optics. *Vision Research*, 1976, *16*, 1145–1149.

Tieman, S. B., and Hirsch, H. V. B. Exposure to lines of only one orientation modifies dendritic morphology of cells in the visual cortex of the cat. *Journal of Comparative Neurology*, 1982, *211*, 353–362.

Tieman, S. B., and Hirsch, H. V. B. Role of dendritic fields in orientation selectivity. In D. Rose and V. G. Dobson (Eds.), *Models of the visual cortex*. New York: Wiley, 1985, pp. 432–442.

Timney, B. N., Emerson, V. F., and Dodwell, P. C. Development of visual stimulus-seeking in kittens. *Quarterly Journal of Experimental Psychology*, 1979, *31*, 63–81.

Toyama, K., Matsunami, K., Ohno, T., and Tokashiki, S. An intracellular study of neuronal organization in the visual cortex. *Experimental Brain Research*, 1974, *21*, 45–66.

Trotter, Y., Gary-Bobo, E., and Buisseret, P. Recovery of orientation selectivity in kitten primary visual cortex is slowed down by bilateral section of ophthalmic trigeminal afferents. *Developmental Brain Research*, 1981, *1*, 450–454.

Tsumoto, T. Inhibitory and excitatory binocular convergence to visual cortical neurons of the cat. *Brain Research*, 1978, *159*, 85–97.

Tsumoto, T., and Suda, K. Laminar differences in development of afferent innervation to striate cortex neurones in kittens. *Experimental Brain Research*, 1982, *45*, 433–446.

Tucker, G. S. Light microscopic analysis of the kitten retina: postnatal development in the area centralis. *Journal of Comparative Neurology*, 1978, *180*, 489–500.

Valverde, F. Apical dendritic spines of the visual cortex and light deprivation in the mouse. *Experimental Brain Research*, 1967, *3*, 337–352.

Van Hof-Van Duin, J. Development of visuomotor behavior in normal and dark-reared cats. *Brain Research*, 1976, *104*, 233–241.

Van Hof-Van Duin, J. Direction preference of optokinetic responses in monocularly tested normal kittens and light deprived cats. *Archives of Italian Biology*, 1978, *116*, 471–477.

Van Sluyters, R. C., and Blakemore, C. Experimental creation of unusual neuronal properties in visual cortex of kitten. *Nature*, 1973, *246*, 506–508.

Vidyasagar, T. R., and Urbas, J. V. Orientation sensitivity of cat LGN neurones with and without inputs from visual cortical areas 17 and 18. *Experimental Brain Research*, 1982, *46*, 157–169.

Villablanca, J. R., and Olmstead, C. E. Neurological development of kittens. *Developmental Psychobiology*, 1979, *12*, 101–127.

Vital-Durand, F., and Jeannerod, M. Maturation of the optokinetic response: genetic and environmental factors. *Brain Research*, 1974, *71*, 249–257.

Vogel, M. Postnatal development of the cat's retina. *Advanced anatomical and Embryological Cell Biology*, 1978, *54*, 1–66.

Walk, R. D. The influence of level of illumination and size of pattern on the depth perception of the kitten and the puppy. *Psychonomic Science*, 1968, *12*, 199–200.

Wark, R. C., and Peck, C. K. Behavioral consequences of early visual exposure to contours of a single orientation. *Developmental Brain Research*, 1982, *5*, 218–221.

Warkentin, J., and Smith, K. U. The development of visual acuity in the cat. *Journal of Genetic Psychology*, 1937, *50*, 371–399.

Watkins, D. W., Wilson, J. R., and Sherman, S. M. Receptive-field properties of neurons in binocular and monocular segments of striate cortex in cats raised with binocular lid suture. *Journal of Neurophysiology*, 1978, *41*, 322–337.

Weiskrantz, L. Sensory deprivation and the cat's optic nervous system. *Nature*, 1958, 181, 1047–1050.

West, M. J. Social play in the domestic cat. *American Zoologist*, 1974, *14*, 427–436.

West, M. J. Exploration and play with objects in domestic kittens. *Developmental Psychobiology*, 1977, *10*, 53–57.

West, M. J. Play in domestic kittens. In R. B. Cairns (Ed.), *The Analysis of Social Interactions*. Hillsdale, N. J.: Lawrence Erlbaum, 1979, pp. 179–193.

Wiesel, T. N., and Hubel, D. H. Single-cell responses in striate cortex of kittens deprived of vision in one eye. *Journal of Neurophysiology*, 1963, *26*, 1003–1017.

Wiesel, T. N., and Hubel, D. H. Comparison of the effects of unilateral and bilateral eye closure on cortical unit responses in kittens. *Journal of Neurophysiology,* 1965, *28,* 1029–1040.

Wilkinson, F., and Dodwell, P. C. Young kittens can learn complex visual pattern discriminations. *Nature,* 1980, *284,* 258–259.

Wilson, P. D., and Stone, J. Evidence of W-cell input to the cat's visual cortex via the C laminae of the lateral geniculate nucleus. *Brain Research,* 1975, *92,* 472–478.

Winfield, D. A. The postnatal development of synapses in the visual cortex of the cat and the effects of eyelid closure. *Brain Research,* 1981, *206,* 166–171.

Winfield, D. A., and Powell, T. P. S. An electron-microscopical study of the postnatal development of the lateral geniculate nucleus in the normal kitten and after eyelid suture. *Proceedings of the Royal Society of London,* Ser. B., 1980, *210,* 197–210.

Winfield, D. A., Hiorns, R. W., and Powell, T. P. S. A quantitative electron-microscopical study of the postnatal development of the lateral geniculate nucleus in normal kittens and in kittens with eyelid suture. *Proceedings of the Royal Society of London,* 1980, *210,* 211–234.

Young, J. Z. The visual system of octopus. *Nature (London),* 1960, *186,* 836–844.

Zetterström, B. The effect of light on the appearance and development of the electroretinogram in newborn kittens. *Acta Physiologica Scandinavica,* 1956, *35,* 272–279.

8

Development of Thermoregulation

MICHAEL LEON

I. INTRODUCTION

Animals live in a world in which the ambient temperature may fluctuate over the course of minutes or seasons. The young of various species and of different ages within species use diverse strategies to maintain their thermal homeostasis. Neonatal thermoregulation may involve physiological and/or behavioral responses, not only by the offspring, but by the parent as well. The development of thermoregulation therefore may be understood best by an analysis of the interactions between mother and young.

The most common approach in a discussion of development is to determine when and how neonates achieve adult competence in a particular aspect of their behavior or physiology. Moreover, the discussion is often framed by analyzing the development of systems of individuals when they are isolated from their parent and siblings, thereby ignoring the importance of the actual social–environmental microhabitat of the young. An alternative perspective allows one to examine the young organism, not as an incompletely formed replicate of the adult, but as an organism whose characteristics have been selected to deal with the specific world in which it develops (Blass and Cramer 1982; Galef, 1981; Oppenheim, 1981). In this review, I will therefore discuss (1) strategies for thermoregulation, (2) temporal and mechanistic analyses of thermoregulatory development, and (3) the long-term effects of early thermal experiences. Although I will discuss the thermal development of species of several different orders, I will focus the discussion on the thermal development of young mammals.

MICHAEL LEON   Department of Psychobiology, University of California, Irvine, California 92717.

297

MICHAEL LEON

The thermal environment in which offspring develop should have important consequences for them starting at the time of fertilization. One may expect, therefore, that the temperature at which embryos develop would be regulated by the physiology and/or behavior of the parent. For example, parents can keep eggs within a specific thermal range by means of heat transfer with direct body contact. This behavior even has been described in an ectotherm, the female Indian python (*Python molarus;* Hutchinson *et al.,* 1966; Van Mierop and Barnard, 1976). The efficacy of this heat transfer behavior is facilitated by a change in maternal physiology. The mothers are normally ectotherms, responding to a decrease in ambient temperature with a decrease in body temperature. When in contact with her eggs, though, the mother python facilitates egg warming by increasing the contraction of her body musculature and thereby elevating her heat production (Hutchinson *et al.,* 1966; Van Mierop and Barnard, 1978). Maternal body temperature thus can be maintained at least 5°C above ambient temperatures from 33° down to 25°C, although both oxygen consumption and maternal temperature decrease at 21°C (Hutchinson *et al.,* 1966; Vinegar *et al.,* 1970).

Warming of the eggs by direct body contact—incubation—is most common among birds. The early importance of parental contact typically is critical for the survival of the young, but as the young grow and develop within the egg, the importance of incubation diminishes (Drent, 1972; Gessaman and Findell, 1979). Many avian species have specialized ventral brood patches that facilitate heat transfer between the parent and the egg (Jones, 1971). Heat transfer to the eggs is maximized both by a loss of the ventral feathers and by an increase in blood flow to the area that comes in contact with the eggs (Jones, 1971). Such changes have been shown to be under endocrine control, with estrogen and prolactin inducing defeathering of the brood patch in a variety of passerines (see Hutchinson, 1975a; 1975b; Steel and Hinde, 1963).

The maintenance of egg temperature is also facilitated by the presence of an insulating and/or shading nest. The nest is particularly important in maintaining the thermal stability of the eggs in the parent's absence (Mayer *et al.,* 1982). As the heat production of the embryo rises over the course of incubation, the importance of maternal contact for warming them is lessened.

Avian parents defend an egg temperature range that varies among species (Drent, 1975; Walsberg and Voss-Roberts, 1983). At the cool end of the range the embryos can slow their growth (Jones, 1971), but even high levels of attentiveness cannot always insure that the young will survive very cold temperatures. White-crowned sparrows *(Zonotrichia leucophrys),* however, are able to keep their eggs relatively warm even in environments that are consistently below the physiological zero of the eggs (25–27°C; White and Kinney, 1974) and are often at lethal, subfreezing temperatures (Webb and King, 1983). The parents achieve this by increasing their contact time with the eggs as ambient temperature decreases (Drent, 1975; Webb and King, 1983; White and Kinney, 1974).

High egg temperatures are also likely to be lethal (Jones, 1971). Parents that incubate in a warm environment, though, can keep the eggs cool by shading them (Maclean, 1967) and/or by serving as a heat sink for the warm eggs. Mourning doves *(Zanaida macroura)* and ring doves *(Streptopelia risoria)* use the latter mechanism, maintaining their temperature below that of nonincubating birds and thereby facilitating heat transfer from the eggs to the parent. Eggs thereby are kept at temperatures much below ambient temperature (Walsberg and Voss-Roberts, 1983; Figure 1).

Not all birds incubate their eggs by direct transfer of body heat to their young. The mallee fowl is perhaps the most fascinating example of egg temperature maintenance without direct parental contact. The males of this species incubate the eggs by constructing and tending a large mound in which decaying vegetable matter has been placed. The heat produced by the decay of the compost heap keeps the eggs warm. The male monitors the egg chamber temperature and, if necessary, reduces the temperature by scratching away the dirt covering eggs, thus exposing them to cooler air. Later in the season, when the fermentation of the vegetable matter is complete, the male will maintain egg temperature by packing the chamber with sand that has been warmed by the sun (Frith, 1962). The eggs are thereby kept warm under changing environmental circumstances without direct parental contact. Indeed, the parent–offspring relationship is such that if the male encounters the emerging hatchling while he is tending the mound, he tosses it aside as he would a stone. As for the mother, her parental duties terminate after egg laying (Frith, 1962).

Mammalian embryos, of course, grow within the stable thermal confines of the womb, their thermal stability ensured by that of the mother. Fetal metabolism can actually be estimated by monitoring the arterial–venous difference in circulating oxygen levels, and the data suggests that the metabolic rate of the fetus is comparable to that of the mother. The metabolism of the offspring, however, increases

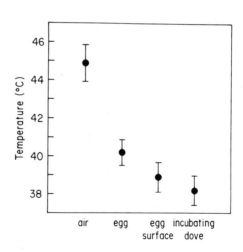

Fig. 1. The relation between the body temperature of incubating mourning doves, their eggs, and the ambient temperature. (After Walsberg and Voss-Roberts, 1983.)

dramatically at birth when the small mammal enters the external world (Meschia *et al.*, 1967; Crenshaw *et al.*, 1968). The embryo therefore seems to have largely passive responses to environmental temperature changes.

B. Sexual Differentiation

While the temperature at which avian and mammalian embryos develop is kept relatively stable by the parents, other species leave their eggs after fertilization to be subjected to varying ambient temperatures. Changes in ambience may induce changes in the *route* as well as the *rate* of embryonic development. The route of development has been shown to be altered dramatically by the temperature at which the young develop under natural conditions. Specifically, differentiation of either male or female characteristics occurs under different ambient temperature, regardless of the genetic sex of the individuals. This phenomenon has been most closely studied in turtles (Bull and Vogt, 1979; Morreale *et al.*, 1982; Pieau, 1982; Yntema, 1976), although similar findings have been reported for lizards (see Bull, 1980) and fish (Conover and Kynard, 1981).

Typically, low temperatures produce entire clutches of male turtles but result in the production of female fish and lizards. Intermediate temperatures allow the appearance of both sexes. This pattern of results has been observed both in the laboratory and under field conditions (see Bull, 1980). The change in sex ratio does not appear to be due to differential mortality (Bull and Vogt, 1979; Conover and Kynard, 1981; Yntema, 1979), nor does embryonic sex reversal appear to occur in these species (Bull and Vogt, 1979; Yntema, 1976). Rather, the change appears in undifferentiated embryos without the production of bisexual forms (Yntema, 1976; 1979; but see Pieau, 1982).

Neither the mechanism nor the function of these permanent changes has yet been determined. Similarly, the importance of these findings for the sexual behavior and physiology of these animals has yet to be investigated. Perhaps the most interesting suggestion in relation to a possible function has been proposed for an atherinid fish, the Atlantic silverside *(Menidia menidia)*. The larvae of this species respond to the cool waters still present in early spring by producing female offspring, while the males are produced in the warmer waters later in the season (Conover and Kynard, 1981). The earlier production of the females allows them to grow to be larger than the males by the time of reproduction, a situation that may be a prerequisite for successful reproduction in that species.

Mammalian sex ratios may also be influenced by ambient temperatures. Hedricks and McClintock (1982) have recently reported that the sex ratio of Norway rats changed with the time of day at which insemination occurred during the postpartum estrus. Given that the body temperature of these females is subject to a daily fluctuation (Kittrell, 1982; Spencer *et al.*, 1976), fertilization may have occurred under different temperatures, thereby inducing the observed sex ratios.

A. Strategies for Thermoregulation

Animals are born with highly different thermoregulatory abilities. The Mallee fowl chicks discussed previously, are buried alive under several feet of hard ground when they hatch. These small feathered chicks burrow to the surface and go off on their own when they emerge (Frith, 1962). Other precocial birds, such as domestic chicks *(Gallus domesticus)* and red jungle fowl *(Gallus gallus),* are much more dependent on parental warmth. While the young eat on their own, they have a limited ability to produce and/or retain heat (Freeman, 1963). These chicks therefore are forced to return to their mother for warmth during periodic brooding bouts (Sherry, 1981).

Mammals are also born along a broad continuum of thermoregulatory competence. At one extreme are animals such as the northern fur seal *(Callorhinus ursinus)* that are born in the Arctic under harsh environmental conditions. The mean ambient temperature during the period of maternal care is 7.5°C and the young seals remain on the beach during this period, continually exposed to strong winds and rain. These neonates are not warmed by maternal contact, and despite their relatively small size (5–6 kg) must deal on their own with the thermal challenge imposed by the harsh weather. These pups are able to thermoregulate extremely well under these circumstances with (1) a very high metabolism; (2) the ability to limit the flow of warm blood from their core to their extremities; (3) a dense pelage, the insulating value of which is kept high by periodically shedding the rainwater with vigorous shaking; (4) an insulating layer of subcutaneous fat; and (5) the ability to produce heat by means of shivering, which seems to be facilitated by specialized adaptations of the body musculature (Blix *et al.,* 1979). The high metabolic rate is supported by the very high fat content in the milk delivered to the pups by the dam (Ashworth *et al.,* 1966).

In contrast, hyperthermia is the dominant thermoregulatory problem of young northern elephant seals *(Mirounga angustirostris),* which are reared on warm beaches off the coast of California and Mexico. These animals throw sand on their backs with their flippers, making a layer of sand that can be several inches thick, thereby shielding the neonate from solar radiation. Moreover, as they throw the sand on themselves, they uncover cool, moist sand that draws off heat from the seal by conduction (White and Odell, 1971).

At the other extreme of thermoregulatory competence are young oppossums *(Didelphis marsupialis virginiana).* These marsupials are born in an extremely undeveloped state and have no measured metabolic response to lowered temperature until they reach 70 days of age (Morrison and Petajan, 1961). Of course, under normal conditions, young oppossums need not deal with falling ambient temperatures within the pouch, and they are thereby able to be reared across the wide range of temperatures in which *Didelphis* lives (see McNab, 1970).

When all other factors that influence heat loss are equal, the size of the animal assumes a critical importance. The amount of heat that is lost by a body is proportional to the amount of surface area. Moreover, the ratio of surface area to the body mass that is losing heat increases as the body mass decreases (McNabb, 1971). Small animals, then, are prone to lose heat easily and to regulate their own body temperature independently. They would be expected to produce large amounts of heat to offset their relatively large heat loss. Indeed, Pearson (1948) calculated that mammals weighing below 2.5 g could not maintain their own thermal homeostasis because they could not find and assimilate food fast enough to produce enough heat to offset their heat loss. While neonates of all endothermic species would be expected to be less efficient than their parents in maintaining thermal homeostasis, one would expect neonates of small endotherms to be at a relatively greater disadvantage in this regard.

Body size, however, is not the only variable affecting heat loss in neonates. Small neonates may be able to maintain their body temperature independently if they have sufficient insulation from fat deposits, fur, or skin that allows them to retain the heat that they produce (McClure and Porter, 1983). Alternatively, as we shall see, young animals may surrender the greater responsibility for thermoregulation to their parents. While one should find thermoregulatory independence to be more prevalent among large neonates, there are examples of thermoregulatory precocity even among animals of small body size.

The importance of body size and body insulation in neonatal thermoregulation can be clearly seen in a comparison of species. Newborn calves are relatively large and furred and are able to thermoregulate effectively without maternal warmth. These animals have a minimum critical temperature (at which they increase heat production in response to a depression in ambient temperature) of 13°C (Alexander, 1961). The newborn piglet is even smaller, has very little body insulation (Manners and McCrea, 1963), and must respond with elevated heat production when ambient temperature is as high as 34–35°C (Mount, 1959). Consequently, piglets are dependent on a continual delivery of calories in the milk to support this heat production, and even brief periods of food restriction may well be fatal for the young (Goodwin, 1957). The domestication of the pig for agricultural purposes may have involved selection for mothers that could provide great quantities of milk under conditions where superabundant food supplies were available. The piglets of such mothers could then afford to devote energy to both growth and thermoregulation, given the great milk supply at their disposal. It would be of interest, therefore, to determine whether the wild ancestors of the domesticated pig, which presumably do not have abundant balanced diets, would have offspring with greater thermoregulatory abilities than those of the domesticated pigs.

Rodents and lagamorphs give birth to still smaller young, amplifying the difficulty of maintaining a balance between heat production and heat loss. Indeed, a critical problem faced by newborn rodent young is whether to invest their limited energy supply into growth or body temperature maintenance and thereby achieve thermal independence. There are clear differences in neonatal thermoregulatory maturity. For example, the guinea pig is born furred, mobile, and able to ingest solid food, while rabbits, though about the same size as guinea pigs, have much less fur at birth, remain in the nest, and are completely dependent on the milk that their mother provides during her daily visit (Zarrow *et al.,* 1965). The rabbits begin to increase their heat production at a higher ambient temperature than the guinea pigs, and though the oxygen consumption of the rabbit is more vigorous, the guinea pig is better able to defend its body temperature, probably because of an advantage in its ability to retain heat (Dawes and Mestyan, 1963). The thermal environment in which rabbit young normally develop, however, is within a very well-insulated nest that normally prevents them from being exposed to low ambient temperatures.

The golden hamster seems to be the only young mammal thus far tested that does not respond to falling ambient temperature with an increase in metabolic heat production (Hissa and Lagerspetz, 1964; Hissa, 1968), suggesting that these young have given the responsibility for regulation of their body temperature to their mothers. Again, litter contact and the maternal nest help to keep the young warm, even in their parent's absence.

A comparison of the development of temperature regulation in cotton rat pups *(Sigmodon hispidus)* and wood rat young *(Neotoma floridana)* revealed that the former, smaller species (113 versus 270 g for adult males) rapidly develops the ability to control their temperature, while the development of the larger is prolonged (McClure and Randolph, 1980). Cotton rats have a linear increase in metabolic rate until they are about 20 g, between days 10 and 12, while the wood rats maintain a relatively low metabolic rate until they are about 26 g. The cotton rats become efficient homeotherms at 10–12 days, while the wood rats reach that stage at 10–21 days; both species are weaned at the time that homeothermy is reached.

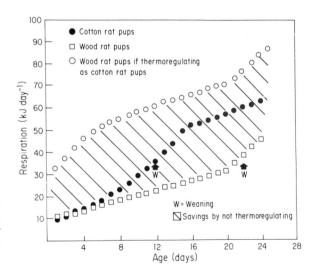

Fig. 2. Estimated energy savings (kJ/rat^{-1} day^{-1}) that might result from deferred heat production by wood rat pups compared to that of cotton rat pups. (After McClure and Randolph, 1980.)

The ability to thermoregulate is dependent both on the young spending a large proportion of their energetic budget on heat production (McClure and Randolph, 1980) and on the density of their insulative fur coat (McClure and Porter, 1983).

McClure and Randolph (1980) suggest that mammals that become relatively large as adults defer the onset of homeothermy to the time when their body mass allows them to produce and retain heat with a relatively low metabolic rate per gram of body tissue. The deferral of body temperature regulation may allow such species to increase their growth efficiency. Neonatal temperature regulation is left to the mother. Species that have a relatively small adult body size may sacrifice growth efficiency for the early independence afforded by homeothermy and use their available energetic resources toward that end. Those young rodents with limited food resources initially may be forced to neglect heat production entirely, because of small size. Alternatively, they may be born thermally independent and remain relatively small.

B. MATERNAL DEFENSE OF NEONATAL TEMPERATURE

Norway rat pups are born small, without fur, and fully dependent on their mother's milk. They are restricted to their nest for the first 2 weeks of life. Even in the first days after birth, though, young rat pups can elevate their oxygen consumption in response to a depression in ambient temperature (Taylor, 1960; Takano et al., 1979), with the lower critical temperature of individual rat pups up to 18 days old about 35°C (Takano et al., 1979). Despite their elevated heat production, though, their body temperature eventually falls as ambient temperature falls (Brody, 1943; Buchanan and Hill, 1947; Gelineo and Gelineo, 1952).

Bruck (1961) calculated that rat pups would have to eat about 1.5–2 times their body weight in milk to maintain their thermal independence. Moreover, their mothers would have to provide more than 80ml of milk/day to a newborn litter of 10 pups. The inability of the pups to thermoregulate independently therefore is necessitated by the energetic limits imposed on them by their small size and poor heat retention. The elevated rate of heat loss in these pups appears to be due to their lack of insulation that is normally provided in the adults by the fur. The pups are born without hair, with a gradual development of fur over the first 4 weeks of life (Kiil, 1949), and this increase in insulation is correlated with a gradual ability of the rats to defend a relatively constant body temperature in cool environmental circumstances (Takano et al., 1979). The pups therefore require great amounts of heat to be produced to maintain body temperature. Moreover, the energy required also exceeds the ability of the mothers to provide it in the form of milk. This species is therefore one that sacrifices early independence from the mother in favor of a prolonged growth period in which their warmth is provided by direct maternal contact.

Rat pups, of course, are born into a litter that remains in an insulated nest. Within that nest, the temperature typically remains between 33 and 35° C, even when the environmental temperature falls to 16°C (Jans and Leon, 1983). The insulating properties of the nest are increased by the dam as the ambient temperature falls, thereby maintaining the temperature within the nest (Kinder, 1927). The young are thereby buffered, to a large extent, from ambient temperature fluctuations during their development.

Along with her nest, the direct heat transfer from the mother during contact bouts keep the young within a 33–35°C range over widely differing temperatures (Jans and Leon, 1983; Leon *et al.*, 1978). Initially, the mothers seek out a relatively warm area (24–26°C) in which to care for the young (Jans and Leon, 1983; Figure 3). When the dams have access to nest material, they construct an insulating nest in a cooler area (about 21°C), but the nest retains sufficient heat to maintain pup temperature between 33 and 35°C. By day 10, mothers no longer maintain high-quality nests and choose instead to care for their young in a warm area, whether or not nest material is available. Mother rats therefore either seek out or construct a relatively warm area in which to care for their offspring.

The mother initially remains in contact with the young for 75–90% of the time (Grota and Ader, 1969; Leon *et al.*, 1978). As the pups begin to grow, develop fur, and increase their ability to thermoregulate, the mother spends progressively less time with the pups (Grota and Ader, 1969; Leon *et al.*, 1978) and gradually allows the insulating nest to fall into disrepair (Rosenblatt, 1969; Rosenblatt and Lehrman, 1963; Figure 4). The pup-warming contact behavior of the mother is quite sensitive to the temperature of the pups, to the surface area–mass ratio of the litter, and to the ambient temperature. Mothers increase their contact time with cool pups, small litters, and litters reared in a cool environment (Leon *et al.*, 1978; Grota, 1973), thereby compensating for thermal challenges to the offspring.

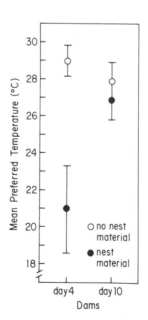

Fig. 3. The mean preferred temperature of mother Norway rats on day 4 and day 10 postpartum with or without nest material available to them. (After Jans and Leon, 1983.)

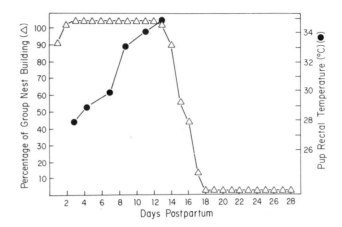

Fig. 4. Percentage of Norway rat females building nests during the first four postpartum weeks is shown, along with the rectal temperature of rat pups taken after 1 hr in a 22°C ambience. (After Rosenblatt, 1969, and Fowler and Kellogg, 1975.)

The dams do not monitor the temperature of the litter and then modify their contact time to bring the pups into a particular (i.e., 33–35°C) temperature range. The evidence for this conclusion is that mothers did not decrease their total daily contact time with pups warmed to 39°C, although they accumulated that time in many short bouts (Leon *et al.*, 1978). Since the pups are never normally kept that warm, the mothers should have spent little time with them if the dams were monitoring pup temperature. In fact, contact time seems to be limited by a rise in maternal temperature; pup temperature is important in influencing contact bout duration only in its ability to influence maternal temperature (Leon *et al.*, 1978). Pup temperature also seems to influence the probability that mothers will maintain contact for more than a few minutes once a bout is initiated. Specifically, mothers are more likely to remain with cool than warm pups for a prolonged contact bout (Jans and Leon, 1983).

As the pups develop some measure of thermoregulatory independence, they begin to move out of the nest in the third week postpartum (Hahn *et al.*, 1956; Rosenblatt and Lehrman, 1963). Under laboratory conditions, mother and young still have contact bouts through the fourth week postpartum (Rosenblatt, 1969), during which time milk is delivered to the young (Babicky *et al.*, 1970). In the wild, however, the young are typically not seen above ground until the fifth or sixth week of life (Calhoun, 1963). The difference in the approximate time of independence by the weanlings may also be due to thermal considerations. Cramer (personal communication) has found that when young rats are cared for in a 19°C ambience, they wean about 10 days later than when they are reared at 30°C. The animals in the wild may be exposed to cooler temperatures than those in the laboratory and consequently may have an extended filial relationship. A cool ambience may allow a mother to remain with the young, while a warm ambience may drive her away from the litter before milk can be delivered to them, thereby inducing weaning.

While defense against a fall in body temperature is a primary concern of many developing mammals, it should also be noted that hypothermia can have adaptive consequences under certain circumstances. While neonatal rats cannot defend their normally high body temperature, they do have an advantage over adults in very cold environmental circumstances. Young rats have the ability to survive after their core temperature falls to 1°C, a temperature that is fatal to adults (Adolph, 1948). This ability may allow them to sustain prolonged maternal absences.

Miller *et al.* (1964) have shown that hypothermic neonatal guinea pigs, domestic pigs, dogs, rats, and rabbits were protected from the effects of severe hypoxia. These authors also reported that cooling hypoxic human neonates improved their survival probability. The cooled organism should have a depressed need for oxygen and may thereby be able to withstand a temporary oxygen deficiency, such as may occur during the birth process, without permanent tissue damage.

III. MECHANISMS OF THERMOREGULATION IN YOUNG MAMMALS

A. PHYSIOLOGICAL MECHANISMS

The mechanisms for heat production have been divided historically into the heat resulting from the contraction of muscles (shivering thermogenesis) and the heat resulting from the metabolic activity of body organs (nonshivering thermogenesis). Shivering may be the principal means of cold-induced thermogenesis in some large mammals such as lambs (Alexander, 1961) and calves (Hales *et al.*, 1968; McEwan-Jenkinson *et al.*, 1968).

Shivering, however, is a relatively inefficient means of heat production, and while small mammals can shiver (e.g., Taylor, 1969), it seems to play a secondary role in endogenous heat production (Bruck and Wunnenberg, 1966). The primary mechanism in these species is nonshivering thermogenesis and the specialized organ producing heat is the brown adipose tissue (BAT; Foster and Frydman, 1979). This tissue is found in discrete fat pads at several locations in the body of neonates, with the bulk of the tissue typically surrounding the major blood vessels in and around the thoracic cavity (Heim and Hull, 1966; Hull and Segall, 1965).

Activation of the BAT is mediated by noradrenergic transmission through the sympathetic innervation of that tissue (Himms-Hagen, 1967). Blockage of noradrenergic activity decreases heat production, while increasing noradrenergic activity, either by experimental administration of the transmitter or by exposure to cold, elevates heat production (Hsieh *et al.*, 1957; Moore and Underwood, 1963). The production of heat is a byproduct of the oxidation of fatty acids (Prusiner *et al.*, 1968; Kornacker and Ball, 1968). Blood flow to the BAT dramatically increases in low ambient temperatures, carrying the heat to the rest of the body (Heim and Hull, 1966).

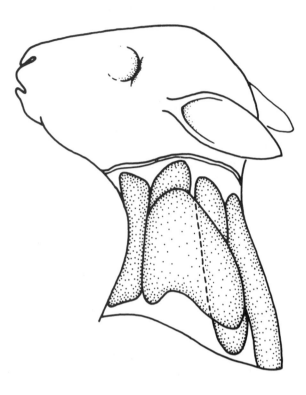

Fig. 5. Cervical and intrascapular brown adipose tissue pads of the newborn rabbit. (After Hull and Segall, 1965.)

B. BEHAVIORAL MECHANISMS

1. THERMAL CHOICE. Young mammals rely on both physiological and behavioral mechanisms for the defense of their body temperature. For example, young pigs (Mount, 1963), kittens (Freeman and Rosenblatt, 1978), Norway rats (Kleitman and Satinoff, 1982), mice (Olgivie and Stinson, 1966), hamsters (Leonard, 1974), gerbils (Eedy and Olgivie, 1970), and rabbits (Klietman and Satinoff, 1981) are able to move toward a warm area and thereby reverse the heat flow from their bodies to the environment.

Animals differ in the developmental course of this ability, and at least in some cases it is possible to find an inverse relationship between the developmental rate of their physiological and behavioral thermoregulatory mechanisms. For example, golden hamsters are able to move rapidly to a warm area soon after birth (Leonard, 1974). Given their inability to elevate their heat production by physiological means (Hissa, 1968), this behavioral competence may be vital for the maintenance of their thermal homeostasis in the first days after birth. Newborn pigs are also quite susceptible to fatal cooling, given the expense of their heat production coupled with their poor insulation (Curtis *et al.,* 1967; Manners and McCrea, 1963; Mount, 1959). These piglets move readily to a warm place (Mount, 1963).

Initially, Norway rat pups have difficulty moving from a cool area to a warm area (Fowler and Kellogg, 1975; Johanson, 1979), but they will go from a warm to a warmer area (Kleitman and Satinoff, 1982). The best strategy for a stray pup may be to remain still in a cool area outside the nest, emitting distress calls (Okon, 1971;

Oswalt and Meier, 1975), because mothers will readily retrieve stray young to the warmth of the nest during the first two postpartum weeks (Wiesner and Sheard, 1933; Rosenblatt and Lehrman, 1963). Once within the relative warmth of the nest, however, the thermal needs of the pups may be best served by moving to the warmer areas of the nest or the huddle formed by its littermates (Alberts, 1978). On the other hand, the brief, infrequent visits of the rabbit doe (Zarrow *et al.*, 1965), coupled with the inability of the young to maintain their body temperature in a cool ambience, may make movement toward the warmth of the nest an imperative for rabbit young (Kleitman and Satinoff, 1981).

2. HUDDLING. While the ability of young mammals to thermoregulate is typically tested in isolates, some neonates normally defend their body temperature within the social circumstances provided by their litter, often within a nest. If one considers the litter as a thermoregulatory unit, one would expect its thermal characteristics to be much different from that of single pups. The litter would have a greater tissue mass that generates heat relative to its surface area that is available for heat loss, allowing increased efficiency of heat retention and a decreased need for heat production (Kleiber, 1961). The individuals in the litter would have both a decreased convective heat loss and an increased insulation on that portion of their bodies in contact with a littermate, both aiding in the retention of heat. Consequently, one would expect that huddling would facilitate the defense of body temperature by neonates and that living within an insulating nest would further facilitate heat retention.

White-footed mice *(Peromyscus leucopus)* have difficulty defending their body temperature as individuals but can withstand a sharp fall in ambient temperature with minimal increase in metabolism if they are tested with littermates in the nest (Hill, 1976). Barnett (1956) also found that litters of young mice *(Mus musculus)* would maintain a nest temperature much above even a very cool ambience. For example, the nest temperature was maintained at 12.7–21.6°C when the ambience was kept between −1.2 and 3.0°C. Huddling in a nest may therefore give the individual pup the thermoregulatory resources of an adult, given both the relatively low surface area/mass of the huddle and the elevated insulation provided by the nest.

Huddling with littermates even without a nest is very effective in facilitating heat retention. Litters of piglets and litters of mice retain heat far more effectively than singletons (Bryant and Hails, 1975; Mount, 1960; Stanier, 1975). Similarly, the heat loss and consequent heat production of Norway rat pups is reduced for litters relative to that of single pups (Alberts, 1978; Figure 6). Moreover, the huddle does not remain a static unit, as pups move to the warm center of the huddle in a cool ambience and toward the periphery of the huddle when ambient temperature increases (Alberts, 1978). These data indicate that individual pups can thermoregulate within the context of the warm huddle by moving toward warmer or cooler areas among its littermates.

3. BEHAVIORAL FEVER. A fever is an elevated body temperature set point that may be accompanied by an elevated body temperature (Snell and Atkins, 1968). The maintenance of this elevated body temperature has been shown to increase

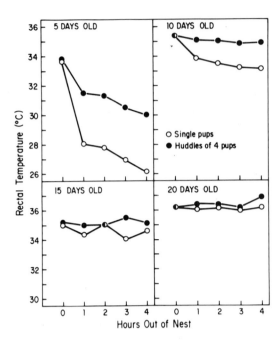

Fig. 6. The body temperature of Norway rat pups of different ages after removal from their nest when they were alone or with three littermates. (After Alberts, 1978.)

the ability of animals to deal with bacterial infections (Kluger and Vaughn, 1978). For example, there is a significant correlation between the extent of the fever developed by adult rabbits and their survival probability following bacterial infection (Kluger and Vaughn, 1978). Survival may be promoted by the ability of fever to stimulate the immune response to bacterial infection (Banet *et al.*, 1982).

While adult mammals are capable of elevating heat output to produce a fever, ectotherms such as reptiles and fish are not able to elevate their body temperatures using physiological mechanisms. Rather, when such animals are infected, they seek out and remain in a warm area, elevating their temperatures behaviorally (Covert and Reynolds, 1977; Vaughn *et al.*, 1974).

Human infants and the young of several other species do not readily react to bacterial infection with a fever (Craig, 1963). Indeed, when the young of a variety of laboratory animals are injected with a dose of pyrogens that induce a fever in adults, the young do not respond with a rise in their body temperature. It is only when the young of some species are given elevated dose of the pyrogens that they produce a fever (Blatteis, 1977; Kasting *et al.*, 1978; Pittman *et al.*, 1974; Szekely and Szelenyi, 1977). Even at the lower adult dose, however, young rabbits can become febrile by moving to a warm area (Satinoff *et al.*, 1976). An interesting aspect of this finding is that the mammalian young can elevate their heat production autonomically in response to a cold ambience, but not in response to a dose of pyrogen that elicits fever in the adult. Instead of using physiological means to elevate body temperature in the presence of a pyrogen, the neonates produce a fever behaviorally.

Rodent young have access to the warmth of the maternal nest that allows them to maintain a high body temperature. Remaining in the warmth of their maternal nest may have an immediate survival value, for the longer mice are kept away from their dam, the more they cool and the more susceptible they are to dying of Coxsackie virus (Teisner and Haahr, 1974). When the pups are kept warm in the absence of their mother, the mortality rate of the pups is greatly reduced. It therefore seems that behavioral thermoregulation may have a critical immediate consequence for pup survival, in addition to having more subtle changes in pup development. Moreover, high body temperatures may have an important adaptive significance for the young under specific circumstances, much as do low body temperatures.

IV. Plasticity in the Development of Thermoregulatory Capacity

A. Effects of Being Reared in Different Ambient Conditions

While the mother, nest, and huddle can combine to buffer the young from exposure to the cold, they cannot buffer the young completely from environmental temperature variation (Barnett, 1956; Leon *et al.*, 1978), particularly given the wide range of ambient temperatures under which animals breed. Consequently, there are differences in the long-term range of temperature to which individuals are naturally exposed during their lives (Hayward, 1965). The young must then accommodate to different thermal conditions when they leave the care of the mother.

The possibility of long-term accommodation to environmental temperature exists among mammals, for those individuals reared under cool ambient conditions are better adapted to such conditions than are those reared in thermoneutral conditions. Gelineo and Gelineo (1951) found that rats reared in a warm ambience had their thermoregulatory development retarded. Similarly, Krecek *et al.* (1957) found that young rats reared at 34°C would experience a more rapid fall in core temperature when placed in 18°C than those reared at 20°C. Since Norway rat pups normally experience a nest temperature of 33–35°C, even at an ambience as low as 16°C (Jans and Leon, 1983), it is likely that the pups reared at the elevated temperature experienced minimal fluctuations in their nest temperature, while those in even cooler ambience experienced the normal, periodic fluctuations in the temperature of their microhabitat. Similar improvements in the ability to deal with low ambient temperatures following neonatal cold exposure have been reported for kittens (Jensen *et al.*, 1980), rabbits (Cooper *et al.*, 1980), and piglets (Ingram, 1977).

As little as 1 hr each day of 0°C exposure of the rats reared in an ambient temperature that remains at about the temperature of the nest (34°C) is sufficient to normalize their thermal development (Hahn, 1956). This type of early exposure to cold has long-lasting effects, for as little as four 1-hr periods of daily cold exposure for the first 2 weeks of life produced rats with an elevated resistance to cold even when they were 19 weeks old (Doi and Kuroshima, 1979).

Human infants also have prolonged responses to different ambient temperatures. Infants had their abdominal skin temperature kept at either 35°C, which is the selected norm for babies in incubators, or at 36.5°C, which is within the zone of thermoneutrality for newborns (Glass *et al.*, 1968). After 2 weeks in these environments, the babies were placed in a 28°C environment for 1 hr. The decrease in body temperature was less among infants reared in the cooler environment, but the rate of body weight gain and body length was greater in the babies reared in a warm environment. The differences in growth could be eliminated if the young reared in the cool environment were given a dietary supplement calculated to support the caloric demands of elevated thermogenesis in the cool environment (Glass *et al.*, 1969).

There are several mechanisms that could account for the changes in thermoregulatory capacity of the developing animals. For example, warm-reared animals may lose heat more easily than those reared in cool conditions. In support of that possibility, piglets reared in a warm (35°C) ambience developed longer extremities than animals reared at 25°C. The resulting increase in surface area available for heat loss would be expected to facilitate heat loss (Fuller, 1965; Weaver and Ingram, 1969). Similarly, Norway rats developed longer tails when kept in a warm ambience (Heroux, 1961). Warm-reared pigs also grew less hair than those raised under cooler conditions, a modification that would be expected to decrease heat retention. Cutaneous water loss did not differ between piglets reared at different ambient temperatures, ruling out the role of that route of heat loss mechanism in producing the difference in thermal response.

A second possible mechanism is that the peripheral and/or central nervous system is altered by early thermal experience. Warm-reared pigs began to pant more quickly than cool-reared pigs when their periphery was warmed, a behavior that would increase their ability to lose heat (Ingram, 1977). These data support that notion that there is a change in the neural response to cold. Norway rats reared either in a warm or a cool ambience, though, had similar peripheral cold sensor responses (Dawson *et al.*, 1981). These data suggest that the fall in body temperature of warm-reared rats is not due to their inability to sense the cold with their peripheral neural responses. Ingram (1977) also heated the hypothalamus of warm- and cool-reared pigs, but found no support for the possibility that the gain of the central thermosensitive neurons was altered.

A third possible mechanism involves the development of different levels of nonshivering thermogenesis mediated by differences induced in the BAT. Warm acclimation in adult rats, for example, has been shown to reduce BAT levels. Moreover, the elevation in heat production in response to noradrenaline injection was not found in warm-acclimated adults (Foster and Frydman, 1979). Similarly, warm-reared rabbits did not show a significant rise in temperature following systemic noradrenaline administration (Cooper *et al.*, 1980). Bertin and Portet (1981) reared Norway rats in either 28°C or 16°C and found that on day 21 postpartum the cool-reared animals had significantly more BAT than the warm-reared animals.

In addition, the glycerol, basal lipolysis, cyclic AMP content, and glycerokinase activity of the BAT was significantly higher in the cool-reared group. These data are consistent with the idea that BAT development may participate in the cold-induced alteration in their thermal development.

Temperature acclimation as adults seems to have different thermal consequences than that developed early in life. For example, intrahypothalamic infusion of norepinephrine (normally producing hypothermia in adults; Metcalf and Myers, 1978) produced a hypothermia in rats that were warm-acclimated as adults, but produced hyperthermia in adults that were reared in the warmth (Ferguson *et al.*, 1981). Similarly, the BAT of cold-acclimated adults has a relatively low glycerol release and cyclic AMP production following noradrenaline infusion (Bertin and Portet, 1976; Dorigo *et al.*, 1974).

B. THERMAL MEDIATION OF EARLY EXPERIENCE

I will consider here the possibility that the thermal consequences of at least some early experiences may mediate the ensuing developmental changes. If, during the first few days of life, infant rodents are removed from their nest and then handled or given mild electric shock, they undergo precocious maturation of some physical and physiological parameters. As adults, these animals have altered behavioral reactivity, altered performance on learning tasks, and altered responsiveness to stressful circumstances (see Daly, 1973; Russell, 1971; Smotherman and Bell, 1980, for critical reviews). Four main theories have been advanced to specify the way in which elevated early stimulation induces long-term changes in these animals. The first theory is that stimulation directly changes the developing nervous system (Levine, 1962). The second suggestion is that the early stimulation is stressful and that this early stress inoculates the animals from additional stress during development (Levine, 1956). The third hypothesis is that pup stimulation changes the stimulus characteristics of the pups, which in turn, affects the maternal care that the young receive (Barnett and Burn, 1967; Lee and Williams, 1974; Smotherman *et al.*, 1977). A fourth hypothesis is that the early experience is mediated by secondary thermal responses to the stimuli (Schaefer *et al.*, 1962). Each of the hypotheses advanced to account for changes in developing physiology and behavior, however, may have a thermal basis.

A factor common to studies in which young rodents received extra stimulation is removal from the warmth of the maternal nest in the course of the experimental procedure. Schaefer *et al.*, (1962) therefore suggested that the effects of handling were mediated by incidental cooling of the young. They showed that cooling of the pups in the nest without handling had the same effects on the adrenal glands as handling during a cold stress. Importantly, handling the young under conditions that prevented the pups from cooling during the treatment prevented the physiological and behavioral changes characteristic of handled–cooled rats (Hutchings, 1963, 1965; Schaefer, 1963). Hutchings (1967) went on to determine that the rate of cooling was the critical stimulus in producing the changes in behavioral reactivity. It would seem, then, that the cooling of the pups may mediate the effects of differential postnatal effects.

The thermal stress hypothesis of Schaefer and of Hutchins was developed using a paradigm quite similar to that used to demonstrate an increased rate of thermal development of pups for daily cooling. Either removal of the dam, or placing the pups in a cold environment, produced long-term changes in development. The principal difference between these experiments is that adrenal ascorbic acid depletion was measured in the early stimulation tests, rather than oxygen consumption or body temperature. In one study, though, Barnett and Burn (1967) stressed young mice that were raised under relatively cool conditions by punching holes in their ears for identification purposes. Mice with ear holes were more cold resistant than mice with no ear holes. Since cold exposure appears to be additive with other stressful neonatal events, these data suggest that early cold exposure and early stressful stimulation may affect a common system.

Changes in maternal behaviors toward cooled young also have been demonstrated with the dams spending more time with cool than with warm pups (Leon, *et al.,* 1978). The rise in maternal body temperature that is normally experienced with warm pups is prevented when the dam is with cool pups. Since it is a rise in maternal temperature that induces the termination of maternal contact bouts, one would therefore predict that the handled, and therefore cooled, pups would receive elevated maternal contact time until they were rewarmed.

If the action of pup stimulation is mediated by its thermal consequences, there may also be a common mechanism that mediates changes in the rate of thermal maturation under the different early experience conditions. Ambient temperature, tactile stimulation, and stress may interact in an additive way by affecting a single mediating system.

One possibility for a common mediating mechanism involves the adrenal hormone corticosterone. Corticosterone is released in young rats following both handling and electric shock (Denenberg *et al.,* 1967; Smotherman, 1983), and it is also released in response to exposure to the cold (Gianutsos and Moore, 1977). Corticosterone may mediate the long-term thermal consequence of both cold exposure and handling by stimulating the development of BAT. Indeed, BAT is a glucocorticoid target organ with a high affinity for corticosterone (Feldman, 1978) and administration of glucocorticoids or ACTH to young rats elevates both BAT weight and fat content (Hahn *et al.,* 1969; Lanchance and Page, 1953). Young rats with elevated BAT reserves should be more able to maintain their body temperature and therefore increase their energetic apportionment to neural and somatic development. The resulting long-term changes in behavior and physiology would follow the early experiences.

C. Thermal Activation

In addition to the thermal state of the pups provoking pup thermoregulatory behaviors, thermal factors may also elicit other responses by the young. For example, the latency of rat pups to attach to their anesthetized dam's nipple is reduced if the young are isolated from the mother in a relatively cool ambience (30°C) before testing at 21°C (Henning *et al.*, 1979). This latency reduction is not seen if the young are kept at 34°C (the approximate temperature of the nest) while away from the dam. No effect of body temperature on the amount of milk ingested by the pups after maternal deprivation was found (Henning and Romani, 1982). When maternally deprived pups are kept at 22–25°C, they have a reduced latency to attach to their dam's nipple at both a warm (34°C) and a cooler (25°C) ambience (Cramer and Blass, 1982). The testing ambience also did not affect milk intake.

It may be the case, therefore, that latency for nipple attachment can be influenced by the metabolic state of the pup. When the pups are in energy deficit, they may have a reduced latency to attach and thereby receive the warmth and calories that should accompany milk delivery, regardless of the environmental temperature at the time of testing. The temperature of the pups may not be an accurate reflection of their energetic state, though, because pups may have identical elevated temperatures that are due either to an elevation in metabolism or to an elevation in heat retention. An energetic deficit acquired during a period of maternal deprivation in a cool ambience may be expressed in a reduction of the attachment latency regardless of the immediate ambient or pup body temperature.

The mechanisms that are involved in the cessation of milk intake when the pups are suckling from the maternal nipple seem not to have a thermal component, for the pups do not differ in intake under high or low temperatures. Ambient temperature, however, clearly affects food intake away from the nipple. Johanson and Hall (1980) presented milk to 3- and 6-day-old maternally deprived pups by means of an intraoral cannula and the pups took in much more of the milk when the ambient temperature was high (34°C) than when it was low (24°C). Activity level as well as mouthing and probing behaviors were also elevated in the high ambience. Neither the elevated milk intake, nor the increase in the other observed behaviors were due primarily to an elevated pup body temperature in the high ambience. Despite the fact that the pups had a depression in milk intake and activity when their body temperatures were depressed, there was an elevation in milk intake, activity, mouthing, and probing in the warm ambience, regardless of pup body temperature.

Adult rats ingest more in a cool than in warm ambience (Brobeck, 1960; Kraly and Blass, 1976). The resulting adult behavior, while not necessarily elicited in response to an internal energetic demand (Kraly and Blass, 1976), seems to compensate for the added energetic needs of body temperature maintenance in the cold. In contrast, the pup behavior may be an adaptation for allowing the young to have food-oriented behaviors in the warmth of the nest.

Indeed, high ambient temperature allows pups to exhibit several kinds of behaviors that appear to be inhibited in a cool ambience. Specifically, 3-day-old maternally deprived pups will approach and begin to ingest milk presented in a puddle, become behaviorally activated, and engage in adultlike behaviors such as face grooming, as well as anogenital and paw licking (Hall, 1979). None of these behaviors are performed either in cooler ambient conditions or in pups that are with their mother.

Warmth may have a combination of permissive and reinforcing properties that allow various forms of learning to occur (Rosenblatt, 1983). Newborn pups will refrain from withdrawing from a cold surface if they received a puff of warm air there (Guenaire *et al.*, 1979). Pups can also be trained to raise their heads or make a spatial discrimination when reinforced with a puff of warm air (Geunaire *et al.*, 1982, 1982b).

Pups can also learn an olfactory discrimination in a warm, but not a cool environment (Alberts, 1981). Alberts and May (1984) went on to show that while olfactory preferences could be acquired in the absence of the mother, those acquired in the presence of the mother were stronger. This phenomenon was not due to the reinforcing effects of nutritive nursing, but could be duplicated with odor experienced while in contact with a warm tube. Their data strongly support the idea that thermal–tactile interactions between mother rats and their young facilitates the induction of filial olfactory preferences. The reinforcing value of warmth may therefore keep the young in the nest where they receive essential maternal care. It then allows critical learning experiences to occur in their mother's presence. The coincidence of maternal odors with maternal warmth may preclude the development of olfactory preferences to irrelevant environmental cues.

V. SUMMARY

Maintenance of the temperature of mammalian young within a range adequate for normal growth and development is accomplished in several ways. Most young are capable of elevating their heat production at birth, in large part, by mobilizing their brown adipose tissue, an organ specialized for heat production. For most small species, however, their heat production is not sufficient to defend their body temperature in a cool ambience for a prolonged period and other mechanisms are utilized to defend pup body temperature. The heat loss of the small young can be minimized by their remaining in an insulating nest and huddling with littermates and the mother. The mother can also serve as a heat source for the pups, warming them during periodic encounters.

Small neonates may put off thermoregulatory independence as an adaptation for rapid growth while larger neonates may have precocial thermoregulatory competence. The relatively greater cost of thermal independence to the small neonates may help to explain the differences in the partitioning of their energetic resources.

Periodic cooling of the young occurs during maternal absences, and the regular cooling episodes may facilitate the development of thermoregulatory competence by the young. The plasticity of the developing thermoregulatory system may allow the newly weaned offspring of geographically widely dispersed species to be ready to meet the environmental contingencies particular to its habitat.

Adolph, E. F. Tolerances to cold and anoxia in infant rats. *American Journal of Physiology*, 1948, *155*, 366–377.

Alexander, G. Temperature regulation in the newborn lamb. III. Effect of environmental temperature on metabolic rate, body temperatures and respiratory quotient. *Australian Journal of Agricultural Research*, 1961, *12*, 1152–1174.

Alberts, J. R. Huddling by rat pups: Group behavioral mechanisms of temperature regulation and energy conservation. *Journal of Comparative and Physiological Psychology*, 1978, *92*, 231–245.

Alberts, J. R. Ontogeny of olfaction: Reciprocal roles of sensation and behavior in the development of perception. In R. M. Aslin, J. R. Alberts, and M. R. Peterson (Eds.), New York: Academic Press, 1981, pp. 322–357.

Alberts, J. R., and May, B. Nonnutritive, thermotactile induction of filial huddling in rat pups. *Developmental Psychobiology*, 1984, *17*, 161–181.

Ashworth, V. S., Ramaian, G. D., and Keyes, M. C. Species difference in the composition of milk with special reference to the northern fur seal. *Journal of Dairy Science*, 1966, *49*, 1206–1211.

Babicky, A., Ostadalova, I., Parizek, j., Kolar, J., and Bibr, B. Use of radioisotope techniques for determining the weaning period in experimental animals. *Physiologica Bohemoslova*, 1970, *19*, 457–467.

Banet, M., Brandt, S., and Hensel, H. The effect of continuously cooling the hypothalamic preoptic area on antibody titre in the rat. *Experientia*, 1982, *38*, 965–966.

Barnett, S. A. Endothermy and ectothermy in mice at −3°C. *Journal of Experimental Biology*, 1956, *33*, 124–133.

Barnett, S. A. and Burn, J. Early stimulation and maternal behaviour. *Nature*, 1967, *231*, 150–152.

Bertin, R., and Portet, R. Effect of lipolytic and antilipolytic drugs on metabolism of adenosine 3':5' monophosphate in brown adipose tissue of cold acclimated rats. *European Journal of Biochemistry*, 1976, *69*, 177–183.

Bertin, R., and Portet R. Effect of ambient temperature on lipid metabolism in brown fat during the perinatal period. *Comparative Biochemistry and Physiology*, 1981, *70B*, 193–197.

Blass, E. M., and Cramer, C. P. Analogy and homology in the development of ingestive behavior. In A. R. Morrison and P. L. Strick (Eds.), *Changing concepts of the nervous system*. New York: Academic Press, 1982, pp. 502–523.

Blatteis, E. M. Comparison of endotoxin and leucocytic pyrogenicity in newborn guinea pigs. *Journal of Applied Physiology*, 1977, *42*, 355–361.

Blix, A. S., Miller, L. K., Keyes, M. C., Grav, H. J. and Elsner, R. Newborn northern fur seals *(Callophinus ursinus)*—do they suffer from cold? *American Journal of Physiology*, 1979, *236*, R322–R327.

Brobeck, J. R. Food and temperature. In G. Pincus (Ed.), *Recent progress in hormone research*. Vol. 16. New York: Academic Press, 1960, pp. 439–459.

Brody, E. B. Development of homeothermy in suckling rats. *American Journal of Physiology*, 1943, *139*, 230–232.

Bruck, K. Temperature regulation in the newborn infant. *Biology of the Neonate*, 1961, *3*, 65–119.

Bruck, K., and Wunnenberg, B. The influence of ambient temperature in the process of replacement of non-shivering thermogenesis during postnatal development. *Proceedings of the Society for Experimental Biology and Medicine*, 1966, *25*, 1332–1336.

Bryant, D. M., and Hails, C. J. Mechanisms of heat conservation in litters of mice *(Mus musculus* L.). *Comparative Biochemistry and Physiology*, 1975, *50A*, 99–104.

Buchanan, A. R., and Hill, R. M. Temperature regulation in albino rats correlated with determinations of myelin density in the hypothalamus. *Proceedings of the Society for Experimental Biology and Medicine*, 1947, *66*, 602–608.

Bull, J. J. Sex determination in reptiles. *Quarterly Review of Biology*, 1980, *55*, 3–21.

Bull, J. J., and Vogt, R. C. Temperature-dependent sex determination in turtles. *Science*, 1979, *206*, 1186–1188.

Calhoun, J. B. *The ecology and sociology of the Norway rat*. U.S. Public Health Service Publication, No. 1008, Bethesda, Maryland, 1963. Conover, D. O., and Kynard, B. E. Environmental sex determination: Interactions of temperature and genotype in a fish. *Science*, 1981, *213*, 577–579.

Cooper, K. E., Ferguson, A. C., and Veale, W. L. Modification of thermoregulatory responses in rabbits reared at elevated environmental temperatures. *Journal of Physiology*, 1980, *303*, 165–172.

Covert, J. B., and Reynolds, W. W. Survival value of fever in fish. *Nature*, 1977, *267*, 43–45.

Craig, W. S. The early detection of pyrexia in the newborn. *Archives of Diseases of the Child*, 1963, *38*, 29–39.

Cramer, C. P., and Blass, E. M. The contribution of ambient temperature to suckling behavior in rats 3–20 days of age. *Developmental Psychobiology*, 1982, *15*, 339–348.

Crenshaw, C., Huckabee, W. E., Curet, L. B., Mann, l., and Barron, D. H. A method for estimation of the umbilical blood flow in unstressed sheep and goats with some results on its application. *Journal of Experimental Physiology*, 1968, *53*, 65–75.

Curtis, S. E., Heidenreich, C. J., and Harrington, R. B. Age dependent changes of thermostability in neonatal pigs. *American Journal of Veterinary Research*, 1967, *28*, 1887–1890.

Daly, M. Early stimulation in rodents: A critical review of present interpretations. *British Journal of Psychology*, 1973, *64*, 435–460.

Dawes, G. S., and Mestyan, G. Changes in the oxygen consumption of new-born guinea pigs and rabbits on exposure to cold. *Journal of Physiology*, 1963, *168*, 22–42.

Dawson, N. J., Hellon, R. F., Herington, J. G., and Young, A. A. Warm-rearing and cold defense in rats. *Journal of Physiology*, 1981, *319*, 51–52.

Denenberg, V. H., Brumaghim, j. J., Haltmeyer, G. C., and Zarrow, M. X. Increased adrenocortical activity in the neonatal rat following handling. *Endocrinology*, 1967, *81*, 1047–1052.

Doi, K., and Kuroshima, A. Lasting effect of infantile cold experience on cold tolerance of adult rats. *Japanese Journal of Physiology*, 1979, *29*, 139–150.

Dorigo, P., Gaion, R. M., and Fassina, G. Lack of correlation between cyclic AMP synthesis and free fatty acid release in brown fat of cold adapted rats. *Biochemistry and Pharmacology*, 1971, *23*, 2877–2885.

Drent, R. H. Adaptive aspects of the physiology of incubation. *Proceedings of the International Ornithology Congress*. Vol. 5. The Hague, 1972, pp. 255–280.

Drent, R. Incubation. In D. S. Farner and J. R. King (Eds.), *Avian biology* Vol. 5. New York: Academic Press, 1975, pp. 333.-420.

Eedy, J. W., and Ogilvie, D. M. The effect of age on the thermal preference of white mice *(Mus musculus)* and gerbils *(Meriones unguiculatus). Canadian Journal of Zoology*, 1970, *48*, 1303–1306.

Feldman, D. Evidence that brown adipose tissue is a glucocorticoid target organ. *Endocrinology*, 1978, *103*, 2091–2097.

Ferguson, A. V., Veale, W. L., and Cooper, K. E. Evidence of environmental influences on the development of thermoregulation in the rat. *Canadian Journal of Physiology and Pharmacology*, 1981, *59*, 91–95.

Foster, D. O., and Frydman, M. L. Tissue distribution of cold induced thermogenesis in conscious warm- or cold-acclimated rats re-evaluated from changes in tissue blood flow: The dominant role of brown adipose tissue in the replacement of shivering by non-shivering thermogenesis. *Canadian Journal of Physiology and Pharmacology*, 1979, *57*, 257–270.

Fowler, S. J., and Kellogg, C. Ontogeny of thermoregulatory mechanisms in the rat. *Journal of Comparative Physiology and Psychology*, 1975, *89*, 738–746.

Freeman, B. M. Gaseous metabolism of the domestic chicken. IV. The effect of temperature on the resting metabolism of the fowl during the first month of life. *British Poultry Science*, 1963, *4*, 275–278.

Freeman, N. C. G., and Rosenblatt, J. S. The interrelationship between thermal and olfactory stimulation in the development of home orientation in newborn kittens. *Developmental Psychobiology*, 1978, *11*, 437–457.

Frith, H. J. *The mallee-fowl: The bird that builds an incubator.* Sydney, Australia: Angus and Robertson, 1962.

Fuller, M. F. The effect of environmental temperature on the nitrogen metabolism and growth of the young pig. *British Journal of Nutrition*, 1965, *19*, 531–546.

Galef, B. G. The ecology of weaning: Parasitism and the achievement of independence by altricial mammals. In D. J. Gubernick and P. Klopfer (Eds.), *Parental care in mammals.* New York: Plenum Press, 1977, pp. 211–242.

Gelineo, S., and Gelineo, A. Environmentl temperature of the nest and the appearance of chemical thermoregulation in rats at the temperature of 21°C. *Bulletin of the Serbian Academy of Science*, 1951, *3*, 149–153.

Gelineo, S., and Gelineo, A. La temperature du nid du rat et la signification biologique. *Bulletin of the Serbian Academy of Science*, 1952, 197–210.

Gessaman, J. A., and Findell, P. R. Energy costs of incubation in the American kestrel. *Comparative Biochemistry and Physiology,* 1979, *63A,* 57–62.

Gianutsos, G., and Moore, K. E. Effects of pre- or postnatal dexamethasone, adrenocorticotrophic hormone and environmental stress on phenylethanolamine *N*-methyltransferase activity and catecholamines in sympathetic ganglian of neonatal rats. *Journal of Neurochemistry,* 1977, *28,* 935–940.

Glass, L., Silverman, W. A., and Sinclair, J. C. Effect of the thermal environment on cold resistance and growth of small infants after the first week of life. *Pediatrics,* 1968, *41,* 1033–1041.

Glass, L., Silverman, W. A., and Sinclair, J. C. Relationship of thermal environment and intake to growth and resting metabolism in the late neonatal period. *Biology of the neonate,* 1969, *14,* 324–340.

Goodwin, R. F. W. The relationship between the concentration of blood sugar and some vital body functions in the new-born pig. *Journal of Physiology,* 1957, *136,* 208–217.

Grota, L. J. Effects of litter size, age of young and parity on foster mother behavior in *Rattus norvegicus. Animal Behaviour,* 1973, *21,* 78–82.

Grota, L. J., and Ader, R. Continuous recording of maternal behaviour in *Rattus norvegicus. Animal Behaviour,* 1969, *17,* 722–729.

Guenaire, C., Costa, J. C., and Delacour, J. Thermosensibilité et conditionnement instrumental chez le rat nouveau-ne. *Physiology and Behavior,* 1979, *22,* 837–840.

Guenaire, C., Costa, J. C., and Delacour, j. Discrimination spatiale avec reinforcement thermique chez le jeune rat. *Physiology and Behaviro,* 1982a, *28,* 725–731.

Guenaire, J., Costa, J. C., and Delacour, J. Conditionment operant avec reinforcement thermique chez le rat nouveau-ne. *Physiology and Behavior,* 1982b, *29,* 419–424.

Hahn, P. Effect of environmental temperatures on the development of thermoregulatory mechanisms in infant rats. *Nature,* 1956, *178,* 96–97.

Hahn, P., Krecek, J., and Krechova, J. The development of thermoregulation. I. The development of thermoregulatory mechanisms in young rats. *Physiologica Bohemoslava,* 1956, *5,* 283–289.

Hahn, P., Drahota, Z., Skala, J., Kazda, S., and Towell, M. E. The effect of cortisone on brown adipose tissue of young rats. *Canadian Journal of Physiology and Pharmacology,* 1969, *47,* 975–980.

Hales, J. R. S., Findlay, J. D., and Robertshaw, D. Evaporative heat loss mechanisms of the newborn calf, *Bos taurus. British Veterinary Journal,* 1968, *124,* 83–88.

Hall, W. G. Feeding and behavioral activation in infant rats. *Science,* 1979, *205,* 206–209.

Hayward, J. S. Microclimate temperature and its adaptive significance in six geographic races of *Peromyscus. Evolution,* 1965, *18,* 230–234.

Hedricks, C., and McClintock, M. K. Regulation of mating and sex ratios during postpartum estrus. Abstract. *Conference on Reproductive Behavior,* Vol. 14. East Lansing, Mich., 1982, p. 20.

Heim, T., and Hull, D. The blood flow and oxygen consumption of brown adipose tissue in the new rabbit. *Journal of Physiology,* 1966, *186,* 42–55.

Henning, S. J., Chang, S-S. P., and Gisel, E. G. Ontogeny of feeding controls in suckling and weanling rats. *American Journal of Physiology,* 1979, *237,* R187–R191.

Henning, S. J., and Romano, T. J. Investigation of body temperature as a possible feeding control in the suckling rat. *Physiology and Behavior,* 1982, *28,* 693–696.

Heroux, O. Climatic and temperature induced changes in mammals. *Review of Canadian Biology,* 1961, *20,* 55–68.

Hill, R. W. The ontogeny of homeothermy in neonatal *Peromyscus leucopus. Physiological Zoology,* 1976, *49,* 292–306.

Himms-Hagen, J. Sympathetic regulation of metabolism. *Pharmacology Reviews,* 1967, *19,* 367–461.

Hissa, R. Postnatal development of thermoregulation in the Norwegian lemming and the golden hamster. *American Zoology,* 1968, *5,* 345–383.

Hissa, R., and Lagerspetz, K. The postnatal development of homeothermy in the golden hamster. *Annales Medicinae Experimentales Biologicae Fennicae,* 1964, *42,* 43–45.

Hsieh, A., Carlson, L. D. and Gray, G. Role of the sympathetic nervous system in the control of chemical regulation of heat production. *American Journal of Physiology,* 1957, *190,* 247–251.

Hull, D., and Segall, M. M. The contribution of brown adipose tissue to heat production in the newborn rabbit. *Journal of Physiology,* 1965, *181,* 449–457.

Hutchings, D. E. Early experience and its effects on later behavioral processes in the rat: III. Effects of infantile handling and body temperature reduction on later emotionality. *Transactions of the New York Academy of Science,* 1963, *25,* 890–901.

Hutchings, D. E. Early handling in rats: The effects of body temperature reduction and stimulation or adult emotionality. *Psychonomic Science,* 1965, *3,* 183–184.

Hutchings, D. E. Infantile body temperature loss in rats: Effects of duration of cold exposure, rate of heat loss, level of hypothermia and rewarming on later emotionality. *Psychology Reports*, 1967, *21*, 985–1002.

Hutchinson, R. E. Influence of oestrogen on the nesting behaviour of female budgerigars. *Journal of Endocrinology*, 1975a, *64*, 417–428.

Hutchinson, R. E. Effects of ovarian steriods and prolactin on the sequential development of nesting behaviour in female budgerigars. *Journal of Endocrinology*, 1975b, *67*, 29–39.

Hutchinson, V. H., Dowling, H. G., and Vinegar, A. Thermoregulation in a brooding female python, *Python molurus biuittatus. Science*, 1966, *51*, 694–696.

Ingram, D. L. Adaptations to ambient temperature in growing pigs. *Pflugers Archives*, 1977, *367*, 257–264.

Jans. J., and Leon, M. Determinants of mother–young contact in Norway rats. *Physiology and Behavior*, 1983, *30*, 919–935.

Jensen, R. A., Davis, J. L., and Shnerson, A. Early experience facilitates the development of temperature regulation in the cat. *Developmental Psychobiology*, 1980, *13*, 1–6.

Johanson, I. B. Thermotaxis in neonatal rat pups. *Physiology and Behavior*, 1979, *23*, 871–874.

Johanson, I. B., and Hall, W. G. The ontogeny of feeding in rats. III. Thermal determinants of early ingestive behaviors. *Journal of Comparative and Physiological Psychology*, 1980, *94*, 977–992.

Johanson, I. B. and Teicher, M. H. Conditioning of an odor preference in 3-day-old rats. *Behavioral and Neural Biology*, 1980, *29*, 132–136.

Jones, R. E. The incubation patch of birds. *Biological Reviews*, 1971, *46*, 315–339.

Kasting, N. W., Veale, W. L., and Cooper, K. E. Suppression of fever at term of pregnancy. *Nature*, 1978, *27*, 245–246.

Kiil, V. Experiments on the hair patterns in rats. *Journal of Experimental Zoology*, 1949, *110*, 397–439.

Kinder, E. F. A study of nest-building activity of the albino rat. *Journal of Experimental Zoology*, 1927, *147*, 117–161.

Kittrell, E. M. W. The circadian temperature rhythm (CRT) in nonpregnant and pregnant and lactating hooded rats: Are there maternal influences in CTR ontogeny? Abstract. *International Society of Developmental Psychobiology*, Minneapolis, 1982, p. 35.

Kleiber, M. *The fire of life*. New York: Wiley, 1961.

Kleitman, N., and Satinoff, E. Behavioral responses to pyrogen in cold stressed and starved newborn rabbits. *American Journal of Physiology*, 1981, *241*, R161–R171.

Kleitman, N., and Satinoff, E. Thermoregulatory behavior in rat pups from birth to weaning. *Physiology and Behavior*, 1982, *29*, 537–541.

Kluger, M. J., and Vaughn, L. K. Fever and survival in rabbits infected with *Pasteurella multocida. Journal of Physiology*, 1978, *282*, 243–251.

Kornacker, M. S., and Ball, E. G. Respiratory processes in brown adipose tissue. *Journal of Biological Chemistry*, 1968, *243*, 1638–1644.

Kraly, F. S., and Blass, E. M. Mechanisms for enhanced feeding in the cold in rats. *Journal of Comparative and Physiological Psychology*, 1976, *90*, 714–726.

Krecek, J., Keckova, J., and Martinek, J. The development of thermoregulation. V. Effect of rearing under cold and warm conditions on the development of thermoregulation in young rats. *Physiologia Bohemoslova* 1957, *6*, 329–336.

Lachance, J. P., and Page, E. Hormonal factors influenced fat disposition in the interscapular brown adipose tissue of the white rat. *Endocrinology*, 1953, *52*, 57–64.

Lee, M., and Williams, P. I. Changes in licking behavior of rat mother following handling of young. *Animal Behavior*, 1974, *22*, 679–681.

Leon, M., Croskerry, P. G., and Smith, G. K. Thermal control of mother–young contact in rats. *Physiology and Behavior*, 1978, *21*, 793–811.

Levine, S. A. Further study of infantile handling and adult avoidance learning. *Journal of Personality*, 1956, *25*, 70–80.

Levine, S. A. The effects of infantile expence on adult behavior. In E. L. Bliss (Ed.), *Experimental foundation of clinical psychology*. New York: Basic Books, 1962, pp. 139–169.

Leonard, C. M. Thermotaxis in golden hamster pups. *Journal of Comparative and Physiological Psychology*, 1974, *86*, 458–469.

Mayer, L., Lustick, S., and Battersby, B. The development of homeothermy in the American goldfinch. *Comparative and Biochemical Physiology*, 1982, *72A*, 421–424.

Maclean, G. L. The breeding biology and behavior of the double-banded courser *Rhinoptilus africanus* (Temmiack). *Ibis*, 1967, *109*, 556–569.

Manners, M. J., and McCrea, M. R. Changes in the chemical composition of sow-reared piglets during the first month of life. *British Journal of Nutrition*, 1963, *17*, 495–513.

McClure, P. A., and Porter, W. P. Development of insulation in neonatal cotton rats *(Sigmodon hispidus)*. *Physiological Zoology*, 1983, *56*, 18–32.

McClure, P. A., and Randolph, J. C. Relative allocation of energy to growth and development of homeothermy in the eastern woodrat *(Neotoma floridana)* and the hispid cotton rat *(Sigmodon hispidus)*. *Ecological Monographs*, 1980, *50*, 199–219.

McEwan-Jenkinson, D., Noble, R. C., and Thompson, G. E. Adipose tissue and heat production in the new-born ox *(Bos taurus)*. *Journal of Physiology*, 1968, *195*, 639–646.

McNab, B. K. Body weight and the energetics of temperature regulation. *Journal of Experimental Biology*, 1970, *53*, 329–348.

McNab, B. K. On the ecological significance of Bergmann's rule. *Ecology*, 1971, *52*, 845–854.

Meschia, G., Cotter, J. R., Makowski, E. L., and Barron, D. H. Simultaneous measurement of uterine and umbilical blood flows and oxygen uptakes. *Journal of Experimental Physiology*, 1967, *52*, 1–18.

Metcalf, G., and Meyers, R. D. Precise location within the preoptic area where noradrenaline produces hypothermia. *European Journal of Pharmacology*, 1978, *51*, 47–53.

Miller, J. A., Miller, F. S., and Westin, B. Hypothermia in the treatment of asphyxia. *Biologia neonatorum*, 1964, *6*, 148–163.

Moore, R. E. and Underwood, M. C. The thermogenic effects of noradrenaline in new-born and infant kittens and other small mammals. A possible hormonal mechanism in the control of heat production. *Journal of Physiology*, 1963, *168*, 290–317.

Morreale, S., Ruiz, G. J., Spotial, J. R., and Standora, E. A. Temperature-dependent sex determination: Current practices threaten conservation of sea turtles. *Science*, 1982, *216*, 1245–1247.

Morrison, P. R., and Petajan, J. H. The development of temperature regulation in the oppossum *Didelphis marsupialis*. *Physiological Zoology*, 1962, *35*, 52–65.

Mount, L. E. The metabolic rate of the new-born pig in relation to environmental temperature and to age. *Journal of Physiology*, 1959, *147*, 333–345.

Mount, L. E. The influence of huddling and body size on the metabolic rate of the young pig. *Journal of Agricultural Science*, 1960, *55*, 101–105.

Mount, L. E. Environmental temperature preferred by the young pig. *Nature*, 1963, *199*, 1212–1213.

Ogilvie, D. M., and Stinson, R. H. The effect of age on temperature selection by laboratory mice *(Mus musculus)*. *Canadian Journal of Zoology*, 1966, *44*, 511–517.

Okon, E. E. The temperature relations of vocalization in infant golden hamsters and Wistar rats. *Journal of Zoology*, 1971, *164*, 227–237.

Oppenheim, R. W. Ontogenetic adaptations and retrogressive processes in the development of the nervous system and behaviour: A neuroembryological perspective. In K. J. Connolly and H. P. R. Prechtl (Eds.), *Maternal development: Biological and psychological perspectives*. Vol. 7. Philadelphia: Lippincott, 1981, pp. 73–109.

Oswalt, G. L., and Meier, G. W. Olfactory, thermal and tactual influences on infantile ultrasonic vocalizations in rats. *Developmental Psychobiology*. 1975, *8*, 129–135.

Pearson, O. P. Metabolism of small animals with remarks on the lower limit of mammalian size. *Science*, 1948, *108*, 44–46.

Pieau, C. Modalities of the action of temperature on sexual differentation in field-developing embryos of the European pond turtle, *Emys orbicularis* (Emydidae). *Journal of Experimental Zoology*, 1982, 353–360.

Pittman, Q. I., Cooper, K. E., Veale, W. L., and Van Pettern, F. R. Observations on the development of the febrile response to pyrogens in sheep. *Clinical Science in Molecular Medicine*, 1974, *46*, 591–602.

Prusiner, S. B., Cannon, B., Ching, T. M., and Lindberg, O. Oxidative metabolism in cells isolated from brown adipose tissue. 2. Catecholamine regulated respiratory control. *European Journal of Biochemistry*, 1968, *7*, 51–57.

Rosenblatt, J. S. The development of maternal responsiveness in the rat. *American Journal of Orthopsychiatry*, 1969, *39*, 36–56.

Rosenblatt, J. S. Olfaction mediates developmental transition in the altricial newborn of selected species of mammals. *Developmental Psychobiology*, 1983, *16*, 347–375.

Rosenblatt, J. S., and Lehrman, D. S. Maternal behavior in the laboratory rat. In H. L. Rheingold (Ed.), *Maternal behavior in mammals.* New York: Wiley, 1963, pp. 8–57.

Russel, P. A. Infantile stimulation in rodents: A consideration of possible mechanisms. *Psychological Bulletin,* 1971, *75,* 192–202.

Satinoff, E., McEwen, G. N. J., and Williams, B. A. Behavioral fever in newborn rabbits. *Science,* 1976, *193,* 1139–1140.

Schaefer, T. Early experience and its effects on later behavioral processes in rats. II. A critical factor in the early handling phenomenon. *Transactions of the New York Academy of Science,* 1963, *25,* 871–889.

Schaefer, T., Weingarten, F. S., and Towne, J. C. Temperature change: The basic variable in the early handling phenomenon? *Science,* 1962, *135,* 41–42.

Sherry, D. Parental care and the development of thermoregulation in red junglefowl. *Behaviour,* 1981, *76,* 250–279.

Smotherman, W. P. Mother–infant interaction and the modulation of pituitary–adrenal activity in rat pups after early stimulation. *Developmental Psychobiology,* 1983, *16,* 169–176.

Smotherman, W. P., and Bell, R. W. Maternal mediation of early experience. In R. W. Bell and W. P. Smotherman (Eds.), *Maternal influences and early behavior.* New York: Spectrum Publications, 1980.

Smotherman, W. P., Wiener, S. G., Mendoza, S. P., and Levine, S. Maternal pituitary–adrenal responsiveness as a function of differential treatment of rat pups. *Developmental Psychobiology,* 1977, *10,* 113–122.

Snell, E. S., and Atkins, E. The mechanisms of fever. In E. E. Bittar and N. Bittar (Eds.), *The biological basis of medicine.* Vol. 2. New York: Academic Press, 1968, pp. 297–419.

Spencer, F., Shirer, H. W., and Yochim, J. M. Core temperature in the female rat: Effect of pinealectomy or altered lighting. *American Journal of Physiology,* 1976, *231,* 355–360.

Stainer, M. W. Effect of body weight, ambient temperature and huddling on oxygen consumption and body temperature of young mice. *Comparative Biochemistry and Physiology,* 1975, *51A,* 79–82.

Steel, E. A., and Hinde, R. A. Hormonal control of brood patch and oviduct development in domesticated canaries. *Journal of Endocrinology,* 1963, *26,* 11–24.

Szekely, M., and Szelenyi, Z. The effect of *E. coli* endotoxin on body temperatures in the newborn rabbit, cat, guinea pig and rat. *Acta Physiologica, Academy of Science, Hungary,* 1977, *50,* 293–298.

Takano, N., Mohri, M., and Nagasaka, T. Body temperature and oxygen consumption of newborn rats at various ambient temperatures. *Japanese Journal of Physiology,* 1979, *29,* 173–180.

Taylor, P. M. Oxygen consumption in newborn rats. *Journal of Physiology,* 1960, *154,* 153–168.

Teisner, B., and Haahr, S. Poikilothermia and susceptibility of suckling mice to Coxsackie B₂ virus. *Nature,* 1974, *247,* 568.

Van Mierop, L. H. S., and Barnard, S. M. Thermoregulation in brooding female *Python molarus bivittalus. Copeia,* 1976, *1976,* 398–401.

Van Mierop, L. H. S., and Barnard, S. M. Further observations on thermoregulation in the brooding female *Python molarus bivittatus* (Serpentines: Boidae). *Copeia,* 1978, *1978,* 615–621.

Vaughn, L. K., Berlheim, H. A., and Kluger, M. J. Fever in the lizard *Dipsosaurus dorsalis. Nature,* 1974, *252,* 473–474.

Webb, D. R., and King, J. R. An analysis of the heat budgets of the eggs and nest of the white-crowned sparrow, *Zonotrichia leucophrys,* in relation to parental attentiveness. *Physiological Zoology,* 1983, *56,* 493–505.

Vinegar, A., Hutchinson, V. H., and Dowling, H. G. Metabolism, energetics and thermoregulation during brooding of snakes of the genus *Python* (Reptilia, boidae). *Zoologica,* 55, 19–55.

Walsberg, G. E., and Voss-Roberts, K. A. Incubation in desert-nesting doves: Mechanisms for egg cooling. *Physiological Zoology,* 1983, *56,* 88–93.

Weaver, M. E., and Ingram, D. L. Morphological changes in swine associated with environmental temperatures. *Ecology,* 1969, *50,* 710–713.

White, F. N., and Kinney, J. L. Avian incubation. *Science,* 1974, *186,* 107–115.

White, F. N., and Odell, D. K. Thermoregulatory behavior of the northern elephant seal, *Mirounga angustirostris. Journal of Mammalogy,* 1971, *51,* 758–774.

Wiesner, B. P., and Sheard, N. M. *Maternal behaviour in the rat.* Oliver and Boyd: Edinburgh, 1933.

Yntema, C. L. Effects of incubation temperatures on sexual differentation in the turtle, *Chelydra serpentina. Journal of Morphology,* 1979, *150,* 453–462.

Zarrow, m. X., Denenberg, V. H., and Anderson, C. O. Rabbit: Frequency of suckling in the pup. *Science,* 1965, *150,* 1835–1836.

Index

Acoustic experience, avian vocal development and, 130–131

Activational responses, milk ingestion and, 110, 111

Adipose tissue, brown, activation of, 307

Adrenergic neurons, embryonic development of, 11

Aggression, 116

Ambystoma, 41

Amygdala
olfactory function and, 176, 177
olfactory input, 193

Aplysia, 152

Area 17. *See* Visual cortex

Auditory feedback, 134

Auditory template, in song learning, 140–141

Auditory–motor learning, 132, 134

Autoradiography, 183–185

Behavior
acoustic, 55
aggressiveness, 116
association with specific odors, 194, 195
catecholamine systems and, 118, 119
development of
factors in, 2
neural determinants of, 99–128
prior to utilization, 112
dopamine and organized, 120
effect of temperature on, 315–316
experience-dependent development and, 283–284
grooming, 60–66, 67
hierarchical expression of, 36
huddling, 106–109

Behavior (*cont.*)
individual variability in expression of adult, 59
induced by electrical stimulation of medical forebrain bundle, 113, 114–115
odor-guided, 180–181, 187–192
patterns of, brain stimulation and, 112–117
sexual, 116, 122–123, 186
social, 76
song, 129–162
suckling, 100–106
thermal mediation of, 313–314, 315–316
thermoregulatory, 308–309
visually guided, 272–275
visual system changes and the development of, 272–281

Bird song, 55–57
development of, 28

Brain
at gastrula stage, 7
space, song behavior and, 139
stimulation, behavioral patterns and, 112–117

Brain stem, monoamine system, 270

Callorhinus ursinus, 301

Canis latrans, 72

Canis lupus, 72

Catalepsy, 118

Catecholamine
behavior and, 118, 119
pathways, development of, 119

Central nervous system, taste responses, 214–218

Cholinergic neurons, embryonic development of, 11

Chorda tympani nerve, gustatory development and, 208, 209, 210, 211, 212

Cistothorus palustris, 153
Clethrionomys britannicus, 71
Communication, 54–59
Conditioning
  cortical, 70
  motor patterns and, 70–72
  paradigm of classical, 152
Cortex
  inhibitory mechanisms in, 121
  olfactory, 195, 196
  olfactory areas, 175, 176
  role in integrating motor capacities, 84
  visual, 27–28, 81, 238, 254–258
Cortical activity, inhibition of, 117
Cortical conditioning, 70
Cossypha heuglini, 137
Cricetus cricetus, 39

Didelphis marsupialis virginiana, 301
Dopamine
  pathways, organized behavior and, 120
  stroking and turnover of, 120

Eisenia foetida, 42
Embryo, thermal environment of, 298
Embryology
  heterospecific transplants in development of, 8–10
  neurobehavioral development and, 2
  as parent discipline of developmental psychology and psychobiology, 3
Environment
  development of cortical inhibitory mechanism and reaction to, 121
  role in behavioral development, 24
Epiglottis, taste buds on, 206
Equifinality, 82
Estradiol, songbird vocalization and, 139
Experience-dependent development programs, 237
  behavior and, 283–284
  implications of, 284

Fat pads, 307
Food intake
  effect of temperature on, 315
  independent, 110–112
  nest experience and, 122
Forebrain
  electrical stimulation of, 112
  role in vocalization, 136
  ventral, lateral olfactory tract spread over, 175
  visual centers of, 241
Fringilla coelebs, 131

Gallus domesticus, 301
Gallus gallus, 301

Gastrula stage, 7
Gene expression, motoneuron activity and, 46
Glomerular complex, modified, 178– 180, 182, 191
  early postnatal activity in, 187–189, 192
  emergence of, 193
  link to olfactory cortex, 195, 196
  location of, 187
  postnatal suckling odors processed by, 186
  processing of odor cues by, 188
Glossopharyngeal nerve
  taste bud innervation by, 221
  taste response recorded from, 224
Grooming behavior
  development of, 60–66, 67
  different phases of, 68
Gustatory development, 205, 236
Gustatory system, human versus sheep, 220–230

Heterospecific transplants, 8–10
Huddling, 106–109
  function, 106
  olfactory cues in, 107, 108
  regulation of size, 107
  tactile cues in, 107
Hyperstriatum ventrale pars caudale, 136, 137, 138, 139, 140, 146, 147, 148, 149, 153, 154, 155, 158
  auditory units in, 140, 141
Hypothalamus
  electrical stimulation of, 112
  olfactory function and, 176

Incubation, 298–299
Ingestion, independent, 109–112
  temperature as a facilitator of, 110–111
Insects, motor development in, 42
Interneuron(s), 80
  in main olfactory bulb, 171
Invertebrates, motor patterns in, 42

Lateral geniculate nucleus (lgn)
  dark-rearing and the development of, 259
  development of, 250–254
  dorsal, cell types in, 242
  monocularity of, 259
  neuronal activity in, 251
  postnatal stages of maturation, 250–254
  preferred stimulus orientation, 245
  relay cells in, 242
  synaptic organization of, 250, 252
  W-cells in, 252
  X-cells in, 242, 243, 245, 252
  Y-cells in, 242, 243, 245, 252
Learning
  auditory–motor, 132, 134
  the development of motor patterns and, 40

Learning (*cont.*)
   perceptual, 134
   vocal, 129–162
Limb coordination, 37
Lordosis, induced by electrical stimulation of
     medial forebrain bundle, 113, 114, 115
Lonchura striata, 131
Lycaon pictus, 72

Magnocellular nucleus of the anterior
     neostriatum, 136, 137, 138, 139, 140,
     142, 145
Maturation, the development of motor patterns
   and, 40
Medial thalamic nuclei, visual system
   development and, 270, 271
Melospiza georgiana, 131
Melospiza melodia, 132
Menidia menidia, 300
Meriones persicus, 39
Messenger RNA, 46
Microtus agrestis, 71
Milk ingestion
   activational responses and, 110, 111
   median forebrain bundle stimulation and, 115
Milk intake, 100
Mirounga angustirostris, 301
Motoneurons, 14
   cortical connections to, 38
   dendritic arborization of, 20
   gene expression and activity of, 46
   interneuron connection to, 80
   in olfactory processes, 196
   spinal cord, 17
   synaptic activity and survival of, 15
Motor activity
   compensatory responses to perturbations in,
     66–70
   elicitation of, 40
   in human fetuses, 47
   limb coordination, 37
   patterns in development, 35–97
   postnatal, 272
   spontaneity in, 41–45
   vision and, 239
   visual system development and, 273
Motor capacities
   arousal and manifestation of, 48
   changes in utilization of, 78
   cortical integration of, 84
   development of, 19, 24–25
     environmental contact and the, 121
   hierarchical organization of, 38
Motor patterns
   candid, 72–76
   conditioning and, 70–72

Motor patterns (*cont.*)
   development of, 35–97
     communication in the, 54–59
     contextual factors in, 78
     control of, 41
     experience and, 53
     genetic "programming" and, 79
     in hamsters, 39
     hierarchical cohesions in, 77
     in human infants, 38
     in insects, 42
     in invertebrates, 42
     learning and, 40
     locomotory experiences and, 82
     maturation and, 40, 41
     in mice, 39
     organizational polarities relevant to, 36
     in rats, 39
     retrograde, 43
     spontaneity in, 41–45
     in squirrels, 39
   elicited, 45
     development of, 45–49
   orientation of, 40
   perserverant activation of, 70–72
   rodent, 39, 60–70
   "social," 74
   social interactions and, 74–76
   socially integrated, 53–60
   vocal, 54–59, 73–74
Motor spontaneity, 41–45
Muscle
   nerves innervating, 14
   skeletal, 13
Mus musculus, 309

Neoepigenesis, 4, 5
Neopreformationism, 4
Nervous system, embryological development of,
     1–33
Nest experience
   dietary preference and, 122
   sexual behavior and, 122
Neural crest, cell differentiation in, 11, 12
Neuroembryology, 1–33
Neurogenesis, 167
Neurons, embryonic versus mature, 18
Nucleus intercollicularis, 136, 137
   vocalization and, 136
Nucleus of the solitary tract, 214, 215, 216
Nucleus robustus archistriatalis, 136, 137, 139,
     142, 145, 146, 147, 148, 153, 154, 155

Odor, behavior associated with specific, 194, 195
Odor discrimination abilities, 180, 181
Odor-guided behavior, 180, 187–192, 194

Olfactory bulbs, 167
 accessory, 175, 176, 177, 192
  autoradiograph, 185
  central connections of, 177
  glomeruli in, 178
  input to the amygdala, 193
  prenatal activity in, 182–187
  vomeronasal nerves in, 186
 main, 169–175, 186, 192
  autoradiograph, 183, 184
  classification of cells in, 171
  electrophysiological analysis of, 190
  glomerular development, 174, 178, 179
  input to the amygdala, 193
  laminar, 169
  neurogenesis in, 171
  postnatal activity in, 190–192
  postnatal suckling odors processed by, 186
Olfactory cortex, 195, 196
Olfactory epithelium, main, 164–167
 maturation of, 164
 neurogenesis in, 167
 receptor cells, 164, 166
Olfactory function
 early development of, 163–203
 neural substrates of, 181
 prenatal evidence of, 164
 suckling and, 163
 role, in newborn, 180
Olfactory marker protein, 166, 169
Olfactory projections, ventral, 175–178
Olfactory tract, lateral, 175–176
Olfactory tubercle, 176
Operant response, acquiring, 113
Optic nerve, 242
 myelination of axons of, 248
Optics, effect on the development of visual
  activity, 247
Oscines, 129–161

Passeriformes, 129–162
Perceptual learning, sensorimotor integration
  following, 134
Periplaneta americana, 42
Peromyscus leucopus, 309
"Plastic song," 132
Poephila guttata, 131
Python molarus, 298

Receptive field
 complex, 243
 retinal, 249
 simple cell, 243
Reflex(es)
 control of, changes during development, 46
 development of, 43, 45
 elicited, 45

Reflex(es) (*cont.*)
 irradiation, 45
 righting, 46
Reinforcement, suckling and, 105–106
Reticular formation, visual system development
  and, 270
Retina
 area centralis of, 248
 cell types of, 241–242
 dark-rearing and the development of, 259
 description of, 241
 ganglion cells of, 242
  conduction velocity of axon, 249
  postnatal morphological changes, 248
  receptive field centers of, 249
 impact of visual centers of forebrain on, 241
 maturation of, 248, 249
 organization of, 241
 overproduction of neuronal elements in, 250
 postnatal stages of maturation, 247–250
 preferred stimulus orientation, 245, 246
 projections to lgn and superior colliculus, 247
 X-cells in, 248
 Y-cells in, 248
Retino-geniculo-cortical pathway, 239–258
 development of, 246–258
Richmondena cardinalis, 133

Saimiri sciureus, 70
Satiety, medial forebrain bundle stimulation and,
  115
Sciurus vulgaris, 39
Sensory system
 development of, 3, 25–26
 prenatal, 28
Sex, temperature and determination of, 300
Sexual behavior, 116
 vomeronasal organ and, 186
Shivering, as a means of heat production, 307
Sigmodon hispidus, 303
Skeletal muscle, differentiation of different types
  of, 13
Skin cells, at gastrula stage, 7–8
Social behavior, defined, 76
Sonograms, 130, 143, 144, 149, 150
Song. *See also* Vocalization
 auditory–motor phase of learning, 132
 behavior, "brain space" and, 139
 development of, 56–57
  auditory feedback and, 132–134, 141, 151
  nonauditory feedback in, 141, 151
  syringeal feedback and, 149
  warbling and the, 131
 learning,
  brain volume changes and, 154–155
  magnocellular nucleus of the anterior
   neostriatum and, 136, 142

Song (*cont.*)
    learning (*cont.*)
        sensitive periods for, 152–154
        testosterone and, 152
    neural control of, 135
    patterns
        crystallization of, 152
        in early-deafened birds, 133
    "plastic song," 132
    species specificity of, 131–132
    subsong, 132
Songbirds
    ontogeny of vocal learning in, 129–161. *See also* Vocalization, in songbirds
    vocal organ of, 135
Speech, 155–158
    development of, 57–59
    motor system in human, 57
    relationship between vocal perception and production of, 157–158
Spinal cord
    morphological and functional development, 16
    motoneurons of, 17
    "retrograde" sequences in, 80
Spontaneity
    human, 43–44
    in motor patterns, 41–45
Steroid hormones, ontogeny of songbird vocalization and, 139
Stripe-rearing, 262–264, 277–280
    behavioral effects of, 277
    effect on orientation selectivity, 278, 279
Stroking, dopamine turnover and, 120
Sturnus vulgaris, 140
Subsong, 132
Suckling, 100
    conditioning and, 102
    internal controls of, 103–105
    nipple attachment, olfactory cue for, 101–102, 103
    nonnutritional benefits of, 104
    odor control aspects of, 187–192
    olfactory function and, 163
    reinforcement and, 105–106
    sensory control of, 188
Synaptic activity
    adrenergic neurons and, 12
    motoneuron survival and, 15
Syrinx, innervation of, 135, 136
"Systemogenesis," 111

Taste
    acquiring preferences in, 220
    development of the sense of, 205–236
    neonatal discrimination of stimuli, 228, 229
    postnatal responses, 210

Taste (*cont.*)
    preferences, 220, 229
    response latencies, 212
    synapse development and development of, 227
Taste buds
    age and number, 230
    appearance of, 206
    in circumvallate papillae, 221, 224, 232
    fetal, 205, 225, 226
    functioning, beginning of, 207–208
    in fungiform papillae, 213, 216, 220, 224, 232
    innervation of, 206
    location of, 206
    number of, 213
Temperatures, cool, effect on neonatal survival, 307
Testosterone
    songbird vocalization and, 139
    song learning and, 152
Tetraeunura fisheri, 57
Tetrodotoxin, as a normal constituent of body fluids, 22
Thalamus, olfactory function and, 176
Thermoregulation
    body size and, 302
    development of, 297–322
    mechanisms of, 307–311
    neonatal, 302–303
    plasticity in development of, 311–316
    role of mother in neonatal, 305–306
    strategies for, 301–304
Timm stain, 177
Triturus vulgaris, 47

Vision
    experience-dependent development of, 239
    extraretinal stimulation, 237
    motor activity and development of, 239
    ontogeny of, 237–295
    postnatal development of 237–238
    retinal stimulation for development of, 237, 239
    sensorimotor coordination and, 239
Visual cortex
    dark-rearing and the development of, 261
    development of, 254–258
    differences in experience sensitivity among cells of, 264–270
    effect of stripe-rearing on, 262
    experience-dependent changes in, 282
    neuronal activity and the development of, 261–266
    orientation selectivity of, 255, 256, 258, 261, 264, 268, 278, 279, 281
    postnatal overproduction of synapses in, 258
    selectivity of, 238
    stimulus orientation for, 242, 255

Visual cortex (*cont.*)
  synaptic density, 256, 258
  synaptogenesis in, 254, 262
  "tuning width" of cells of, 256
Visual system, 246–258
  activity-dependent development of 258–270
  extraretinal factors in the development of,
      270–271
  eye alignment and development of, 247
  monoamine system and development of, 270
  motor activity and the development of, 273
  neonatal, 246–247
  neuronal activity and development of
      postnatal connections in, 258
  optics and the development of, 247
  orientation selectivity of, 271
  retinal organization and, 241
  sensory stimulation and the development of,
      283
  stripe-rearing and the development of, 277–
      280
Vocalization, 54, 55. *See also* Song
  bird song, 55–57
  in birds, 27
  birds versus humans, 155–157
  central nervous pathways for, 135–136
  deafness and, 156
  development of, steroid hormones and, 139
  forebrain and, 136
  human, 155–158

Vocalization (*cont.*)
  motor activity and, 55
  neural lateralization of, 136–137
  nucleus intercollicularis and, 136
  preliminary phase of, 156
  sex differences in, 137
  in songbirds, 129–162
      acoustic experience and, 130–131
      auditory–motor phase of learning, 132
Vomeronasal organ, 167–169
  anatomy, 167
  cross section, 168
  glands in, 169
  receptors, 180, 186
  reproduction and, 186

Warbling, 131
Water intake, 109–110
W-cells, 241, 242, 249, 252, 255
Weaning, effect of mother on, 122

X-cells, 241, 242, 248, 249, 252, 259, 260
  receptive fields, 243
Xenopus laevis, 19, 20

Y-cells, 241, 242, 248, 249, 252, 259, 260
  maturation of, 248–249
  receptive fields, 243

Zonotrichia leucophrys, 131, 298